# THE EVOLUTION
## OF THE
# ENGLISH FARM

# THE EVOLUTION
# OF THE
# ENGLISH FARM

## M. E. SEEBOHM
F.R.HIST.S.
(M. E. CHRISTIE)

*Revised Second Edition*

Republished by
EP Publishing Limited
1976

First published 1927


This is a reprint of the second
revised edition of 1952, published
by George Allen & Unwin Ltd., London


Republished 1976 by
EP Publishing Limited
East Ardsley, Wakefield
West Yorkshire, England
by permission of the
original publisher


ISBN 0 7158 1173 8


**British Library Cataloguing in Publication Data**
Seebohm, Mabel Elizabeth
    The evolution of the English farm. — 2nd revised ed.
    Bibl. — Index.
    ISBN 0-7158-1173-8
    1. Title
    338.1'0942      S455
    Farms-England-History

Please address all enquires to EP Publishing Limited
(address as above)


Printed in Great Britain by
The Scolar Press Limited
Ilkley, West Yorkshire

CUM HUMILITATE

ILLI

QUEM IN BOVARIO NATUM

BERCARII PRIMI ADORAVERUNT

# FOREWORD

LONESOME must be the book which comes into the world unattended by the willing help and furtherance of friends; and herein is recorded my gratitude to all those who have watched over the growth and final launching of this work: first to the one who worked that I might have time to write, then to those, severally, who made for me a translation of the "Gerefa," took photographs of the ox teams, latinized the dedication, read the manuscript chapter by chapter and made useful suggestions, and who helped in many other ways.

To the late Professor Lilian Knowles I am deeply indebted for her good-heartedness in finding time in the midst of a busy life to go through the manuscript and give much valuable advice; also to Mr. B. M. Headicar for his kindly furtherance, and to Mr. Annesley Voysey for his criticisms of the final chapter.

Thanks are also due to the following for their kind permission to reproduce illustrations: the Trustees of the British Museum, the Society of Antiquaries, the Royal Anthropological Institute, the Director of the Municipal Museums, Hull, and Messrs. A. Bulleid and H. St. George Gray.

For the second edition I am most grateful to Mr. Findlay Stratton for his help and advice on the last chapter. Thanks are specially due to Dr. E. C. Curwen for his great kindness in allowing me to reproduce a number of his illustrations. I am indebted also to Mrs. Quennell and her publishers, Messrs. B. T. Batsford, Ltd., for permission to use an illustration, and to Mr. O. G. S. Crawford, C.B.E., F.B.A., for his photo of the Soay sheep.

# CONTENTS

# ILLUSTRATIONS

---

I am indebted to the authorities of the British Museum for their kindness in allowing me to reproduce many illustrations from their Guides to the Stone Age, Bronze Age, Early Iron Age, and Roman Antiquities (Figs. 2b, 3a, 3b, 4, 7, 9, 12, 13, 14, 19a, 20, 26, 27, 28); to the Society of Antiquaries for leave to reproduce Figs. 11 and 25 from the article by Professor T. McKenny Hughes, Figs. 30, 31 and 32 from Cædmon's *Paraphrase*, and for all those from the Luttrell Psalter in Chapters VI and VII. To the Royal Anthropological Institute I am indebted for Fig. 6 from the article by Professor E. B. Taylor; and to the Curator of the Municipal Museums, Hull, for the bronze implement shown in No. 15. Messrs. A. Bulleid and H. St. G. Gray have very kindly allowed me to reproduce Figs. 8, 21 and 22 from their work on the Glastonbury Lake Village.

# CHAPTER I

## INTRODUCTION. THE NEOLITHIC FARM

THIS book is written for those who love a farm-yard: such a farm-yard as may be seen pictured at the beginning of this book and in many other places still in pleasant reality. There is the comfortable farm-house, with its roomy, stone-flagged kitchen, and open door into the sunny yard; the dairy adjoining, the cow-byres beyond, the stables, the cart-sheds, the piggeries, the henhouse and other buildings completing the homely quadrangle, with the cat sunning herself on the steps of the granary, and the dog regarding the whole with placid possession. Nearby is the rick-yard, and the barn with its great doors and dim raftered roof; the cock crowing lustily without and the hens industriously picking up their livelihood; the ducks and geese breaking into sudden clamour round the pond. On the sunny side of the old house is the gay little garden, and the orchard, where such young things as chicks and lambs find shelter, and the busy bees from the row of hives ply their work. Beyond and around all lie the fields, green pastures and corn-lands, the immemorial fields from which our forefathers won sustenance for man and beast, and in the distance the old water-mill where once the corn was ground into flour.

All this should you love, the sights, sounds and smells—particularly the warm and comfortable smell of the cow-byre—the wholesome contact with the fruitful earth, unhurried in her seasons, if you are to take pleasure in what is written here.

The object of this book is to inquire how this most primitive of institutions—the farm—one so fundamentally important to the existence of the human race, was developed. We shall endeavour to show how that wild and skin-clad creature "homo sapiens" became a farmer; how in time he collected round him his dwelling place, his various stock, his farm-buildings and his implements; what crops he grew and how he set about it; through what phases of organization his agricultural endeavours passed, what he ate, how he lived, and in what ways his ever-busy wife was employed.

Many of these things have come and gone and changed. The great plough-ox, for instance, the mainstay of our forefathers, the unit upon which the Domesday Record was constructed, has vanished, even from the Sussex Downs; even his successor, the steady, ponderous farmhorse, is now largely—though not entirely—displaced by the noisy tractor. The great barn too is passing, and with it many other ancient customs and institutions which it will be our pleasant duty to record, and perhaps the pastime of some to read.

Long ages ago, when the early races of mankind were drifting out of Asia, moving slowly from East to West with the sun, England was a little outpost in the north-west, the last spot which successive waves of migration reached. Hence each fresh race brought to England the accumulated results of its long experience: such great steps in human progress as the domestication of animals and the discovery of the use of metals did not have to be slowly evolved on the spot, but for the most part were brought over in an advanced stage of development from the Continent—Neolithic man, Bronze-Age warriors, Celts, Romans, Saxons, Danes and Normans each brought their own contribution. For this reason the progress of civilization in England falls into much more clearly defined stages than would otherwise have been the case, although we must not lose sight of the fact that change or development was always going on, that ancient customs long survived the introduction of new ones—in fact many different stages of civilization continued to exist side by side, the old types long preserved in the mountains of Wales and the North, and the whole of Britain never being conquered simultaneously. All that we can really do is to note the arrival of fresh elements, and record the general standard of development reached in each successive period.

Up to about 6000 B.C. Britain was still joined to the Continent, but the sea-level was gradually rising, and not long after that date the North Sea and the Channel began to join and this country became an island, separated at first merely by a swampy depression where the Straits of Dover now are. Under a warm, damp climate the coast-level continued to sink little by little until the present coast-line was reached sometime between 2000 and 1500 B.C. and the present climate pretty well established.[1]

The first inhabitant of these islands, Palaeolithic man, was a hunter and an eater of roots and berries, leading a precarious existence among strange beasts. It did not occur to him to domesticate any of them—many indeed would have been singularly unsuited for the purpose—neither did circumstances lead him to think of tilling the soil, therefore with him we are not concerned. After him came the men of the Mesolithic

[1] D. Jerrold—*Introduction to the History of England*, 38-61.

Culture, users of very small, finely-worked flint implements; but they too were merely hunters and gatherers of food.

The history of the farm-yard begins with the arrival about 2800-2500 B.C. of Neolithic men (i.e., folk who had arrived not only at chipping out implements of flint, but polishing them to a better shape), men whose forebears had learned far back in the Middle East to domesticate animals and cultivate a few cereals. At some period in the dim past the women of the tribe began to cultivate little patches of grain-producing plants, scratching the ground with a rude hoe of flint or wood, and thus laying the foundations of a new mode of life. Thus woman became the partner of man the hunter, and sometimes even matriarch of the tribe, and the fruitful soil was deified as Mother Earth the producer of crops and fruits. But for the birth of settled farming livestock also was necessary, and so man, as hunting grew more difficult, learnt to tame and breed beasts suitable for food. By far the earliest animal to be domesticated was the dog, not so much as an article of food as a helper in hunting, and later as guardian of the flocks and herds. By degrees oxen, sheep, goats, and eventually pigs and horses were domesticated, and man became a wandering pastoralist, seeking fertile pastures for his flocks and herds, and fighting for them when necessary, thus often developing a tough warrior stock, like the ancient Israelites. When these two elements of stock-breeding and cultivation were fused together, when man learned to settle in a spot where crops could be grown, then we get the beginning of true farming as it came to Europe thousands of years B.C. Man ceased to be merely a wandering herdsman and turned his energies to the labour of serious cultivation, while his wife took up her realm indoors and presently developed the arts of spinning, weaving and pottery making.[1]

So in course of time, about 2800-2500 B.C. these men of the stock that peopled the Mediterranean shores wandered up to the English Channel, bringing with them their domestic animals and their grain. They are described as small, dark men with long-shaped skulls, oval faces and aquiline noses, of the average height of 5 ft. 6 in. The watery obstacle to their advance would have been far narrower and shallower than at present, and when the family and the smaller stock were embarked in a long hollowed-out tree trunk, or a coracle of hide-covered wickerwork, it should have been possible on a calm day to swim the larger beasts across. To cross the Straits, even though narrow, in a log canoe burdened with struggling and

[1] D. Jerrold, op. cit., 42, 43.

protesting animals would seem to us a most hazardous proceeding. Possibly the immigrant selected his stock for the voyage
on the same principle as Noah, a pair of each sort, of not too
unruly deportment, and securely roped with thongs or twisted
fibres. But the later comers of this period were much more
versed in seamanship, sailing up from the Mediterranean to
the south-west of Britain in what must have been sizable ships.[1]

The first comers, now referred to as the Windmill Hill folk,
from the settlement of that name near Avebury, with its
characteristic pottery, found the climate of Britain far more
humid than at present, and consequently the land was covered
very largely by dense forests and wide stretches of morass. In
this wild and untamed land therefore the first farmers with their
inadequate tools had to choose for their settlements the spots
that Nature had left fairly clear. They needed dry uplands
with pasture for their little flocks and herds, and sheltered
slopes for their homes, not too far from the water in the valley
and the food and fuel of the woods. Hence their traces are
found on moors, downs and wolds, in groups of pit-dwellings,
clusters of little round stone huts, or sometimes in convenient
caves. The farmer's first house was doubtless a cave but the
first he made for himself seems to have been a pit in the ground,
an inconspicuous form of habitation which did not catch the
eye of an enemy. Some of these underground dwellings were
quite elaborate structures and remained in use for long ages
in the more troubled parts of the country. They were
approached by a sloping trench, sometimes roughly paved and
lined with stones, roofed with slabs, and terminating in a round
chamber, seven to ten feet deep, which might be connected by
short passages with others. In the middle of the chamber
there was usually a fireplace, with a hole in the roof above to
let out the smoke, which could be closed at will. Any sudden
warning of an enemy's approach must have caused an inconvenient thickening of the atmosphere, but these dwellings would
at any rate have the advantage of being warm in winter and
cool in summer. In the Isle of Portland subterranean beehive
huts have been found, roofed with overlapping stones and
having a hole about sixteen inches in diameter at the top. These
apartments are about eight to ten feet high and ten to twelve
feet across.[2] In the opinion of some such rooms were only used

---

[1] G. Clarke: *Prehistoric England*, 78. D. Jerrold, op. cit , 62.

[2] Addy: *Evolution of the English House*, 12. B. C. A. Windle: *Remains
of Prehistoric Age in England*, 257, 267.

as storehouses, but they may well have been inhabited in times of stress.

But the more courageous soon worked their way towards the surface. They still preferred to dig to some depth before beginning their walls, and these little houses are known as "pit-dwellings." A round or occasionally oval hole was excavated, perhaps three feet deep or more, and varying from six to twenty-five feet in diameter, and dry stone walls were built round it, four to six feet thick. The entrance, low and narrow, often had upright stone doorposts and another stone for lintel. Inside, in the single room, there was usually a low dais at one side for use as a seat by day or a bed by night. There was also a sunk hearth for a fire, but the cooking was more often done in a hole outside, at any rate in dry weather. Most of the huts would have a tree trunk set up in the centre to support the roof, probably formed of large boughs meeting in the middle, filled in with interlaced smaller branches and covered with turf or heather. Others, especially in the stony districts of Cornwall and Dartmoor, were beehive-shaped with layers of stones overlapping towards the top, and almost completely meeting except for a smoke-hole at the apex.

The simple type of round hut or "pit-dwelling" remained in use for many centuries, and was found up to Roman times. They were characteristic of the very early Neolithic culture of the so-called "Peterborough" folk, found in the eastern part of Britain, a people now thought to be survivors of the Mesolithic period who had absorbed the new Neolithic culture, brought over to them probably by immigrants from the Baltic.[1]

Another equally early type of dwelling, used by the contemporary people of the Windmill Hill culture in the West, was roughly rectangular, though broader at one end than the other, sometimes as much as twenty feet long and probably gabled. These huts, the first examples of the angular building, had a basin-shaped cooking-place of baked clay hollowed out in one corner.[2] The settlement at Skara Brae in the Orkneys, characteristically Stone Age though actually later in date, had these long houses, with closets in the thickness of the walls, furniture made of sheets of flagstone, wall shelves, shelved dressers, beds with heather mattresses and skin canopies, and a central hearth. They also had covered passages from hut to hut and an efficient sewerage system.[3] The Neolithic farmer

[1] D. Jerrold, op. cit., 65.
[2] G. Clarke, op. cit., 29, 31.
[3] V. Gordon Childe, *Prehistoric Communities of the British Isles*, 84, 85.

B

had thus acquired a home for himself, but there is no indication that he had as yet any outbuildings for his beasts, and his barns and storehouses, if any, continued for many centuries to be underground pits. The smaller animals such as pigs and calves may of course have shared his hut, but for the most part they were probably penned within a stone wall or earthen rampart at nights to protect them from the wolves. On the bare uplands, where the cattle and sheep ranged, these men had strongholds into which to drive the animals in time of danger, or to round them up before the winter, and perhaps as a retreat for themselves. Usually there was a dewpond in one corner. Living on the heights as they mostly did, the water supply must have needed thought. They might be able in normal times to drive their animals down to the nearest stream to drink, but on chalk Downs these are not very frequent, though the water-level then was higher than now. The farmer had no buckets in which to carry the water home, though he may have used water-skins. Probably he had little use for water domestically except for drinking; skin clothes would not require laundry work, nor was the scrubbing brush invented. The "dew-pond on the height" was probably his main standby, and this was made by digging a saucerlike depression in the chalk and lining it with straw covered by a layer of clay, thus cutting off the heat of the earth and causing water to condense in the hollow at night. It was protected round the edge with a chalk rim and a layer of loose flints was put in the bottom.

The first British farmer was, as we have seen, fairly well stocked, though as yet without the resources of the poultry-yard or the assistance of a cat to guard his grain from rats and mice.

The earliest *dog* whose bones have been discovered, belonging to the Windmill Hill folk, seems to have been not unlike a large fox-terrier, long-legged, short-backed, with a small head and a particularly wide chest.[1] But larger dogs were known on the Continent at this time, notably a breed on the Baltic coast. descended from the wolf, and the ancestor of later mastiffs and shepherd dogs, and it is very likely that these were brought over to the East Coast of England with other wares to the men of the Peterborough culture and thus introduced to the country. These great shepherd dogs were needed to guard the flocks and herds from robbers and wild beasts, and to watch over the pigs venturing into the forest for acorns and beechmast in

[1] G. Clarke: *Prehistoric England*, 23.

Fig. 1.—Soay Sheep (*Photo by O. G. S. Crawford, C.B.E., F.B.A.*).

autumn. They must not be confused with the sheep-dog of later and more peaceful times, whose office was merely to take the flocks to and from their pasture.[1]

The *ox* of the Neolithic farmer was a medium-sized, robust animal, a cross between the wild urus then still surviving in Britain, and the little short-horned *Bos longifrons* of the Continent, some of which were presumably brought over by the first comers. The result was an animal smaller than the urus but long-horned and broad of forehead, a breed which is now represented by the rough Highland cattle of Scotland.[2] They were plentiful during this period, the people being more concerned with cattle-breeding than with agriculture, and the ox being the primitive standard of value. How early the ox was used for draught purposes we do not know, but his first use in this respect may well have been to haul away the tree trunks laboriously felled by stone axes and wedges.

The Neolithic *sheep* were small and slender-legged, with long narrow skulls and rather erect horns like a goat, hence they are called goat-horned or turbary sheep. Their tails reached to the hocks. Sheep probably descended from this early type are still found in the Faroes, Orkneys and Shetland, and on Soay in the St. Kilda group of islands. These animals to-day are a dark blackish brown with light underparts and have very short wool. The primitive sheep were not shorn but moulted naturally in large blanket-like masses during the summer, when they were rounded up and plucked by hand. The sheep with heavy fleeces needing to be shorn were a later development.[3] However in Neolithic times *goats* were far more plentiful than sheep. The use of wool was little understood, goats gave more milk and were hardier and could pick up a livelihood more easily in difficult times. Indeed this must have been the heyday of the goat in England. Goats of a similar species to the Neolithic are still to be found in Crete.[4]

The early *pig* seems to have been a species domesticated from the European wild boar, and would be a fair-sized, rather ferocious animal. The similarity between the wild boar and the domestic pig is greatest during this period, the difference becoming more marked later, when they were crossed with the smaller turbary pig such as were found in the Swiss Lake

[1] *Antiquity*, vi. 24, pp. 415-16.
[2] G. Clarke, op. cit., 22.  E. C. Curwen: *Plough and Pasture*, 33-35.  V Gordon Childe, op. cit., 34.
[3] *Antiquity*, x. 38, pp. 203-5.  J. M. Tyler: *New Stone Age in Northern Europe*, 78.
[4] *Antiquity*, xii. 46, p. 142.

Dwellings.[1]  Even as early as this they were much valued for food, and on their woodland diet would be able to get through the winter better than the rest of the stock.  But all the poor beasts must have fared badly in winter, for the art of making and storing hay was as yet unknown, and the supply of straw from the few scrops would be very scanty.  This and the meagre winter pasturage, with acorns, nuts and browse from the forest, would be all they could get.  If penned within the village stockade their litter might be of rushes, bracken or brushwood. Poor feeding combined with much milking would tend to keep the already undersized stock small.

Whether the Neolithic farmer was provided with *horses* is a matter of doubt, but it seems very improbable, and some archæologists consider that they were not domesticated in Europe until the Bronze Age.  Prof. Boyd Dawkins, however, records the finding of bones of the domestic as well as the wild species in British Neolithic deposits, and they were found in the Peterborough settlement, very early in the period.[2]  But

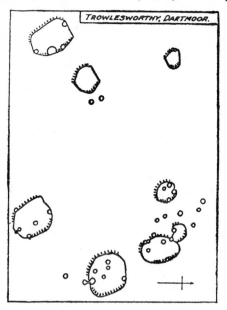

Fig. 2a.—Ancient Corn Plots
*The smaller circles represent huts.*

[1] Ibid.

[2] G. Clarke, op. cit., 22.  *Antiquity*, xii. 46, p. 143.  W. Boyd Dawkins: *Early Man in Britain*, 261-262.

in any case it would be many centuries before these rare animals came to be used on the farm.

*Agriculture* was of necessity still in its infancy throughout this period, for early farming was much handicapped by lack of implements; indeed it is difficult to realize how it was carried on at all without the help of any form of metal. Yet we have reason to believe that little plots of corn were grown, irregular in shape, since the idea of a straight furrow had not yet occurred to the farmer or his wife. They were roughly oval or circular enclosures, about half an acre in average size, surrounding a group of huts, the stones cleared from the surface being collected and heaped round the edge. A slight lynchet formation can sometimes be seen on sloping ground.[1]

The natural course would be to choose a favourable spot that was not too steep and did not require too much clearing, burn off the rough growth and dig up the ground as best they could. The greater dampness of the climate might help them by making the ground softer, and the ashes of the burnt surface growth would fertilize the soil to a small extent.

For *crops* we know the Neolithic farmer grew two sorts of wheat, the short-eared club-wheat and the variety known as emmer, and a certain amount of barley, mainly the naked sort.[2]

For *tools* his only resources were stone, wood and bone or horn. His most characteristic implement was the celt or axe-

FIG. 2b.—JADEITE CELT (Canterbury)

head, made usually of flint, but sometimes of other hard stone, and fashioned in three stages. The celt was first blocked out in the rough, and then chipped to a finer shape and edge, and finally polished, either along the cutting edge only or over the entire surface if time, patience and the desire for an elegant implement prompted.

[1] E. C. Curwen: *Plough and Pasture*, 54.
[2] E. C. Curwen, op. cit., 46, 48. G. Clarke, op. cit., 21.

The form of the axe-head showed a certain development as time went on.  At first it was almond-shaped, sharpened to a blunt edge all round, and with the broad end polished to a cutting edge either straight or rounded, the pointed end being driven through a hole in the wooden haft.  The next stage shows the axe-head longer, narrower and flatter but still chipped to an edge along the sides, a cross-section giving a parallelogram with slightly rounded ends.  Lastly it became like a thick chisel blade, with straight, squared sides, giving a complete parallelogram in section.

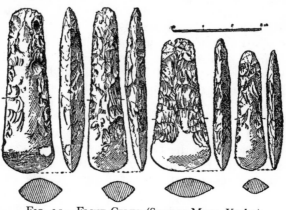

FIG. 3a.—FLINT CELTS (Seamer Moor, Yorks.)

The method of fitting these axe-heads into a handle was to insert them at right angles through a hole in a stout wooden haft.  The idea of boring a hole for the handle through the axe-head itself, i.e., a broad axe-head or "hammer-axe," did not evolve until very late in the period, or perhaps not until after the introduction of metal.  The boring of the hole weakened the stone and would necessarily be a work of much labour and patience.

FIG. 3b.—METHOD OF FITTING CELT TO HANDLE (Solway Moss)

Besides the ever useful axe the Neolithic farmer fashioned from flints "grubbing tools," adzes, borers, drills, wedges, and sometimes even finely cut saws for working wood or bone, and he had also sharp-edged flints of various shapes and sizes which, fitted into handles at an acute angle, could have been used as hoes and picks for breaking the soil. Picks formed from deer

FIG. 4.—ANTLER PICK (Grime's Graves)

antlers we know were used in the flint mines of the period, so there seems no reason why they should not also have been used for agricultural purposes. The blade bones of oxen, mounted as spades, were also used in the mines, and would be useful for shovelling soft soil in the fields.

Another early form of pick, which survived until recent times in Sweden, and which owing to its simple nature was doubtless of universal application, was the "hack." A stout stake of some hard wood was selected, with a projecting branch at a right or slightly acute angle. This was cut short, trimmed and sharpened into a fairly efficient implement.[1] But the most primitive of all tools was the "digging-stick," a sharpened stake with the point hardened in the fire, and sometimes weighted

FIG. 5.—SWEDISH WOODEN HACK

with a stone, perforated and fitted on. It would be used first to grub up edible roots of plants, and hence by a natural transition to prepare the ground for their sowing.

This stake would undergo various developments; the end could be made broader, flattened, and slightly hollowed, and

[1] E. B. Tylor: *Journal of Anthropological Institute*, x. 76.

used as a rude spade for already broken ground; or it could have a flint point bound to it to give it greater efficacy. Possibly even some enterprising spirit thought of pushing it in front of him in a slanting position, with a thong tied round the middle of it for his wife to pull, going before it, thus forming a rudimentary plough. Flint implements large enough to be used as rude

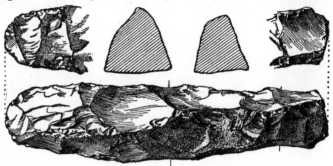

FIG. 6.—"THAMES PICK"

plough-points have been found.[1] But with the development of that most fundamental instrument of agriculture we must deal later, its day was not yet.

The ground thus scratched, sown, and fenced in some way against the incursions of beasts, the harvest had ultimately to be dealt with. The imagination quails before the thought of reaping an entire field of corn with a sharpened flint, yet somehow it was done. A delicately chipped flint sickle, about ten inches long, shaped like a knife-blade with a curved and taper-

FIG. 7.—FLINT SICKLE (Grovehurst, Kent)

ing point, is preserved in the British Museum, and implements such as this wielded by practised hands doubtless accomplished much. Later in the period serrated flint flakes were set in a sickle-shaped wooden frame and used for reaping.

The ears of corn, carefully collected and dried, were stored in some way, probably in a dry pit or storehouse, until needed.

[1] W. G. Smith: *Journal of Anthropological Institute*, x. 84.

They could be threshed in a rough and ready manner by beating with sticks or rubbing between the hands, and the grain was then either parched, or bruised and pounded into meal with stones by the women of the household. The earliest method of dealing with it was to pound it with a round stone crusher in the same way that nuts and roots had been pounded in still earlier times. But as time went on various improvements were devised. The corn-crushers were used, either upon the flat surface of a rock, or in a suitable hollow in the face of it, if one were available. In any case the hollow would come in course of time with use, and in this way the idea of a mortar gradually developed. The crusher was at first globular, and then became flattened on one side, and later was elongated to fit into the mortar, until a pear-shaped, rudimentary pestle was evolved.

When some woman began to use the stone with a circular rubbing motion, instead of merely pounding, the idea of true

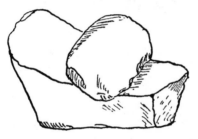

Fig. 8.—Saddle-stone (from *Glastonbury Lake Village*)

grinding began to arise, and a rude saucer-quern was formed.[1] Later still, rubbing the crusher backwards and forwards in a shallow hollow she came upon the idea of the saddle-quern, the primitive form of mill which, side by side with the mortar, remained exclusively in use almost until the Christian era. The saddle-stone was so called from the resemblance of its concave surface to the seat of a saddle. The woman working it knelt behind it, pushing the stone forwards with both hands, hence the Roman name "mola trusatilis," the thrusting mill.[2] This particularly monotonous work was always left to the women and slaves of the household. The coarse meal thus produced was made into a sort of porridge, or kneaded into little flat

[1] V. Gordon Childe, op. cit., 35.

[2] Hence also the Biblical expression "the maidservant that is *behind* the mill" (Exodus xi. 5). The Hebrews used mortars also, for manna was "ground in mills or beaten in a mortar" (Num. xi. 8). R. Bennett and T. Elton: *History of Corn Milling*, i. 8-67.

unleavened loaves and baked on the hot stones, to form what must have been a rather luxurious article of food. More usually the farmer drank milk and ate the various forms of meat prepared by his womenfolk, which included besides game—venison, wild boar, beavers, rabbits, hares and the like—all the domestic animals indiscriminately from horseflesh to superfluous puppies, and any of the stock, including dogs in time of scarcity, which were past usefulness.

A common method of cooking was to put the flesh into a pit in the ground and cook it by means of red-hot stones or charcoal. It could also be put into rough clay pots of water and boiled by dropping in hot stones, as the pottery was not sufficiently well fired to stand putting on the fire; or again it could be roasted or broiled in small pieces over the open fire, or seethed in a bag formed from the hide. Fire would be produced with flint and iron pyrites, or by a fire drill-stick.

For light we know that the flint miners had little hollowed-out chalk lamps, presumably burning fat with a rush or flax wick, and these may have been used in dwellings, but as fat in those lean times cannot have been very plentiful, artificial light other than firelight would be rare.

To the farmer's wife probably fell much of the work of agriculture, since it is thought to have been her invention; but besides this she was very likely the first to fashion rude pots of clay and bake them at the fire. At first she imitated the receptacles of other materials which she already had; thus the housewife of the Windmill Hill culture made her pots like their leather bags, round containers stiffened by withy hoops at shoulder and lip. The wife of the Peterborough district on the other hand made hers like baskets.[1] She ornamented them with patterns incised with the fingernail, but oddly enough it did not occur to her to flatten them at the bottom so that they would stand securely. Little spoons of pottery were also made, and there were hollowed out bowls and scoops of wood.

It seems unlikely that the Neolithic farmer's wife knew the art of spinning and weaving, but possibly she was able to make some sort of felt from the hair of her goats.[2] Their ordinary clothes however were made of skins, cleaned with flint scrapers, and sewn together. No doubt she learnt to do this skilfully, with a bone needle threaded with fibres or thin strips of hide, and sometimes fastening them with buttons of bone or horn.

[1] G. Clarke, op. cit., 47. V. Gordon Childe, op. cit., 36.
[2] Déchelette: *Manuel d'Archéologie Préhistorique*, 334, 579. J. M. Tyler, op. cit., 84

Here then are the beginnings of the farm-yard as far as we can discover them. The farm-house at best a round or oblong hut, the flocks and herds of little beasts with scant shelter or foddering, no barn or rick-yard beyond an underground pit or a little pile of straw, and no domestic fowls. Yet here is indomitable man making the very stones scratch him a hard-won sustenance from their parent soil, wresting a little homestead from the waste and guarding it as best he can from the wild life of the forest. Here is the root of the matter firmly planted; we have only to watch it grow and spread.

# CHAPTER II

## THE BRONZE AGE

DURING the next 1400 years, from approximately 1900 to 500 B.C., there came to Britain from the Continent many successive waves of invaders all of whom had learnt the use of bronze (a mixture of copper and tin) and brought that art with them, together with many other advantages, as their civilization developed.

To the little dark Neolithic farmer, laboriously chipping his way through life with a sharpened flint, arrived about 1900 B.C. a fresh race of men, some crossing from Brittany to Wessex and others from Holland to the East Coast. They were a taller and more muscular folk of Alpine stock, of fairer complexion and rather rugged features, and were distinguished by their comparatively broad skulls, which gave them a rounder shape of head than the long-skulled Neolithic inhabitants. By a curious coincidence too they preferred to bury their dead for the most part in round barrows, instead of the long barrows of their longer-skulled predecessors. They were apparently descendants of the Mesolithic people of Spain, who had acquired the art of agriculture, the use of metal and the habit of trade. They were known as the "Beaker folk," from the shape of their characteristic pottery.

Implements made of copper came into use on the mainland of Europe during the second millennium B.C., but it is doubtful whether Britain ever had a true Copper Age, for most of the copper ore from British mines, notably those of Cornwall, contained a small percentage of tin, thus forming naturally a poor kind of bronze, before that composite metal was intentionally manufactured. The introduction of true bronze (10 per cent. tin to 90 per cent. copper) now took place, and successive improvements in the implements made from this valuable metal were evolved, or brought over by fresh arrivals from time to time. The tall, round-skulled newcomers of the second millennium successfully subdued the Neolithic inhabitants and amalgamated with them, thus producing a considerable variety of types in colouring and shape of skull, but the small, dark earlier race is thought to have long survived in the Silures of South Wales, and perhaps in the tributary villages of "taeogs" mentioned in the Welsh Laws. The invaders, like their predecessors, at first naturally preferred the open uplands, unencumbered by forests and morass, for even after the precious bronze came into the hands of their chiefs, it might be some time before it was common enough to be used habitually for wood-felling and clearing. Indeed no great amount of forest clearing was attempted before the time of the Romans. Bronze implements once established brought with them many possibilities of enlargement in man's horizon, but meanwhile habit was established and it is in such places as the wolds of Yorkshire and the Midlands, the moorlands of Devon and Cornwall and the downs of the South Country that most remains of the Bronze Age are found, although after the coming of the Celts in the first millennium B.C. valley sites began to become more popular.

The "Beaker-folk" were joined a century later by a more warlike company of people known as the "Beaker-battleaxe folk," a fusion of Alpine and Nordic races, crossing from the Rhineland about 1800 B.C. and from Brittany about 1700 B.C. All these men of the early Bronze Age were pastoral rather than agricultural people, paying more attention to their animals than to their crops. After them, in the Middle Bronze Age (1400 to 1000 B.C.), came a time of great barbaric chieftains, worshipping in the vast temples of Stonehenge and Avebury and having their tombs in great burial mounds along the Wessex heights. This was a time of great trading activity in gold, bronze and amber, but in spite of fresh invasions from the Continent little progress was made in agriculture until the last millennium B.C. But with the coming of the first wave of the Celtic peoples about 750 B.C. (known as the Urnfield culture, from their urn burials) a real change took place, for these folk brought not only the "leaf-shaped sword" but the first light plough, thus creating something like a revolution in agriculture and laying the foundation of farming as we know it.[1]

Such is the rough outline of the period; we must now consider in detail the gradual developments on the farm.

Little change took place in the dwelling-house. Caves and pit dwellings were still inhabited for various reasons by those who desired to be inconspicuous rather than comfortable, but the normal type of dwelling was the round hut. These, adopted from the earlier Neolithic men and improved upon, were now on the whole larger and better built. The lower walls were sometimes as much as three feet thick, and occasionally such refinements as special rooms partitioned off by slabs of

[1] D. Jerrold: *Introduction to the History of England*, 69-75.

stone, or paved floors are to be found. These huts were usually about 15 to 20 feet in diameter, having two large upright stones as doorposts, the approach and entrance often roughly paved with stones to prevent a morass forming in the doorway—an early beginning of the flagged path and stone doorstep of the later farm-houses so familiar to us. Some of the smaller huts

Fig. 9.—Hut Circle, with Compartments (Ty Mawr, Holyhead)

were roofed beehive-fashion with overlapping stones, but a roughly thatched or turfed roof supported by a central pole, as before, was probably more usual. These clusters of little dwellings, ranging in number from five or more to occasionally as many as fifty, were generally situated in positions sheltered from the north-west winds, and were sometimes surrounded by thick stone walls. Some houses, as at Grimspound, developed a sort of porch, with the outer door at right angles to the inner, which would certainly add to the comfort of the inhabitants, and might serve to shelter an animal or two at night. This

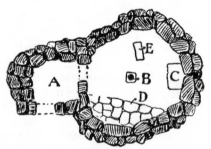

Fig. 10.—Plan of Hut, Grimspound, Dartmoor. (a) Porch. (b) Stone for Central Post. (c) Hearth. (d) Raised Dais. (e) Cooking Hole.

particular group of twenty huts was surrounded by a wall 5½ feet high and 10 feet thick.[1]   But as these Bronze Age men were

[1] Ibid., 76. T. Rice Holmes: *Ancient Britain*, 154.

mainly a pastoral people many would probably live in bothies built of boughs, and tents of skins, of which no trace would remain.

The farmer's *stock* during this period did not change much more than his house. As more pasture was cleared by the bronze axe the condition of his animals would tend to improve, and more cultivation would mean more straw. Even the bronze dagger and spear-head, though not at first sight connected with agriculture, would be of the utmost value to the farmer in protecting his flocks and herds from the ravages of wolves and other wild beasts.

His *oxen* remained the same mixed breed until late in the period, after 1000 B.C., when the Celts brought the small short-

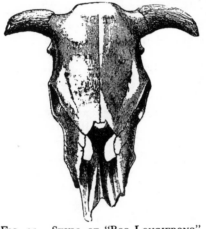

FIG. 11.—SKULL OF "BOS LONGIFRONS"

horned *Bos longifrons* with them in larger numbers. This animal had a small broad skull and short, stout downward curving horns, its characteristic being the length of forehead from the orbits to the top of the skull. This breed superseded the long-horned crossbreed and remained the normal variety until the coming of the Romans.[1]

The same type of small, goat-horned *sheep* continued, but as conditions improved they tended to supersede the hardier goat on account of the greater usefulness of their wool, for it was during this period that the art of weaving was developed. The wool was still plucked off by hand, or the operation sometimes assisted by driving the sheep through a narrow passage,

[1] E. C. Curwen: *Plough and Pasture*, 35.

presumably between bushes, so that much of the wool was detached or at least loosened.[1]

The climate had become warm and dry, and some writers think that herds of cattle and sheep were run on ranches, bounded by long banks and ditches, along the high ground in Wessex. Small pear-shaped bronze bells have been found, which probably adorned the necks of the leaders of the herds, and thus the first tinkle of sheep-bells on the downs may date from this time.

*Swine* underwent some alteration by the introduction of the smaller long-legged turbary pigs from the Continent, well known to the Swiss Lake-dwellers, and which were crossed with the type domesticated from the wild boar. They were perhaps brought to England in large numbers by the Celts, the later comers in the Bronze Age. A curious legend of the introduction of some new kind of pig is preserved in the *Mabinogion* and possibly it might not be too great a flight of the imagination to suppose that it had its foundation in this event. The theft of these animals from Pryderi by Gwydion ap Don is represented as the cause of much envy and strife. "Lord," says Gwydion in describing them, "I have heard there have come to the south some beasts, such as were never known in this island before . . . they are small animals, and their flesh is better than the flesh of oxen . . . and they change their names. Swine are they now called . . . and still they keep that name, half hog, half pig."[2] Lord Avebury considers that the Bronze Age pigs although slender in build were about the same size as at present.

The *horse* was certainly known in Britain during this time, but was scarce until the late Bronze Age, when it became important with the evolution of wheeled vehicles and particularly chariots, but although sometimes an article of food its occupation was mainly warfare, and not the humble uses of agriculture.[3]

This Age was naturally one of great advance in implements. At first the precious metal was used only for the weapons of the chiefs, but as it became more plentiful it quickly spread downwards, and even though many stone implements were used contemporaneously with bronze to the very end of the

[1] Ibid., 32. W. Youatt: *The Sheep*, 33. V. G. Childe: *Prehistoric Communities of British Isles*, 182.

[2] *Mabinogion*, Part IV: "Math, the Son of Mathonwy." See Magnus Maclean: *Literature of the Celts*, 229 n.

[3] D. Jerrold, op. cit., 77.

period, the metal tools which came into general use, especially
during the last millennium B.C., worked a revolution when
applied to agriculture.

The coming of the bronze axe or celt marks a great change,
and this implement shows three main stages of development,
with of course many intermediate varieties. At first the wedge-
shaped stone axe, driven at right angles through its handle,
was merely imitated in copper or poor bronze, but by the time
true bronze appeared the hammering of the cutting edge had
produced a fan-shaped expansion of the blade at that end. It
was soon found more convenient to fit these celts into a cleft
stick by way of handle, and this was obtained by cutting a
branch from a tree with part of the trunk still attached, thus

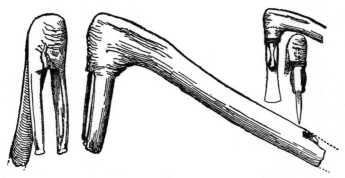

FIG. 12.—METHOD OF FITTING CELT TO HANDLE

giving a wide V-shaped piece of wood, the shorter limb of which
was cleft for the insertion of the axe-head, which was then fixed
securely by lashings, probably of raw hide. To make the celt
fit more securely slight ridges or flanges came to be hammered
up at each side, and next a stop ridge was made about half-way
up to prevent the axe-head from splitting the cleft handle when
in use. Hence the development of the *palstave*, about the
middle of the Bronze Age.

The next stage was to hammer over the wings or flanges
on each side to hold the two prongs of the cleft handle, and
from this was developed the *socketed celt*, with a single cavity
into which the handle was inserted without being cleft. A
bronze loop was made at the side of the celt by which to bind
it to the haft. This latest form of celt was brought by the
Urnfield folk about 750 B.C. together with chisels, gouges and
winged axes. The idea of a transverse socket, through which

C

the handle could be passed at right angles to the blade, as in modern axes, did not occur to them, although horn and stone axe-hammers were pierced in this way at the very beginning of the period.

Other implements followed the axe, notably a sickle. This was at first a crescent-shaped piece of bronze, fastened to the

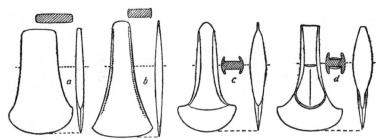

FIG. 13*a*.—STAGE 1.   DEVELOPMENT OF BRONZE CELT

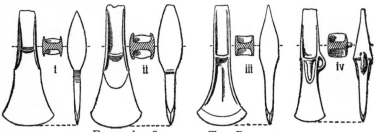

FIG. 13*b*.—STAGE 2.   THE PALSTAVE

FIG. 13*c*.—STAGE 3.   THE SOCKETED CELT

handle by rivets, but this form was little used by the Britons, who presently invented a socketed sickle, peculiar to Britain and Northern France.[1]   This tool must have been of inestimable value at harvest-time to the proud farmer who owned one.

Various forms of hoe and digging-stick were doubtless used now shod with bronze, and quite possibly some form of hand-

[1] G. Clarke: *Prehistoric England*, 22.

plough resembling the caschrom described in the next chapter may have been evolved, but at last, about 750 B.C. the Celts, a race of peasant agriculturists from the Lower Rhine, brought to Britain that triumph of the late Bronze Age, the light bronze-tipped plough, drawn by oxen.

FIG. 14.—BRONZE SICKLES

One of the earliest of British ploughshares has been found in Holderness, a beak of bronze about six inches long, with an oval socket to receive the end of the wooden plough-beam, and a rivet-hole on each side.[1] Various legends attend the coming of this all-important implement. Ancient Welsh mythology

FIG. 15.—BRONZE IMPLEMENT, PROBABLY A PLOUGHSHARE, EXCAVATED AT HOLDERNESS

states that the ancestors of the Cymri were instructed in agriculture by Hu Gadarn—Hu the Mighty—who was the first to draw a furrow with a plough on British soil. If this chieftain, known as one of the three national pillars of the Isle of Britain, ever existed in the realm of fact it is of course quite possible that he imported a bronze ploughshare, and the substratum of truth is that the Celts of the late Bronze Age did actually introduce it.

[1] *The Naturalist* (Huddersfield), May 1917, p. 157.

One theory of the origin of the plough is that it was evolved from the digging stick, held in a slanting position and pulled by a thong fastened to the middle. The cord would in time be replaced by a pole or beam, and a yoke attached to the end of it for the oxen. Another view is that it evolved from the hoe. A heavy bronze hoe dragged backwards through the ground would make a continuous light furrow, and the convenience of having it pulled in some way, instead of perpetually walking backwards, must soon have been obvious. This simple hoe-like form, larger and heavier in construction, occurs in ancient Egyptian, Syracusan and Etruscan drawings, and was known in Sweden. For a long time it was pulled through the ground by man himself, or more probably by his womenfolk,

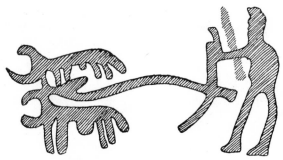

FIG. 16.—PRIMITIVE PLOUGH

but eventually by his cattle. As Tyler justly observes: "The first man who made his cow instead of his wife draw the plough was a great benefactor."[1]

But the need for additional weight and means of guidance made itself felt, and hence the plough came to be made in two pieces—the forward pointing pole ending in the ploughshare was continued backwards and upwards to form the handle or "plough-stilt," and a cross-piece was fastened to this horizontally for purposes of pulling, and was known as the "plough-crook." The ploughman, besides putting his weight on the handle, could press with his foot on the projecting end of the cross-pole when necessary, though it must have been an awkward proceeding to anyone not possessed of three feet. This was known as the "over-treading" plough, and was still found in England in Romano-British times. Thus we arrive at the simple form of plough drawn by a single yoke of oxen shown

[1] E. B. Tylor: "On Origin of Plough and Wheel Carriage," *Journal of the Anthropological Institute*, x  77 on.

on the early Scandinavian rock carvings of the Bronze Age. Here we have the advent of the plough-ox, that useful beast which for so many centuries was to be the central figure of the farm-yard economy. The man who discovered that the fierce bull could be converted by emasculation into the patient and tractable ox was the bringer of a great gift to mankind.[1]

A rude form of cart also began to make its appearance, and we must consider how man arrived at this. The first form of vehicle is thought to have been a sort of rough sledge. This idea might easily occur to a man using his oxen to haul away the tree trunks he had laboriously felled, and it is possible that the Neolithic farmer had advanced to using a primitive sledge. The next step forward was made when a weight was encountered too heavy to be easily moved, and rollers were introduced underneath, thus giving rise to the idea of wheels. Rollers, however, were not easily attached and were at best clumsy and awkward, and the middle of the trunk was therefore whittled down to the thickness of a pole, leaving a rough disc at each end. Here we have the primitive wheels and axle all in one piece. Stops were then fixed under the body of the sledge to keep the axle from shifting, but otherwise leaving it quite loose. But the advantage of larger wheels soon became apparent, and large discs were then cut from a tree trunk and connected by a pole which passed through a hole in the middle of each and was fixed by a pin or nut, so that the whole turned together. This was the form used in the primitive ox-cart, drawn by a pair of oxen with pole and yoke. The next idea was to lighten these solid wooden wheels by cutting out four large holes, and if the arms of the cross thus formed proved too weak they were strengthened by lashing on spokes. Presently these spokes took the place of the arms of the cross and were set strongly in a rim, as at present, and increased to six or eight. The rim could be enclosed in a rawhide tyre which would shrink tight as it dried. Another variety, instead of cutting out a complete rim, was to make a rough wheel of several pieces of wood nailed together, which was less likely to split. This form survived in ancient Roman farm carts.

The body of the cart or sledge probably began as a hide stretched between poles, and this would be replaced by rawhide laced across in strips, and later by a wooden bottom. A simple early form was the V-shaped cart, two poles being fastened together at one end, and held apart at the other end

[1] J. M. Tyler: *New Stone Age*, 111.

by a cross-beam. Oxen could easily be yoked to the pointed end, and a pair of wheels attached underneath. Such a cart, with side hurdles, is seen in the Saxon Calendar (see Fig. 33). On a square frame a dashboard of wattle in front would form a kind of chariot, or hurdles at the side would bring it very near to the mediaeval farm cart.[1] It is impossible to say how far the Bronze Age farmer had advanced in this direction, but there is no reason why he should not have got as far as this, and waggons are thought to have been in use before the beginning of the Iron Age.

All early forms of cart survived side by side; indeed sledges were still habitually used on the rough land of Dartmoor farms, and throughout Devon, as late as the eighteenth century, wheeled vehicles being unknown there until about 1750.

With regard to agriculture, it is easy to imagine what strides the farmer could now make with a gleaming bronze plough-point to break the stubborn soil, drawn by a yoke of oxen, and with a sharp-edged bronze sickle to reap the fruits of his labour. The light plough brought by the Celts merely dug a straight furrow in the soil, without turning the sod over, no mould-board being yet attached, and it was therefore necessary to cross-plough the plot of ground at right angles, hence the squarish plots of what is called the Celtic Field system, now revealed in great numbers by air photography. It was still in the main the bare uplands which were inhabited and cultivated, and a glance at a contour map of England will show how all the main ridges of high ground—the South Downs, North and Hampshire Downs, and southwards the Dorset Downs, the line of the Chilterns and their eastward extensions, and the Cots-wolds connecting with the Northampton heights and up to Lincoln Edge—all converge on Salisbury Plain. It is difficult to realize that this wide, open stretch of sparsely inhabited country was once the hub of British civilization, but so it was, from the dawn of our history down to the first century B.C. There were the great temples, and there vast stretches of corn-land, and from that centre the great pre-Roman trackways ran in all directions. Husbandry assumed a real importance and spread and flourished as the population increased.

Of the farmer's *crops* barley seems to have been predomi-nant, mainly the beardless variety, with some of the six-rowed sort; and also the long-eared common wheat, which was grown

[1] E. B. Tylor: *Journal of the Anthropological Institute*, x. 79-81. Roger Pocock: *Horses*, 105-7. *Antiquity*, ix. 34, p. 142.

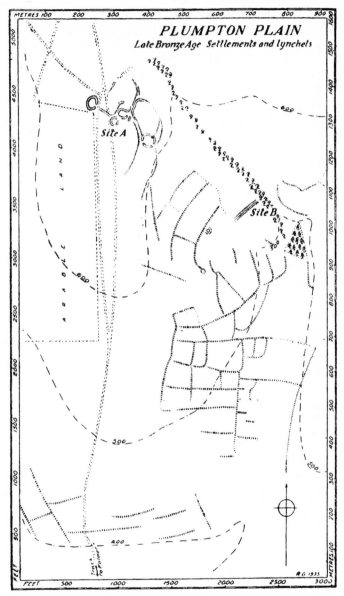

FIG. 17.—CELTIC FIELDS

as far north as Yorkshire. Flax also was cultivated, especially from the middle period onwards.[1]

The idea of manuring the fields by hand, instead of merely grazing the cattle on them, was arising too, for in the Shetland hamlet of Jarlshof, in the late Bronze Age, the cattle stalls were carefully arranged for the collection of dung—the first indication known of such a practice.[2]

Corn was still ground throughout this period on a saddle-stone or in a mortar, as we have described in the last chapter. From all these implements a certain amount of grit wore off and mingled with the meal, a disadvantage which is shown by the particles adhering to the teeth of Bronze Age skeletons, and which implies that the flour was not sieved or bolted in any way. It also suggests incidentally how general a form of food bread, gruel and other preparations of meal had now become. The so-called bread would be flat unleavened cakes of wheat- or barley-meal, accompanied by perhaps more "home-killed meat" and less game than formerly. Ponies were still eaten but were not plentiful. Food altogether would be less hardly won and life a little less laborious. The skeletons of this Age show less disparity of stature between men and women, and this is sometimes taken as evidence that the hard lot of the farmer's wife was a little improved. At any rate she may have been better fitted to cope with all that she had to do.

Her cooking was done, as before, sometimes on open fires outside the hut, and sometimes in cooking holes lined with stones. One comparatively luxurious hut has been found on Dartmoor with a separate kitchen just outside the door, provided with a hearth and cooking hole,[3] but this was evidently an unusual and possibly prideful innovation, and was not imitated. The plentiful supply of cooking stones scattered about these settlements shows that these were the most common means of cooking food, and hide-boiling would probably long survive also. One solitary instance of the use of coal as fuel (the earliest in the world) is found during this period, but it was no doubt an extremely unusual practice.[4] But towards the end of the Bronze Age the metal industry was developed to such an extent that it came to be used for domestic and industrial utensils, which were manufactured in Britain and distri-

[1] E. C. Curwen: *Plough and Pasture*, 48. G. Clarke: *Prehistoric England*, 21. T. Rice Holmes: *Ancient Britain*, 151. V. G. Childe, op. cit., 157.
[2] V. G. Childe, op. cit., 182.
[3] V. C. H. Devon, i. 354.
[4] V. G. Childe, op. cit., 157.

buted up and down the land by itinerant traders, so that the
farmer and his wife benefited greatly. Pots and cauldrons
were made, and wooden buckets with hoops of bronze, though
possibly these were only found in the wealthier households.
Since the coming of the Beaker folk the housewives had flat-
bottomed earthenware vessels which would stand steadily, but
they were still made by hand without a wheel, and baked at the
fire—such simple forms as bowls, jars, pipkins, cups and
strainers, the last possibly for wild honey. They attained to
a good deal of skill in the art, ornamenting their pottery as
before with incisions of the finger-nail, or impressions of twisted
fibres, or with a pointed tool.

A lathe for turning wood had been invented early in the
period,[1] and they had ladles of horn.

Spinning with distaff and spindle, and weaving on a simple
loom now became definitely an occupation of the farmer's wife.
Flax was woven into linen, and in addition woollen cloth was
made, especially in the more civilized parts, and sometimes
of fine quality. The Celts brought in improved spindle whorls
and cylindrical loom-weights to hold the weft taut.[2] In the
wild villages of Dartmoor leather scrapers were more plentiful
than spindle-whorls,[3] which shows that skin garments were
still commonly worn there. That useful little article the safety-
pin made its appearance, and also buttons of jet and other
materials.

The chief advance, therefore, that we see in the Bronze Age
farmer is in his implements and equipment, and the consequent
development of agriculture made possible by the application of
metal to the tilling of the soil. His stock remains practically
unchanged, his housing and even his general level of civilization
do not show any great advance, but the coming of the plough
marks a great step forward in the history of mankind.

---

[1] G. Clarke, op. cit., 48.

[2] V. G. Childe, op. cit., 157, 193.

[3] The spindle-whorl was the little wheel or ring, usually of stone, through
the hole in the centre of which the spindle (of wood or bone) was fitted. It
gave a rotatory motion to the spindle, which was twirled between finger or
thumb by the tapering end projecting below the whorl, the thread being
wound round the upper and longer part.

# CHAPTER III

## PART 1.—THE EARLY IRON AGE

WE have seen how the first wave of Celtic folk to reach these shores (the Urnfield people) had laid the foundations of true agriculture by bringing with them the light plough drawn by a pair of oxen. About 500 B.C. they were followed by a second invasion of Celts (formerly known as the Brythonic Celts) who brought with them the use of a new metal of the greatest importance. These men, coming from the Rhone, were familiar with the use of iron, a metal much easier to work and simpler to prepare than bronze, since it needed no alloy. They belonged to what is known on the Continent as the Hallstatt culture (so called after the village in the Austrian Lakes where characteristic discoveries were made), and landing on our south and east coasts they founded what is known here as Iron Age A, spreading their culture as far north as Scarborough and westwards to Exeter. Settling down, they lived in open villages, and sometimes in single isolated farms, but were remarkable for their immense fortifications on the hilltops. The average size of the little single farms seems to have been about fifteen acres, but their vast earthworks, built presumably to preserve themselves and their flocks and herds, covered anything from six to eighty acres. One hundred and fifty or two hundred years later, in the third century B.C., they were joined by a third wave of Celtic invaders, bringing with them the later Iron Age B. These last were a cultured and artistic people, deriving their inspiration from contact with the Grecian cities of the Mediterranean, and evolving from it the best traditions of Celtic art in skilled metal and enamel work. They spread all over England, westward to Cornwall and northwards over Scotland, coming to terms with their predecessors but not exterminating them. They were the people who built the marsh village of Glastonbury, from the excavation of which remains, preserved in the peat, we learn so much of their standard of living and equipment. But during this middle period the centre of commerce and civilization was gradually shifting from the high tableland of Salisbury Plain eastward to the more sheltered Thames Valley, Sussex and Kent.

Lastly in this momentous period came an invasion of the people called Belgae from Picardy; a Celtic folk with a strain of Teutonic blood, not artistic but more developed in the technique of agriculture than their predecessors. Arriving in 75 B.C.—the first definite date in English history—they established themselves first in the south-east, especially in Kent and Hertfordshire, under their chief Cassivellaunus, and were joined by a second influx shortly after the brief visit of Julius Caesar. This second wave settled in Hampshire and Wessex, making their capital at Winchester, and presently destroying the peaceful settlement of Glastonbury. They established themselves in valleys and open spaces beside river crossings, their superior implements enabling them to clear forests and prepare the heavier soils for their great contribution to agriculture, the heavy wheeled plough drawn by a team of oxen.[1]

Thus we shall be able to trace a great advance in many directions. The farmers' *house* began to develop, the arrival of iron implements making the felling and working of timber much easier. At first the round shape of dwelling continued, but larger and with more elaboration. At Little Woodbury a house has been found 50 feet in diameter, with an inner circle of posts, possibly marking the partitioning off of stalls and sleeping places. Smaller ones also have two circles of posts, either as an extra support for the roof, or perhaps to form a sort of clerestory to give extra light and ventilation.[2]

At Glastonbury, in Iron Age B, we find the ancient circular huts constructed of wattle and clay with thatched roofs supported by a central pole, but placed upon an island artificially formed in the middle of a shallow lake, like the crannogs of Ireland. The island was formed of brushwood, stone and timber, laid on the bed of the lake and kept in place by a palisading of piles. A floor of clay was superimposed on this within the huts, the natural subsidence of the understructure necessitating the renewal of the whole floor with its central hearth stone from time to time. A number of these huts have five or six fresh floors, and some as many as nine or ten. The hearths were made in most instances of baked clay but sometimes of stones and occasionally of gravel or marl. The whole village was stockaded round the confines of the island, and the superior defence provided by the surrounding water no doubt made up for the somewhat unstable nature of the foundations.

There were also some rectangular houses showing the advance made in the use of timber. These had ground plates of oak beams with mortise holes to receive the posts of the

[1] D. Jerrold, op. cit., p. 69 et seq.
[2] G. Clarke, op. cit., 39. J. Hawkes: *Early Britain*, 30. Kendrick & Hawkes: *Archaeology in England and Wales*, 153 et seq.

hurdles forming the walls, quite in the mediaeval style, and these again fitted into a wall-plate at the top, forming vertical walls about six feet high.    Door sills were of timber, and there were pathways of rubble from hut to hut.[1]

Another type of rectangular house, which seems to have been developed by the Celts, was the simple booth or bothy, which survived in Wales for many centuries.    It was constructed

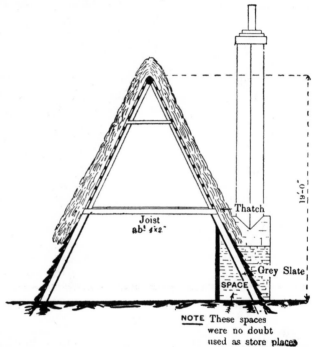

FIG. 18.—BOOTH FORM OF DWELLING, WITH MODERN CHIMNEY

by taking two pairs of young trees, either with a suitable natural bend, or with the branches so lopped as to leave one bending inwards; a pair of these was then set into the ground at each end of the proposed dwelling in the form of a Gothic arch, connected by a ridge pole from point to point and if necessary strengthened by other ties.    These pairs of trees were known later as gavaels, forks or columns, and the house was easily enlarged by adding another pair of trees and ridge pole at either end.    The sides were formed usually of wattle daubed with clay, or, in rather later forms, of planking or turf.    A door-

[1] A. Bulleid and H. St. George Gray: *Glastonbury Lake Village*, i. 14, pp. 55-60.    Hereafter referred to as G.L.V.

way was left at one end, the door being also of wattle in many cases, and there was a hole in the roof for the smoke to escape; rushes were strewn thickly round the walls inside and covered with coarse cloth, and there the family slept, in their day clothes, with their feet to the fire on the central hearth, receiving as Giraldus Cambrensis says, centuries later, "much comfort from the natural heat of the persons lying near them . . . they expose alternately their sides to the cold and to the hardness of the

FIG. 19.—BOOTH FORM OF DWELLING, SHOWING EVOLUTION
OF STRAIGHT WALLS

bed." [1]   This simple house of one "bay" would at first naturally be made of any size that came convenient, but later the standard size of the bay became fixed at sixteen feet between each pair of crucks or forks, a size transferred to the dwelling-house from the ox-house, where this was considered suitable room for four oxen, standing in their stalls as they did under the "long-yoke in the field."

Occasionally during this period there are found curious habitations of a courtyard type; oval enclosures of massive dry masonry, with oval or round chambers built as recesses in the solid surrounding walls. [2]

[1] Addy: *Evolution of the English House*, 17-28, 34, 35.   Anglo-Saxon gaefles=forks, hence gable. G. Cambrensis: *Description of Wales* (Everyman's Library), 170.

[2] Kendrick & Hawkes: *Archaeology in England and Wales*, 177.

As regards the farmer's *stock*, there was little change until the coming of the Belgae towards the end of the period.

*Oxen* would probably increase as more land was cleared for cultivation and as the population became more numerous. The little dark Bos longifrons (or Celtic shorthorn) remained the usual breed, and there is a theory that the black breeds of Celtic districts, Galloways, Aberdeen Angus, Kerry, Dexter and Welsh black are descended from this.[1] Bones of long-horned cattle are seldom found among the remains of the Iron Age.[2] Cows among a primitive pastoral people such as the Celts were of the greatest importance, not only on account of their milk, flesh, hides and powers of traction, but as the recognized standard of value. In the absence of coinage (until the coming of the Belgae) the value of everything was still expressed in cows, and the tribesman's wealth was centred in his herds. An awkward form of currency in the shape of long iron bars was introduced in the middle period; they were used at Glastonbury and in other parts of the south and Midlands until superseded by true coinage, but even as late as the sixth century A.D. we find bequests of horses being made of the value of five cows or of three cows, according to their quality. The Brigantes of the north, a people who gave considerable trouble to the Romans, are said to have possessed herds of snow-white cattle, but there is no evidence of larger cattle at Glastonbury. The Belgae of the south-east certainly had plenty of cattle, of whatever sort, at the coming of the Romans, and were skilled stockraisers and dairy-farmers.

Pliny describes how the Gauls, and therefore probably their kinsmen in Britain, dealt with murrain among their cattle and swine. The herdsman had to go fasting and seek the flower of the water pimpernel, and having found it he must pluck it with the left hand and carry it without looking back, straight to the drinking troughs of the cattle and put it therein.

*Swine* were much valued by the Celts and seem to have been plentiful. At Glastonbury all the remains are those of the turbary pig,[3] but the type descended from the wild boar and crosses of the two must have continued in many parts. The result of the crossing of these two strains seems to have been a large and fierce breed daunting to foreigners, for Strabo, writing about 30 A.D., reports that "the swine live abroad and are remarkable for their height, strength and swiftness—indeed it is as

[1] F. H. Garnier: *Cattle of Britain*, 10.
[2] E. C. Curwen: *Plough and Pasture*, 35.
[3] G.L.V., ii. 659.

dangerous for a stranger to approach them as the wolf." The inhabitants of the south fed them in the great forests of the Weald, guarded by fierce dogs.

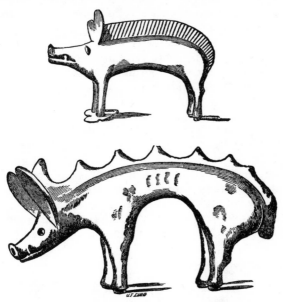

Fig. 19a.—Bronze Boars of Early Iron Age.

*Sheep* in the pre-Roman village of Glastonbury were, with lambs, the most common form of animal food. Three different types are found there: (i) a species with long thin legs and upright goat-like horns, akin to the ancient type of the Swiss Lake Dwellings; (ii) a kind with large horns, more like the Highland sheep and those of St. Kilda; (iii) a few without horns at all. This is the first appearance of the hornless sheep in Britain, and it is thought that they may possibly have been females of the ancient turbary species with the horns reduced.[1]

The *goats* found there were small and short-horned.

Sheep must have been plentiful in the country, for from an early period of the Iron Age the Celts were skilled weavers of tartans and other woollen cloths, which testifies to a good supply of wool.

*Horses* though still small were improving, and a good horse was much valued, some attention evidently being paid to breed-

[1] G.L.V., ii. 657-9.

ing.   Excavations of the early Iron Age show that ornamental horse trappings were much thought of, but judging from the absence of horseshoes among these finds they do not seem to have been shod before the time of the Roman occupation.   In the early Iron Age and at Glastonbury they were from 11½ to 12½ hands high, of the Exmoor pony type, but some with more slender limbs.[1]   It is doubtful how much they were used for farm work, but they were certainly used for food.

*Dogs* early became of importance in Britain, and the Belgae actually exported three kinds of sporting dogs—spaniels, greyhounds and buckhounds.   At Glastonbury the largest dog found was rather less than two feet at the shoulder, while the smallest was about the size of a fox-terrier.[2]

From the Belgae the British farmer at last received his *poultry yard*, for this people introduced fowls and geese shortly before the time of Caesar.   At first it was not considered right to eat them, but they were kept for amusement, probably cock-fighting, and doubtless for their eggs.[3]

The climate during the Iron Age became after about 400 B.C. cold and wet, so that the beasts probably had to be stalled in winter.

FIG. 20.—IRON SOCKETED CELT (Walthamstow, Essex)

The introduction of iron naturally brought about a great improvement in the farmer's *implements*. He now had a slasher or leaf-knife for cutting leaves and grass for fodder, and also a pruning hook and bill hook.   Iron sickles were made with the wings hammered over to form a socket for the handle.   His spade could have an iron tip, probably not unlike an ancient form which long survived in the Highlands of Scotland, a primitive spade of wood broadening out with a notch for the foot on the right side only.   The blade of this was covered with iron about 6 or 8 inches up in the front and 2 or 3 inches up at the back.   His light plough and his hand plough could now

[1] G.L.V., ii. 649-51.   G. Clarke, op. cit., 23.
[2] G.L.V., ii. 660.   G. Clarke, op. cit., 36.
[3] *Caesar's Commentaries*, trans. T. Rice Holmes, 135

have good iron points, and at Glastonbury an oak hand-plough was found, the pole about 5 feet 6 inches long, curving downwards to a wedge-shaped projection 11 inches long.[1] This implement has a strong resemblance to the Highland caschrom, which some consider a step between the spade and the plough. This has a long wooden handle about 2½ feet long, shod at the end with an iron tip. A peg is inserted at the angle for pressure by the foot. The point of this instrument is pushed well under a clod of earth, which is levered up and turned over to the left. It is said to do the work of four spades, and a number of them wielded by gangs of ten men or women turn a very good furrow.[2]

FIG. 21.—OAK HAND PLOUGH OF EARLY IRON AGE
(*Glastonbury Lake Village*)

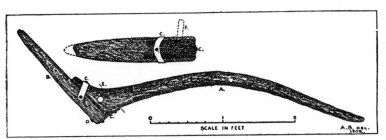

FIG. 22.—WEST HIGHLAND HAND PLOUGH

But with the coming of the Belgae in 75 B.C. and onwards an immense step forward was taken, for these people were provided with a deep-cutting effective plough which really turned the sod over. The wheeled plough, according to Pliny, was invented in Rhaetia "at a comparatively recent period." Instead of being a mere "lever furnished with a pointed beak" it was provided with a coulter, the knife-like blade fastened in front of the share which, in Pliny's words, "traces beforehand by incisions the future furrows which the share is to open out, and

---

[1] G.L.V., i. 348. Iron saws were found at Glastonbury and a high level of turning and joinery was reached. These early saws had the apices of their teeth set in the opposite direction to modern ones, so that wood was cut by drawing the saw *towards* the worker.

[2] G. L. Gomme: *Village Community*, 279-82.

D

cuts up the dense earth before it."[1]    Whether he was right or not about its country of origin the heavy wheeled plough was known in Denmark about 400 B.C., and there the mould-board was strengthened against wear with pebbles inserted into drilled holes in the side.[2]    This much more efficient plough, drawn by several pairs of oxen, made it possible for the British farmer to cultivate the heavier lowland soils, the lighter uplands being left to the users of the light plough, which continued in use side by side with the other.

This was the beginning of the eight-ox plough, which doing away with the necessity of cross-ploughing and the consequent rectangular plots of the Celtic field system, found most convenient a long narrow strip of land on which the cumbrous team did not have to be turned too often.    Thus we have the foundation of the mediaeval land system which we shall presently see developing.

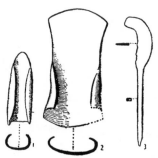

FIG. 23.—1. IRON AGE PLOUGH-SHOE.    2. SHARE FOR HEAVY BELGIC
PLOUGH.    3. COULTER FOR SAME.

Four-wheeled carts were found at Glastonbury, the wheels used having in some instances twelve well-turned spokes, although solid wheels 17¼ inches in diameter are also found there.    They could now be strengthened with iron tyres or have the edges studded with large nails.

*Agriculture* therefore during this period was able to undergo great development as more and more clearing of the land was done.    About 330 B.C. Pytheas reported that abundant wheat was grown in the south-east of Britain, and in Caesar's time much wheat and barley was exported from Kent under the skilled cultivation of the Belgae, who raised plentiful crops from the newly-cleared virgin soil.    The more northern and western

[1] Pliny: *Natural History*, xviii., ch. 48.
[2] E. C. Curwen: *Plough and Pasture*, 67-8.

parts of the island were behindhand in their development, though probably not so much as early Roman writers, largely depending on hearsay, would have us believe.

The farmer's *crops* at this time received important additions, though wheat was the principal crop grown. Oats, originally a weed in summer wheat, were brought over by the Celts in Iron Age B and were suited to this northern climate, then going through a phase of cold and wet. Rye, a weed in wheat and barley, also appeared early in the period. Barley continued to be grown, the husked variety now predominating. Emmer was the most important variety of wheat in the early Bronze Age, but Spelt was also grown. Beans and peas are now met with for the first time, being grown at Glastonbury, where the inhabitants rowed across to the mainland in dugout canoes to cultivate their crops.[1] Millet is mentioned by Pytheas as forming part of the food of the inhabitants of Britain in the fourth century B.C. but we hear no more of it, and possibly he was mistaken. Flax must have been grown to a considerable extent to furnish linen.

Manuring and dressing of the land was early resorted to by the Celts. The more advanced Belgae relied much on marling and chalking, according to the nature of the soil, but did not chalk the same field twice. Pliny in the first century B.C. notices this practice as particularly British. All marl, he says, is laid on the ground immediately after ploughing, and "acts as a fat in relation to the earth." Fine white chalk was "taken from a considerable depth in the ground, the pits being sunk in most instances as much as a hundred feet. These pits are narrow at the mouth, but the shafts enlarge very considerably in the interior as is the case in mines. It is in Britain more particularly that this chalk is employed. The good effect is found to last full eighty years."[2] Here we have an accurate description of those mysterious pits, now known as "Dene holes," the use and age of which was so long in dispute. The "civilized Gauls" of Britain had discovered that chalk from a depth forms a much more efficacious dressing for the soil than that on the surface. Pliny, in addition to white marl and chalk, mentions as a third variety "fuller's chalk mixed with unctious earth," which he says promotes the growth of hay, the effects lasting thirty years.

Early in the Iron Age the Celts introduced new ways of dealing with the harvested corn. Pytheas in the fourth century B.C. related how the natives of Britain "collect the sheaves in

[1] E. C. Curwen: *Plough and Pasture*, 47-8. *Antiquity*, xii. 46, p. 141. G.L.V., 1. 16, 629.

[2] Pliny: *Natural History* (trans. Bostock & Riley), xvii. 4, pp. 153, 454.

great barns and thresh out the corn there, because they have so little sunshine that our open threshing-places would be of little use in that land of clouds and rain," the climate having deteriorated, as we have seen, since Bronze Age days. This difficulty of the dampness of the climate was overcome by the Celts by drying the corn on racks before it was threshed. The fire was built upon a stone hearth and at a suitable distance above, slight rafters were placed across on posts, and straw laid evenly upon them, the corn to be dried being spread upon this. Another plan was to make a large round sieve of interwoven twigs to hold the corn and place it above the fire or over red-hot flints.[1] After drying it could either be threshed, or stored in the ear.

Diodorus Siculus (44 B.C.) describes how the Britons, having cut off the ears of corn and stored them underground, "cull therefrom daily such as are old, and dressing them, have thence their sustenance." This ancient method of dressing the corn by "burning off" the husks, survived so long in the Western Isles of Scotland that it was described by Martin in 1703, and was done in the following way. A woman holding a handful of corn in the left hand by the stalks set fire to the ears, and with a stick in the right hand, which she managed dexterously, beat off the grain at the very instant that the husk was burnt. Corn may thus be dressed, winnowed, ground and baked in an hour.[2] A small farmyard of the early Iron Age has been excavated at Little Woodbury, a holding of about twenty acres. Here the corn was dried on frames supported on pairs of posts six to eight feet apart, and the grain was then threshed in oblong hollows in the ground under a rough shelter. The seed corn was selected and stored in a small rectangular granary raised on piles, and the rest roasted to destroy the germ, and stored in subterranean silos, circular pits three to five feet across and eight or nine feet deep and lined with a receptacle probably of plaited straw. At Glastonbury these pits were so large that the farmer descended into them by a runged ladder or a notched tree trunk. They had to be discarded every five years or so, and were filled up with rubbish.[3]

The corn thus dried was much easier to grind, and although the ancient saddle-stone long continued in use the rotary quern had found its way to these shores by the middle of the period, and at Glastonbury outnumbered the earlier implement. These

[1] Jones: *Myvyrian Arch. of Wales.* Glossary. Adamnan: *Life of St. Columba,* ed. Reeves. 88 n.
[2] Martin: *Description of the Western Isles of Scotland,* 204.
[3] G. Clarke, op. cit., 24-5.

Fig. 23a.—Quern

early querns were conical or beehive in shape, with handles projecting sideways and steeply sloping grinding surfaces.[1]

The Celtic farmer lived simply, after the manner of pastoral
people, mainly upon the produce of his herds—milk, butter,
probably cheese, and according to Strabo, much swine's flesh,
both fresh and salted. Gauls on the Continent supplied Italy
with smoked and salted bacon, so that it was evidently a Celtic
art[2] and would have been brought to Britain by the Belgae, if
not before. It is sometimes stated that cheese was not known in
Britain before the coming of the Romans but this seems almost
incredible among pastoral tribes—cheese-making vessels are
found on the Continent in the last millennium B.C.—and the
statement of Strabo (A.D. 30) that *some* tribes had not the art
of making it, is surely an argument that others had it. Pliny
speaks of the "barbarous nations" who were "either ignorant of
the merits of cheese or have totally disregarded it," but he may
have been referring to the more backward parts of the country.
"Yet," he says, "they knew how to thicken milk and form an
acrid sort of liquid with a pleasant flavour, as well as a rich
butter."[3]

The people of Glastonbury were skilled fishers, having nets
with lead sinkers, and doubtless other villages round the coast
were equally accomplished. Oysters were much eaten, where
obtainable, from the beginning of the period.[4]

With the introduction of oats the housewife would be able to
grind them and make porridge, and hens and geese would provide her with eggs. She could mix rye flour with her wheat or
barley meal to make a fresh variety of bread. In many places
housewives had dome-shaped clay ovens in which to do their
baking, presumably not unlike the cloam ovens still surviving in
Cornwall. Firedogs and spits have also been found for roasting,
and skillets and cauldrons for boiling, so that with all these
resources cooking must have been greatly improved. Early in
the Iron Age she had pots of a warm glossy red. However, the
Belgae brought with them wheel-made pottery, plates and jars of
graceful design in a lustrous black, a great advance on the simple
handmade pots hitherto known.[5] The carpenters of Glastonbury were skilful craftsmen in wood, making tubs of staves and
small turned vessels. Buckets were elegantly hooped and

[1] E. C. Curwen, op. cit., 109. G.L.V., ii. 609-10.
[2] Mackenzie: *Ancient Man in Britain*, 114.
[3] Pliny: *Natural History*, xi. 96. E. C. Curwen, op. cit., 36.
[4] Kendrick & Hawkes, op. cit., 174.
[5] G. Clarke, op. cit., 37, 39. *Archaelogical Journal*, Vol. 87, pp, 259-61.
J. Hawkes: *Early Britain*, 30.

mounted with bronze at first, until iron came into common use.

Weaving also had much advanced and had become quite a household industry. The Celtic housewife was a skilled weaver of tartans, which involved incidentally much skill in dyeing: she had a good timber loom-frame, with triangular loom weights of baked clay to stretch the warp, and a long-handled comb to press the weft into place.[1] Her spindle whorls were of many materials, but her needles seem to have been still of bone.

The Celtic farmer had four great days in the year upon which he made holiday, with bonfires on the hills and much drinking of mead or metheglin. These were:—

(1) Beltane, May 1st, the joyful celebrating of the waking of the earth from its winter sleep; later to become "May Day" with its revels.

(2) Midsummer Day, or the triumph of the sun; afterwards celebrated as St. John the Baptist's day.

(3) The Feast of Lugh, in August, when the turning point of the sun's course was reached; afterwards "Lammas," August 1st.

(4) Samhain, the last day of October, the farewell to the sun for the long dark winter, and the beginning of the Celtic year. This later became "Hallowe'en," the "holiday of dead and evil spirits."[2]

His drink on these festive occasions was mead made from wheat and honey, or a kind of beer made from barley and later described by Dioscorides as "headachy, unwholesome, and injurious to the nerves."

It is clear then that during the Iron Age great strides were made in the practice and organization of farming. The three stages of the period continued side by side, the early Hallstatt peasant culture lingering in the north and west, with even Bronze Age survivals, and the La Tène culture with its high artistic attainment flourishing particularly in Cornwall and stretching up through Britain, from Weymouth to Bristol, and along the ridge across Northamptonshire to Lincoln Edge.[3] But in the south and south-eastern parts peopled by the Belgae the farmer, before the coming of the Romans, had acquired his poultry-yard, improved his other stock, learned the use of the eight-ox wheeled plough, added immeasurably to his implements, and was supplied with the quern and the potter's wheel.

---

[1] V. G. Childe, op. cit., 201.

[2] C. Squire: *Mythology of the British Islands*, 406-8.

[3] D. Jerrold, op. cit., 113-114.

# CHAPTER III

## PART 2.—DEVELOPMENT OF THE CELTIC FARM

### REFERENCES.

V.C. = Venedotian Code ⎫
D.C. = Dimetian Code ⎬ *Ancient Laws and Institutes of Wales*
G.C. = Gwentian Code ⎭
A.W.L. = Anomalous Welsh Laws (Record Commission)

# CHAPTER III

## PART 2.—DEVELOPMENT OF THE CELTIC FARM

With the establishment of the Celts in Britain we stand, historically speaking, on firmer ground, for they are a people still surviving in Scotland and Wales, and much of their ancient literature and tradition has been handed down. There survived in South Wales in particular a body of tribal custom and law which, assembled and written down in the tenth century A.D. by Howell Da, shows not only the development which the Celtic village system underwent during some twelve centuries, acquiring and assimilating additional advantages from successive neighbours, but also the body of immemorially ancient tradition underlying it. It seems appropriate therefore to describe it here, following on the record of the arrival of the Celts in this country, even though some features of it belong to a later date.

From these Laws we get an extraordinarily detailed picture of village life on the outskirts of Britain, where the early Celtic folk took refuge from subsequent waves of invasion.

They were a simple, pastoral people, living like the Swiss in the sheltered valleys during the winter, and moving with their flocks and herds to the hill pastures for the summer. Their dwelling was the simple booth or "bothy," built as described earlier in Chapter III (see p. 45) of two pairs of trees connected by a roof-pole. According to early Welsh Law the "three indispensables" of a bothy were a roof-tree, roof-supporting forks, and wattling for the sides.[1] The winter houses were more carefully and permanently constructed with auger holes and wooden pegs, and were valued at 20d. a bay instead of 12d. for the summer houses. This dual system of dwellings was still in use in the twelfth century, but the tendency would be for the family to remain more and more in the permanent and elaborated winter houses, and for the herdsmen only to migrate in summer to the booths in the hills. The single-bay house was easily enlarged by the addition of another pair of trees at one end, thus doubling the length, or increasing it to any extent desired, and from this evolved the Celtic patriarchal farmhouse in which three generations of the same family lived together in common. The central hall was formed of a series

[1] A.W.L., 676.

of pairs of tree trunks as before, but with the addition of aisles at the sides formed by lean-to buildings or penthouses. These were partitioned off into bedrooms or private compartments for the various branches of the family, and the hall formed the common living- and dining-room, and in early times doubtless the kitchen also, but this came to be a separate building among the outhouses.[1]

The ordinary homestead or "tyddyn" consisted of a house of "six columns and penthouses," with chamber, cowhouse, barn, kiln, sheepcote and pigsty—a very representative farm-yard. This manner of living in the primitive patriarchal household, which must have had its drawbacks when the patriarch became too old to rule with a firm hand, was not however the only one. Another and possibly later method was to form a group of "tyddyns" or homesteads, still confined to a kindred, by giving a number of the young men separate households, each with his own dwelling, cattle-yard and corn-yard, a cow and a yoke of oxen, and a share in the common fields. These groups originally consisted of nine homesteads forming a "trev" or hamlet, with a smithy and a bath house for common use, prudently placed seven fathoms from the nearest house for fear of fire, and roofed with shingles, tiles or sods. But not every trev had a smithy of its own, for a smith in his smithy together with a book, and a teacher versed in song, were later considered as the three ornaments of a trev. The trev was also at the outset provided with a plough, an oven, a churn, a cat, a cock, a bull, a herdsman and a dog, for the common use.[2] Each little house was surrounded by a fence of wattles or thorns to keep the cattle from eating the thatch, and the whole trev would probably be enclosed in a stockade.

A certain number of tyddyns grouped together were under obligation to pay to their chief a food rent or "gwestva," consisting in the Gwentian district of a horseload of wheat-flour, an ox, seven threaves of oats, a vat of honey, and twenty-four silver pence; but otherwise they were free. Even the "taeogs" of the tributary villages only had a few building and billeting obligations, and for the rest paid from each group, in winter a sow, a salted flitch, threescore wheaten loaves, a tub of ale, twenty sheaves of oats and pence for the servants, and in summer

---

[1] For a detailed description of an unsavoury example of this kind of hall, see the "Dream of Rhonabwy" in the *Mabinogion*. In this instance the cattle evidently wandered about the house at will.

[2] F. Seebohm: *Tribal Custom in Anglo-Saxon Law*, 36-7. To this day in Breton villages the oven is commonly situated in the middle of the village green.

a tub of butter, twelve cheeses and bread.[1]   These taeogs, who were probably the subdued remnants of a more ancient race, were grouped in villages but not in kindreds like the free tribes-men.   They had homesteads of much the same sort, with swine and cattle of their own, and had summer settlements in the hill pastures in the same way as their masters, but except that they paid tribute as described, and might not sell their stallions, swine or honey without leave, do not seem to have been in a very dependent condition.

As regards *stock*, their swine were evidently the vigorous crossbreed of the rest of the country, and remained turbulent animals, for the Welsh Laws provided that if swine entered a house and scattered the fire about so as to burn the house the owner of the animals should pay for the act.   It became customary to send them to the woods for mast and acorns from the feast of St. John[2] until January 15th, between which dates the woods were preserved for their use.   A sow was valued at 30d. and a boar at 90d., a very high value compared with other animals.   The legal herd was twelve sows and a boar.[3]

*Cows*, as we have seen, were the recognized standard of value.   The "normal cow" as used for the payment of fines, was one in full milk and until she had had her fifth calf.   To be tested she was taken to untouched pasture, and was expected to fill the standard milk measure twice daily.   The Venedotian measure was three thumbs' width at the bottom, six in the middle and nine at the top, and nine thumbs diagonally in depth.   The Dimetian measure was slightly smaller, being seven thumbs diagonally in depth.   When the Laws were codified in the tenth century the money value of an ox or cow in their prime was fixed at threescore pence, which was equivalent to three ounces of silver.[4]   The legal herd was supposed to con-sist of 24 cows, contributed by the various households of the hamlet and tended by the common herdsman, the hamlet being provided with one bull as we have seen, but doubtless these numbers increased and varied as time went on.   In winter the cattle were fed on the lowland pastures near the homesteads and each man's cattle were penned at night in his own cattle-yard, and here also his plough-ox or oxen would be housed practically all the year round.   For May, June and July the

[1] F. Seebohm: *English Village Community*, 197-8. A threave in later times was 24 sheaves.

[2] Probably the Decollation of St. John the Baptist, Aug. 29th.

[3] V.C., 127, 134.   D.C., 274.   A.W.L., 408.

[4] V.C., 128, 133.   D.C., 277.   G.C., 388.

greater part of the community migrated to the hills to feed their cattle on the upland pastures, returning at the beginning of August for the harvest. Later no doubt this migration was confined to the herdsmen, whose "three indispensables" were a bothy, a dog, and a knife, with which to cut the materials for the bothy and also perhaps for defensive purposes. The milk of the herd went into the common churn, and the produce was divided among the owners of the cattle according to the test milking of their several cows on a certain fixed day.[1] A simple but interesting system of co-operative dairy-farming.

White cattle with red ears appear in the Laws as of special value, and these were probably the larger breed introduced by the Saxons.[2]

*Sheep* and *goats* were not highly valued, being equally priced at 4d. a head, but the former at any rate must have been universally kept for every household seems to have been provided with a sheepcote,[3] and a plentiful supply of wool would be necessary for the weaving of tartans and other garments. The milk of both animals was probably used for butter and cheese-making, whenever that art was introduced, if not for drinking; a "milch sheep" in mentioned in the Dimetian Code. In the *Mabinogion* we read of a flock of goats being kept in a house at the extremity of the forest. Celtic shepherds counted their sheep by the score, a custom which was handed down by the unchanging peasantry to succeeding generations, for in Cumberland and the North, and parts of the West, sheep are counted by the "rhyming score" to this day, and this count has been shown to be a corruption of the ancient Welsh numbers, now become a meaningless jingle. The legal flock in the Laws consisted of thirty sheep and a ram.

*Horses*, according to the ordinary Celtic practice, when in use on the farm were kept at home to feed or train, but the mares when not in a state to work were turned loose in the forest to foal. The mare was afterwards recaptured, but the foal left to grow up wild until needed. Stallions also were turned loose in the forest.[4] This custom is exemplified by the Domesday record for Cornwall and some other parts, where "equi sylvestres" or "indomiti" are frequently mentioned among the manorial stock. A good horse, worthy of bequest to a monastery, was worth in the sixth century three to five cows,

[1] F. Seebohm: *Tribal Custom in Anglo-Saxon Law*, 38.
[2] V.C., 3. D.C., 168.
[3] V.C., 135-8.
[4] J. Thrupp: "Domestication of Animals in England." *Trans. Ethn. Soc.*, N.S. iv. 167.

but these proud animals were not for farm use. The "working horse," now a recognized member of the farmyard, was valued in the tenth century at threescore pence, i.e., the same value as a "normal cow." Its recognized work was to "draw a cart uphill or downhill," to carry a burden or to harrow.[1] The pack-horse was used to carry many things, notably firewood and victuals. In Adamnan's *Life of St. Columba* (sixth century) there is a pleasant description of the white horse, "obediens servitor," which carried the milk pails in Iona from the byre to the monastery and was beloved by the Saint. These pails were wooden, with covers secured by a crossbar passing through two holes in the sides.[2]

The legal stud, according to the Dimetian Code, was fifty mares. Any man selling a horse was to warrant it against the staggers, black strangles, farcy and restiveness.

The *herd-dog* was all important to the Celtic farmer. One thoroughly well-trained, which would go before the herd in the morning or behind it in the evening, and would make three turns round it during the night, was worth the best ox, and was one of the herdsman's "three indispensables." Dogs were used to guard cattle, sheep and swine, but for the last-named perhaps mastiffs were used, as they would need protection against the beasts of the forest. The three sorts of "curs" classi-fied in the Laws are the mastiff, the shepherd-dog and the house-cur, the last of which was valued at 4d. like the cat. A vicious dog was not tolerated, for it was decreed that "if a dog accustomed to bite persons bite three persons and is not killed by its master, then it is to be tied to the leg of its master with a rope of two spans in length and there killed." Also it was enacted that the scowl of a man at a dog which attacked him was one for which no reparation need be made.[3] Evidently at other times one's expression must be controlled.

The domestic *cat* was much prized by the Celts, though she was apparently of Roman introduction, as we shall see. As guardian of the barn from rats and mice she was a very impor-tant person; if anyone dared to kill the proud puss that guarded the King's barn the penalty was considerable. The corpse was suspended by the tail with the head downwards "upon a clean, even floor," and wheat was then brought by the offender and heaped about it until the very tip of the tail was covered. If

[1] V.C., 128, 129.
[2] Adamnan's *Life of St. Columba.* Ed. Reeves, 361.
[3] V.C., 138. D.C., 244, 245. G.C., 382. A.W.L., 691.

so much corn could not be produced the luckless man had to
forfeit a milch sheep with its lamb and wool.   The worth of
an ordinary farmer's cat was 4d.; a kitten "from the night it is
kittened until it shall open its eyes 1d., and from that time until
it killed its first mouse 2d., after that event 4d."   Anyone selling a cat was "to answer for her not going a-caterwauling every
moon, that she devour not her kittens, that she have ears, eyes,
teeth and nails, and be a good mouser."[1]   Each hamlet of nine
homesteads was, as we have seen, provided with one tom-cat.
The number of cats in each household was not restricted, for
if a man and his wife had occasion to separate and divide their
property the husband was entitled to one cat and the wife to
"the rest of the cats."

*Poultry,* that useful contribution of the Belgae to the farm-
yard, became so universal in Wales that little children, dogs
and cocks were considered the three signs of inhabitancy of a
country.   Hens and geese were then valued at 1d. each, and
cocks and ganders at 2d.   Goslings, of which there should be
twenty-four in a nest, were worth 1/2d., chickens 1/4d. until they
roosted, and then 1/2d. until they laid an egg or crowed.   Poultry
might be taken or impounded if found upon the cornfields during the first fortnight after sowing, and after the grain was
formed.   It was evidently not customary to clip their wings,
for it was considered impossible to keep hens and geese out of
fenced gardens since they could fly.   However, anyone who
found a hen trespassing in his flax-garden or his barn might
detain her until she or the owner compensated him with an
egg.   Geese that trespassed met with very short shrift.   Anyone finding geese in his corn might cut a stick "as long as from
his elbow to the end of his little finger and as thick as he may
will" and kill the geese with it.   If they were found damaging
corn in the corn-yard or barn they fared even worse; "let a rod
be tightened round their necks and let them remain there until
they die."[2]   The poultry-yard does not seem to have yet become
the particular property of the farmer's wife, for if the household goods were divided on separation the husband took the
poultry with his one cat.

Everyone, from the King to the taeogs of the tributary
villages, kept hawks for taking game.   Wild ducks were sometimes taken with these but were not yet domesticated.

*Bees* at the beginning of this period were probably not
domesticated; but honey was greatly valued, being the only

[1] V.C., 136.   D.C., 282, 283.   G.C., 355.
[2] V.C., 136, 138, 158.   D.C., 275

method of sweetening,[1] and the wax being needed for candles for the chief's household, and later for the altar. Honey was much used also for making mead. The collection and preparation of honey was an important work, and perforated earthenware vessels were used for straining it   In the Laws, and again in Domesday Book, we find sextars of honey a large item in the tribute paid by groups of trevs to their overlord. By the tenth century bees were not only domesticated but almost canonized. "The origin of bees is from Paradise" states the Gwentian Code, "and on account of the sin of man they came from thence and God conferred His blessing upon them, and therefore the mass cannot be sung without the wax." Bees of old stock were valued at 24d., a first swarm at 16d., and the subsequent swarms decreasing each time, until a swarm after August was worth only 4d. A swarm of bees on a branch was considered a "free hunt."[2]

In that remote and hilly country the hoe and the foot plough probably long continued to be the usual way of cultivating the soil, but the Welsh had various traditions about the introduction of the plough—Hu Gadarn the Mighty, coming with the Cymri to Britain as we have seen is said to have been the first to draw a furrow with the plough,[3] but another tradition has it that Elldud introduced the plough in the time of Theodosius[4] (A.D. 379-395). If so this later one may have been the heavy wheeled plough known in the rest of the island 400 years earlier. In any case by the time the Laws were codified a heavy eight-ox plough had evidently been long used and well established, but with a variation of their own for instead of yoking the oxen in pairs in the usual way they were yoked four abreast in two rows, which must have been a most intricate matter to yoke, guide and turn. The parts of the plough enumerated in the Laws are the plough-head (the draught-iron at the end of the beam), wheels, bars and stilt (the second handle was a later addition), sock (or share), beam (to which the oxen were attached), coulter, and cleansing hurdle, cleansing spud and goad for the ploughman and driver, and yoke and bows for the oxen.[5]

The oxen were usually yoked four abreast, the "long yoke" measuring 12 feet (16 Welsh feet of 9 inches) being stretched

---

[1] Unless perhaps birch-sap was used for this purpose.

[2] V.C., 138. G.C., 360.

[3] C. Squire: *Mythology of Ancient Britain and Ireland*, 53.

[4] E. C. Curwen: *Plough and Pasture*, 75.

  V.C., 150.

athwart the four with "bows" of wythes passing round their necks. There were, however, shorter yokes of nine feet, six feet and three feet, which shows that they were also yoked in threes or pairs, or even sometimes singly or in tandem. These four yokes remained in use down to about A.D. 1600.[1] Some of them would no doubt be for carts, but it seems that lighter ploughs were sometimes used, for Giraldus Cambrensis writes in the eleventh century that they "seldom yoke less than four oxen to the plough," which implies that as few as four were quite often used. A team of six, used by one David, is mentioned in the Laws.

The ploughman and driver were responsible for the team and plough delivered to them while in use, and had to pay for any damage that they did. The ploughman's duty was to guide the plough, assist the driver in yoking the oxen, but loosen only the two "short-yoked." He also had to know how to make a plough and nail it from the first nail to the last. He carried a "cleansing spud" with which to clear away the soil clinging to the plough. For some curious reason the thickness of his nail when he had been a ploughman for seven years was used as a measure for the thickness of the gold plate which was rendered to the King on certain occasions, the alternative measure being the shell of a goose's egg.[2]

The driver was to yoke the oxen carefully "so that they be not too tight nor too loose and drive them so as not to break their hearts." He had also to "furnish the bows of the yokes with wythes, and if it be a long team the small rings and pegs of the bows." He carried a long rod with which to goad on the oxen, "mitigating their sense of labour by the usual rude song."[3] The unfortunate man walked backwards in front of the advancing team, and when he not unnaturally fell down was "frequently exposed to danger from the refractory oxen." But at any time, according to Giraldus Cambrensis, the Welsh farmer preferred warfare to husbandry and "rushed eagerly from his plough to arms."

*Agriculture* to the Celtic tribes of the interior long remained secondary in importance to pastoral pursuits, but they nevertheless evolved a definite form of co-operative tribal husbandry on the open field system. The fields of the kindred were

[1] V.C., 90. Rhys and Brynmor Jones: *The Welsh People*, 249.

[2] V.C., 154, 155. D.C., 168.

[3] Giraldus Cambrensis: *Itinerary Through Wales* (Everyman's Library), 29, 166, 184. This was continued as late as 1806 and consisted of a "sort of chaunt of half or even quarter notes."

divided into long narrow strips called "erws," separated by an unploughed balk of turf two furrows wide, and representing the usual amount ploughed by an ox team in a day. The length of the erw was a "furrow long" or furlong, i.e., originally the distance that an eight-ox team could conveniently draw the plough without stopping to take breath. Its width was settled in various ways[1] but the total area was less than that of the present English acre. To each free tribesman was apportioned five of these strips for his support, but by co-operation with his neighbours and by licence of the chief he might also cultivate portions of the waste land of the community. The rules for this co-aration were most carefully laid down in the Laws. Twelve erws were ploughed with a team of eight oxen contributed by the co-tillers, and the strips and their produce were apportioned in the following manner. The first strip to the ploughman, the second to the man who contributed the plough-irons (i.e., the share and the coulter), the third to the owner of the "exterior sod ox," the fourth to the owner of the "exterior sward ox," the fifth to the driver, the strips from six to eleven to the owners of the remaining six oxen of the team in order of worth, and the twelfth or "plough erw" was for the maintenance of the woodwork of the plough.[2] The return received by the husbandman was thus in direct relation to his contribution to the outfit. Evidently ploughing had become a very thorough matter and the plough a heavy and cumbrous implement, of which the timber frame (and wheels) was common property and only the cherished irons a private possession. On separation the ploughshare went to the woman and the coulter to the man. Few tribesmen would have possessed more than a single yoke of oxen, so that some co-operative arrangement for ploughing was a necessity.

Neither horses, mares, nor cows might be put to the plough, but oxen in their prime were used. Ploughing started legally on February 9th, and for the first two years that an ox was put to the plough he was expected to work from morn to eve; in the third and subsequent years he was let off at noon. There were three ploughings in the year, one in autumn or winter for wheat, one in spring, from February to April, for oats and barley, and a third in summer, presumably of fallow ground.[3]

[1] According to the Venedotian Code "16 feet = the rod in the hand of the driver. As far as he can reach with that rod stretching out his arm are the two skirts of the erw," i.e., its breadth. V.C., 90.

[2] V.C., 153. The "sward ox" would be the one nearest to the balk, and the "sod ox" that furthest from the balk.

[3] V.C., 133, 156. Giraldus Cambrensis, 166.

Harrowing was done with horses, but the provision of a horse for this purpose was not a co-operative obligation, each man evidently being responsible for the harrowing of his own land. Two sorts of harrows are mentioned in the Laws; the harrow proper, presumably with teeth, worth a legal penny, and the thorn harrow (composed of bunches of cut thorn bushes) worth only a curt penny.[1]

Reaping was done by cutting off the ears high on the stalk with a sickle or "reaping hook." In the eleventh century, however, Giraldus Cambrensis states that "instead of small sickles in mowing they make use of a moderate-sized piece of iron formed like a knife with two pieces of wood fixed loosely and flexibly to the head which they think a more expeditious instrument." Harvesting was not a co-operative affair but the community turned out at the sound of a horn,[2] each household to reap their own strip in the open fields.

The harvested corn was treated, as we have seen, in various ways, but was commonly dried in a kiln. An improved form with perforated tiles was no doubt borrowed from the Romans. A "piped kiln" is mentioned in the Laws as being more valuable than the ordinary one. The kiln was evidently in some cases a large building, for in Iona the kiln was large enough for the corn to be threshed there also, as soon as it was dried, and the King's kiln in the Welsh Laws was used as a lodging-place for his huntsmen; they would doubtless lie warm on the dry straw from the threshing. The stones of the kiln appear in the Laws as one of the three things which "preserve a memorial of land and homestead";[3] possibly these were the stones on which the fire was made, but it seems more probable that they were the quern-stones, for the place where the corn was dried and threshed would obviously be a convenient place for grinding it also. The kiln of the hamlet, like the smithy, had to be situated nine paces from the nearest house, for fear of fire.

A flail and a winnowing-cloth are mentioned in the Laws, also an iron spade and a wooden shovel, a pickaxe, billhook, weeding hook, fork and rake, iron saw, ropes of hair and of elm-bark, shears for the sheep's wool, and a good assortment of carpenter's tools, also a barrow. The farm-yard cart by the tenth century had become universal, for every household was provided with a "car and yoke," in the Gwentian Code speci-

[1] V.C., xxii. 249-50.
[2] Giraldus Cambrensis, op. cit., 184. This practice of summoning and cheering reapers with a horn is still noted by Tusser in the 16th century.
[3] V.C., xvi. 9. D.C., ii. 16, etc. A.W.L., 655.

E

fied as an ox car, which suggests that it was commonly drawn by oxen; yet in the provisions relating to horses it was one of the duties of a working horse to draw a car, so that evidently both animals were used for the purpose. In the sixth century St. Columba drove about Iona, to overlook the brethren working on the farm, in a cart the wheels of which were secured to the axle by some sort of bolt or pin (Lat. *obex*). These bolts, which seem to have been easily moveable, were sometimes forgotten, with disastrous results to anyone riding in the cart.[1]

But to return to the harvest. The barns were left open to air from the entering of the first sheaf until the Calends of winter (Nov. 1st) and were then closed in the following manner: "with three eatherings on the sill and a wattle upon the doorway with three bands thereon, two on the back and one on the front."[2]

To grind the corn, the revolving quern, as we have seen, was introduced about the middle of the Iron Age and soon superseded earlier forms. By a curious provision of the Venedotian Code if a husband and wife separated the husband took the upper stone of the quern and the wife the lower one, a proceeding which would seem to render them both useless. One can only infer that the querns of a tribe were made of much the same size and pattern, so that if either man or woman entered into a fresh union they could fit their stone to that of their fresh partner. The quern was often kept in a quern-shed, and bolting of the meal was evidently carefully attended to, as the household sieve might not be lightly lent out. It is recorded that St. Columba ground his own corn in a quern, commencing the hymn *Altus Prosator* as he put the first feed into the hopper of the mill and finishing the hymn and his grinding simultaneously.[3]

The watermill, possibly introduced by the Romans, but certainly adopted by the Saxons, penetrated slowly to Wales, for in the tenth century a mill was sufficiently rare there to be considered an "ornament of the kindred" possessing one.[4]

The Welsh seemed to have relied entirely on farm-yard manure for dressing their land, and fixed rules were laid down about it. If a man manured a piece of land with the permission of the owner it gave him a claim on that land for a number of years according to the manure he used. Gardens had to be

---

[1] G.C., 366. Adamnan's *Life of St. Columba*. Ed. Reaves, 172, 228.
[2] V.C., xxv. 12. Eathering=a binding or hedge of flexible rods.
[3] V.C., 38, 87, 146. Adamnan, op. cit., 362.
[4] V.C., 87.

manured every year, so could only be claimed for one year at a time.   Fallow land was ploughed and sown for two years without manuring, but if manured the lessee worked it for four years consecutively.   Land dressed with "rotten dung" (i.e., "where cattle are accustomed to be without folding") was worked for two years; if dressed with "yard dung" for three years, and if with "car dung" for four.   Newly-cleared woodland was also used for four years before it returned to the owner.[1]  A "manuring pannier" occurs in the list of implements in the Laws.

Burning of the woods, to clear fresh land, was a co-operative service rendered by the community on request.   Heath was burnt in March under the Venedotian Code, and under the Dimetian Code from the middle of March to the middle of April.   The most valued trees were the oak and the yew for their wood, and the crabtree and hazel for their fruit.[2]

Their usual crops would have been wheat, barley and oats. Rye, by tradition, was not well thought of.   According to the Welsh bard Taliesin's version of the Creation it originated in the following way: Adam was given a spade and some wheat seed; Eve also was given seed, but hid a tenth of it, and this portion consequently turned to black rye to shew the mischief of thieving.   Again in the *Mabinogion* Elphin's wife is said not to have kneaded rye-dough since her marriage, which implies that it was a rather despised form of food, not used by the prosperous.

Peas and beans oddly enough are not mentioned in the Laws. Every household seems to have had its flax-garden to provide the necessary linen, and it was protected by law against the incursions of scratching hens and mousing cats.   Hemp does not as yet appear.   Cabbages and leeks were grown in the garden, and also orchard produce, which until Roman times at any rate would be almost exclusively apples, although Giraldus Cambrensis declared in the next century that the Welsh had neither gardens nor orchards.   An orchard in the Laws was one of the three "ornaments of a kindred," together with a mill and a weir, so would probably be a rarity not very long introduced.

The cattle of the hamlet, except during May, June and July, when they were taken to the hills, were fed upon the common pasture lands and the waste ground, to which each household had a by-road.   The meadow land selected for hay was reserved from the feast of St. Patrick (March 17th) to November 1st, that

[1] V.C., xvi., 9.  G.C., xxxii. 14, gives car dung three years, cleared woodland five years, fold dung two years and freshly broken-up land two years.

[2] V.C., 127, 141, 142.  D.C., 218.  Crabs would be used for making verjuice and nuts were crushed to extract the oil.

it might yield two crops in the year. It was enclosed by a fence and swine were never allowed on it.[1] Besides this the cattle were herded on the corn-fields after harvest and on the fallow ground in order to manure them. Each household was allowed two reserves of grass—a fenced close, and a meadow fenced for hay for the allotted period. Materials for thatch were also regularly harvested and would probably be reeds, broom or heather, according to the locality, as straw would be needed for littering or feeding the cattle.

Hunting was entirely free, and every tribesman might take what game and wild animals he could, unhampered by the restrictions later imposed by manorial lords.

Horses were used for food until Christian times, when the eating of horseflesh was forbidden on account of its connection with heathen sacrifices to Odin and Thor, and the British farmer gave it up very unwillingly even then. But probably the introduction of Christianity released him from the superstitious objection of his ancestors to eating the flesh of geese, fowls and hares, so that he was not without compensation. Even as late as the eleventh century Giraldus Cambrensis represents the Welsh as eating flesh in larger proportion than bread. It was then the custom to have one meal a day, in the evening, at which the dishes were few and not highly seasoned, consisting often of chopped meat and broth. The humble dwellings he describes were not provided with tables or cloths, but the dishes were placed upon the ground on rushes and fresh grass, three persons to each trencher in Christian times in honour of the Trinity. Bread was made daily in broad thin cakes, probably upon the "baking girdle" mentioned in the Laws, and the first piece broken off each loaf was given to the poor.[2] With this girdle, a brass or iron pan, a trivet, cauldron and flesh fork, a baking board and bowl, some other pieces of crockery and a "house-knife" the farmer's wife dealt with her cookery. She would now be able to make oatmeal porridge.

By the tenth century the loom was improved by the provision of a reed to press down the weft, in place of the former hand comb.[3] The woven cloth was anciently cleansed by "waulking" —working it backwards and forwards upon a board between two rows of women, with the feet or hands, with the help of some saponaceous substance and a great deal of song. After

[1] V.C., xxv. 27.
[2] Giraldus Cambrensis, op. cit., 169.
[3] V.C., 146.

the Conquest this process was gradually superseded by the fuller and his fulling-mill.

The life of the farmer's wife was no doubt a very busy one, but among the Celts woman had a high standing. Her property was legally defined, and if she wished to leave her husband she could do so within three nights of the seventh year of their union, taking with her her share of the household goods, children and stock.[1] The wife of a freeman was permitted by law to give away her mantle, shift, shoes, head cloth, meal, cheese, butter and milk without asking the advice of her husband, and could if she were so disposed lend all the household furniture, which must have given her considerable control over his comfort. The wife of a taeog might give away nothing but her headgear, and lend her sieve, but that only for the distance her call could be heard with her foot upon the threshold of her house,[2] her generosity therefore being restricted by her lung power.

Such was the life of the ancient Celtic farms, carrying on their separate way of life in Wales, but gathering from Roman and Saxon what advantages they offered and weaving them into the daily life of the tribe. But now we must turn back.

[1] V.C., 36. This was a sort of probationary union which became a legalized one if the seven years were completed.
[2] D.C., 253.

# CHAPTER IV

## THE ROMAN OCCUPATION

HAVING traced the development of the Celtic tribal farm as shown in its survival in Wales and the West right through to the tenth century A.D., we must now retrace our steps nine hundred years to the coming of the Romans, and consider what effect this entirely new influence had upon the arrangements of the British farm-yard.

The greater part of Britain was subdued by the Romans in the first century A.D., but the uplands, the wilder and more inaccessible parts, were not conquered until the second century. Indeed, the northern and western parts, although nominally subdued, penetrated by roads and formed into military districts, were never really Romanized. The typical Roman villa is not found in them and their Celtic institutions survived almost untouched, as we have already seen. The rest of the country, as far north as York, and westwards to the Welsh border and to Exeter, became a well-ordered and prosperous civilian State, and by the third century country houses and farms of the Roman type were plentiful, except in the Midlands, where they were comparatively few, probably owing to the still vast extent of forest and marsh-land.

Britain as the Romans found it was an even damper country than it is now, but with their coming the resources of civilization were brought to bear upon it for the first time. Morasses were drained, great tracts of forest cleared, and low-lying districts reclaimed by means of embankments, and thus large stretches of fertile soil were added to the land available for cultivation.

What was the condition of Britain in the first century as the Roman conquerors found it? In the southern and eastern districts occupied by the Belgae, who had arrived a century or two earlier, we are led to believe that agriculture, stock-raising and dairy-farming were well developed: marling and chalking of the land was understood, large farm-houses "like those of the Gauls" abounded, the farmer had his fowls and geese (though not for the larder), and probably understood the making of cheese. The rest of the country was backward in varying degrees, particularly in the Highlands of Wales and the North, where many still lived in primitive round huts and cultivated irregular plots of ground where they could and with what implements they had. But in the belt of country between these districts and the Belgic South-east the old "Celtic field-system" continued, with the use of the light plough. Some lived perhaps in simple booths which have disappeared, but the majority in round huts, rather better constructed than before; indeed this early form of dwelling continued right through the Roman period, as is shown by such settlements as the villages of Cranbourne Chase,[1] a circumstance which suggests that the life of the poorer and more outlying population was not much affected by the Roman occupation. In these villages, however, there was an interesting variety of dwellings, showing the penetration of Roman influence. Besides the primitive round huts of daub and wattle where the poorer villagers lived, there were square and flat-sided houses of timber, fastened together with iron nails and clamps, and plastered and painted inside. Some of these even had British imitations of the Roman hypocausts for warming their rooms by flues beneath the floor; in fact, in the village of Woodcuts there was a well-to-do quarter containing quite a number of these better-class houses.

What effect the coming of this fresh and much more developed civilization had upon the arrangements of the British farm in general has still to be carefully explored. In many parts it seems that the culture imposed by the Roman occupation merely formed a fertilizing stream, which influenced its surroundings but did not alter the general aspect.[2] But the fact is undisputed that Britain became one of the most important grain-producing countries of Europe. Much fertile land was cleared, drained and brought under cultivation. The wide spaces of Salisbury Plain and Cranbourne Chase and the open South Downs remained great stretches of corn-land until about the fourth century A.D. By that time, partly because the centre of British civilization had shifted to the Thames Valley and the South-east but more probably because of the growing importance of the wool industry, they were going out of cultivation and becoming wide grazing grounds.[3] The Roman taste for urban life was providing the farmer with market towns to which he could take his produce for sale.

By the end of the Roman occupation a distinct Romano-

---

[1] The chief of these, which we shall have occasion to refer to again, were Woodcuts, Rotherley and Woodyates. The Chase is situated on the borders of Wilts and Dorset.

[2] Vinogradoff: *Growth of the Manor*, 83, etc.

[3] Collingwood and Myres: *Roman Britain and the English Settlements*, 2nd Edit., 178-9.

British type had evolved—sturdy, muscular men rather short of stature, their heads fairly long, flattish on the top but projecting at the back, the forehead upright, square and rather low, with much the same brain capacity as at present.[1]

Under the Romans the land seems to have been organized for purposes of agriculture in four different ways.

The first of these was the typical *Roman Villa,* which introduced an entirely new feature into the life of Britain. Here we have the first appearance of the "lord," the rich man in a large country house, farming an estate by the work of slaves and dependents. The large Belgic homesteads, judging by the continuity of location, seem to have transformed themselves easily into this type of estate, and many fresh ones, small and great, grew up, inhabited mainly by Romano-Britons. The Roman villa of the full-blooded type was a large farm managed by a "villicus" or bailiff, on behalf of the lord whose country estate it was, and under this official were slaves arranged in groups of ten called decuriae, who worked such of the land as was not let out to coloni or freemen. These coloni cultivated their portions as separate farms and paid tribute in cattle or produce, their dwellings probably forming a little hamlet outside the walls of the main homestead.

The buildings of this type of villa usually consisted of a homestead with one or two courtyards. If the establishment was a grand one with two courtyards, the inner one contained, besides the dwelling-house, the ox-stalls, the stables for the horses, and the quarters for the other livestock. In the outer court, with the abode of the villicus at the entrance, were the cellars, granaries and storehouses, and the quarters of the slaves, also the common kitchen where the slaves did all their indoor work, and the "ergastulum" underground, where refractory slaves worked out their punishment in chains. The Roman villicus was not too exalted to dine with his slave-labourers, as later farmers did with their farm-servants.

The homestead or villa-house[2] itself showed, of course, a great advance architecturally, for the Romans considered the booth a barbarous and obsolete form of dwelling. The lower portions of the walls were built of stone or brick, but the main part was often of beams of wood placed near together and the interstices filled with clay mixed with chopped straw. Inside they were plastered with stucco and often painted. The roof was of shingles or tiles, roofing tiles being a speciality introduced

[1] Ibid., 17.
[2] The "villa" was the whole estate, not the house only.

by the Romans.[1]  For poorer houses and farm buildings a
turf roof laid on a foundation of heather was often used,[2] with
a plank along the eaves to keep the turves in place until they
grew together.  Yellow ochre was sometimes used to colour
the walls.

As regards plan, the type which came to be most highly
developed was the "corridor" house.  This was originally a
simple row of rooms opening on a corridor, which in time
became enlarged at right angles until the building ran round
a square court.  If the establishment was not so large as the
one described above, this courtyard included the dwellings of
the household, the quarters of the slaves and perhaps the
stables, and was called the "villa rustica," while the second
and outer courtyard, the "villa fructuaria," contained the
barns, granaries and storehouses.  The apartments of the household were warmed by hypocausts, an arrangement of flues
carrying hot air beneath the floor, and a separate bath-house
was included in the inner court.[3]  These grand residences combined the functions of country house and farm-house, but the
second type, the "basilical," was of a humbler nature suited to
a farm-house only.

The latter form seems to have been a survival of the ancient,
typically Saxon type common to all Aryan peoples at an
early stage of their development, and we have already seen a
variety of it in the Celtic homestead of "six columns and
pent-houses."[4]  This was a long building with two rows of
pillars down the centre like a nave, and apartments partitioned
off in the aisles for family or other uses.  Originally the family
apartments were across the end, with the hearth in the open
space in front of them, and the side aisles were given up to
stalls for the beasts.  The horses were placed down one side
and the oxen down the other, with their heads towards the
centre of the building so that they could be foddered from
the main floor.  The maidservants slept in the loft over the
cow-stalls and the menservants over the horses.  The corn
and hay were stored on boards among the joists of the roof,
and the corn threshed on the central floor.  A pig-sty and
calf-house were arranged on each side of the main doorway.

[1] H. M. Scarth: *Roman Britain*, 171.

[2] This was later called "dovet."

[3] We may, perhaps, see in these courts the origin of the square farm-yard
so often met with in England, with the barn and stack-yard beyond, but
according to some it was typically Saxon.

[4] Whether they evolved it from the booth or adopted it from the Romans
or the Belgae we do not know.

This extremely self-contained establishment, combining the whole farm-yard under one roof, would seem to us to have its disadvantages, but we are assured that "so long as the smoke of the great hearth fire, which had no chimney, permeated the whole building, insects and the bad stench were driven away."

Before the Romans arrived in Britain, however, they had assimilated this type of house, and becoming more particular had relegated the beasts to the courtyard, their stalls becoming

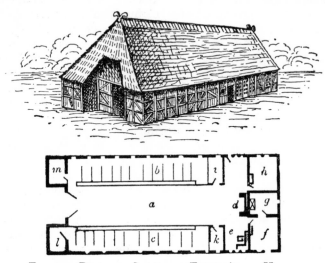

FIG. 24.—PLAN AND SKETCH OF EARLY ARYAN HOUSE

a, Floor.  
b, Stables.  
c, Cattle stalls.  
d, Hearth.  
e, Space for household.  
f, Best room.  

g, Store-room.  
h, Living-room.  
i and k, Chambers.  
l, Calf-house.  
m, Pigsty.  

small rooms. The central floor remained in the Roman house as the "atrium," with the family hearth in the centre and a hole above to allow the smoke to escape. Behind the hearth the "tablinum" originally contained the bed of the master of the house, and the rooms on either side were arranged in the same manner as in the early form described. The pillars rested upon stone blocks with recesses in their tops to receive the posts. Hypocausts and sometimes baths were included even in this humbler type of house, but baths were more usually semi-detached or quite separate.[1]

[1] For particulars of Roman housing see J. Ward: *Roman Era in Britain* 73-89, etc.

As regards situation, the favourite site chosen by the Roman farmer for his house seems to have been on the slope of a hill, backed by a wood, from which, if possible, a brook should flow, the whole overlooking a pleasant stretch of meadow-land, and facing south or east.

The second method of dealing with the land was to form *Military Colonies* by settling bodies of veterans upon the land. These men, we must remember, would not necessarily be Romans in blood, but, the Roman army being entirely cosmopolitan, might well have been Britons who had served their time in it. The land of the colony would, however, be laid out according to the rules of the Roman Agrimensores.[1] The territory apportioned was first divided by two perfectly straight roads crossing each other at right angles, one being called the Decumanus Maximus and the other the Cardo Maximus. The village or little town was usually situated at the cross-roads, and the four quarters or "regiones" were divided into neat rectangular holdings of arable land, as far as possible in rows, numbered and lettered according to their position in relation to the two main roads, and connected with them by lanes and cartways also running at right angles to the one and parallel with the other like a modern American town. There was no occasion for anyone to lose their way. Each holding was an estate, called a centuria, of 200 to 240 jugera (Roman acres), or a half-centuria of 100 to 120 jugera, and was cultivated quite separately by the owner. Those portions of land not suitable for cultivation, either in the midst of the centuriae or between the rectangular limits of the mass of the holdings and the irregular boundaries of the district, formed the common waste and woodland of the colonists.

Thirdly there were the *Agri occupatorii*, single or double holdings for little companies of veterans, planted in any suitable spot. These men were given their holdings together with an outfit of a pair of oxen and 50 modii[2] of each of two kinds of corn (wheat, oats or pulse), and cultivated them according to "common sense and the custom of the country." These men, whether they lived together in a little village or on scattered farms, could therefore adopt either the Belgic open-field system of strips of land, ploughing by co-operation, or could cultivate their square holdings independently according to the Roman method.

[1] For a detailed description of this system see Coote: *Romans of Britain*, 53 onwards. Actually there were only four of these in Britain—Colchester, Gloucester, York and Lincoln.

[2] The Roman modius was slightly less than a peck

Lastly there were the *tributary cultivators,* i.e. the remains of the Celtic tribes in their trevs, and probably a certain number of "Laeti," bodies of imported subject races, such as the Alamanni, planted upon foreign soil after the custom of the Romans.  The districts, probably very large in extent, where these tribal organizations were suffered to remain were divided into cantons, according to tribes, and the trevs were allowed to survive "with a modest but definite measure of self-government," the native chief becoming the Roman magistrate.[1]  Here the Celtic system of agriculture was not interfered with, and very wisely too.  One can imagine how disconcerting the uncompromisingly rectangular methods of the Romans must have been to the Celt with his natural bias towards curves, his irregular enclosures for cultivation, and his recent emergence from round huts.  However, even in these villages it seems that the Romans made some attempt to impress their methods, for in the "Anomalous Welsh Laws" a passage occurs providing that there shall be "two lawful roads in every trev, a road along and another across"[2]—an echo surely of the rules of the Agrimensores as described above in the Military Colonies.  The "lawful roads" were to be 1½ fathoms wide,[3] and the by-roads, which were to run from every habitation to the common waste, 7 feet wide.

We must next consider what alterations took place in the Romano-British farmer's stock and outhouses.

That all-important animal the *ox* remained throughout this period the little dark-hued *Bos longifrons,* slightly smaller than the present Kerry cow, but short-horned.  It is probable, however, that the Romans modified and improved the native breed by the introduction of some of

FIG. 25.—ROMAN OX

their own larger cattle with lyre-shaped horns.  Mr. McKenny Hughes deduces from excavations that the result was a somewhat larger breed with stouter horn cores, the horns having a tendency outwards and upwards instead of the forward curve of the original *Bos longifrons.*  Many intermediate sizes and

[1] Vinogradoff: *Growth of the Manor,* 51.  Haverfield and Macdonald: *Roman Occupation of Britain,* 186.

[2] *Ancient Laws and Institutions of Wales* (Rec. Com.), 525.

[3] A fathom equals 6 feet.

shapes are found.[1]   At Rotherley (Cranbourne Chase) the
average size of the oxen measured was 3 feet 3½ inches at
the shoulder; at Woodcuts it was 3 feet 3¼ inches, and at
Woodyates 3 feet 5½ inches.   The largest found was 4 feet
2 inches, which must have been an exceptionally fine animal.[2]
Palladius preferred red or brown oxen and black cows.[3]

The ox now began to be shod, to help him in his work on
rough ground, with iron shoes made in two crescent-shaped
pieces, and later called "cues": a practice which was continued
in Sussex until about forty years ago.   The Romans put their
oxen to the plough at three years old, and if refractory kept
them fasting in the yoke for a day and a night.

The plough oxen were accommodated in special houses,
8 feet of standing room being allowed for each yoke or pair of
oxen.   Four oxen therefore required a space of 16 feet, and this
became the recognized width of the "bay" or space between
each pair of crucks or forks used in erecting a simple house.
This booth-like form of building, dear to the Celts, was used
by the Romans for outhouses, and occurs in the writings of
Palladius, a Roman of the fourth century A.D., particularly
interesting to us both because he wrote at the time when Roman
culture was at its zenith in Britain, and also because an unknown
monk of Colchester in the early fifteenth century thought it
worth while to make an English translation of his work on
farming.[5]   A shed for beasts, he says, should be made of forks,
boards and boughs (in striking resemblance to the Welsh booth),
roofed with shingles, tiles, broom or sedges, and the animals'
hoofs should be kept dry, strong planks being put under horses'
feet.   His opinion was that stables should look south with a
north light, and should have a fire in winter, presumably in the
centre of the floor, which seems very considerate.   Vitruvius,
writing in 11 B.C., says that the ox-stalls should be united with
the hall, with the cribs facing the fire and the east, which seems
like a relic of the self-contained farmhouse with everything
under one roof, as described earlier in the chapter.   He says,

[1] T. McKenny Hughes: "Breeds of Cattle in the British Isles," *Archaeologia*, lv. 135 onwards. He also considers, with Professor Owen, that the typical Chillingham wild ox is the representative of this improved Roman breed, instead of being, as is usually held, the descendant of the wild Urus.

[2] Pitt Rivers: *Excavations at Cranbourne Chase*, ii. 219.

[3] Palladius: *On Husbandrie*, 129, 130.   E.E.T.S.

[4] See chap. iii. p. 51. Addy: *Evolution of the English House*, 21, Palladius, 19.

[5] It is an interesting conjecture as to whether Palladius had an earlier connection with Britain.   Could a MS of his work have survived in Britain to be handed down to the monk of Colchester?

however, that the stables, though in a warm place, should not face towards the fire.

The Roman herd was fifteen cows to one bull.

*Horses* also were small, being rarely larger than the present Exmoor pony, viz. 11 to 12 hands,[1] and were used extensively for food as well as other purposes. Some larger ones, of 13 hands, have been found, and at Woodcuts one of 14½ hands, which must surely have been a specially imported treasure belonging to a rich man. At Silchester their bones are not nearly so common as those of the ox, sheep and dog, and they

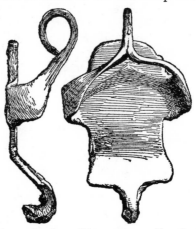

FIG. 26.—IRON HIPPOSANDAL (London)

are all very small.[2] They are thought to have been shod first with "hipposandals," a sort of shoe having a complete flat sole of iron with turned-up flanges, the front and back ones having hooks or eyelet holes by which to attach it. They were sometimes made of rope, and are supposed to have been used on stony or snowy ground, or to protect an injured hoof. From the time of Valentinian II (375-92) they had shoes not unlike modern horse-shoes, but without the turned-down heels. Evidently the Romans made some kind of "sandal" for their oxen also, for Palladius directs that the floor of the dovehouse should be scattered with "old broom (spartea) with which shoes have been made for oxen," as a charm against weasels.[3] Both horse- and ox-shoes are found in the Romano-British settlements of Cranbourne Chase. Small and well-finished bells are found, which were probably worn by the horses.

[1] A "hand"=4 inches, and the height is taken at the shoulder.
[2] *Archaeologia*, liii. 287.
[3] Palladius, 20. E.E.T.S.

It is commonly stated that the Romans were the first to bring hornless *sheep* to Britain, but we must not forget the hornless skulls found at Glastonbury which may have belonged to a genuinely hornless breed. This new breed was, perhaps, chiefly confined to the Roman villas and more favoured districts, for the horned variety are more abundant in Roman refuse-heaps than the hornless, and the former seem to have remained the dominant breed.[1] The sheep bones found at Cranbourne Chase still show the long-legged, slender-boned Highland breed, elegant and almost deer-like animals with brown wool. In size those found were on an average slightly smaller than the present Highland ewe, and of a slenderer build. The largest found in the Cranbourne Chase villages measured 2 feet 1 inch at the shoulder, and the smallest 1 foot 7 inches, the average being about 1 foot 11 inches. There were a few traces of a shorter-legged, stout-built breed, but in very small numbers.[2] Sheep with a true fleece, which needed to be cut off with shears, were now introduced, hence the growth of the wool industry in the fourth and fifth centuries.[3]

Palladius liked his ewes large-boned and with long, soft wool, and his rams tall, broad and white-wooled, with long tails and large foreheads.

*Goats* were also kept, a variety with simple recurved horns being present in many excavations of the period, especially in the North. Not many, however, were found at Cranbourne Chase. They were evidently popular as mutton.

Both cattle and sheep were exported from Britain under the Romans.

It must be remembered that all animals of which these measurements are given were found in outlying districts not of the most advanced civilization; it is quite possible that on the estates of the rich Roman villa-farmers the animals were of larger and finer breeds.

*Asses* may possibly have been introduced at this time, for the Romans used them for field work. A Romano-British lamp has been found, decorated with what is thought to be a corn-mill turned by an ass, but this is rather slender evidence to go upon.

*Pigs* continued comparatively large, those found at Cranbourne Chase being of the species akin to the wild boar, with long legs and large tusks. The average shoulder height

[1] W. Boyd Dawkins: *British Pleistocene Mammalia* (Prelim. treatise), xiv
[2] Pitt Rivers, op. cit., ii. 220-2.
[3] Antiquity x., 38, p. 205.

of those found at Woodyates is 2 feet, and at Woodcuts 2 feet 4¼ inches.[1] Palladius considered six piglings enough in a litter, and thought black pigs did best in cold countries and white pigs in warm ones.[2]

*Dogs* were kept in considerable variety, the bones of those found at Cranbourne Chase and Silchester varying from the size of a mastiff to that of a terrier, but all were slender in build.

The domestic *cat* seems to have been first introduced to Britain by the Romans, but probably first as a pet and not as an assistance to the farmer against rats and mice in his granary, a capacity in which it came to be much valued by the Celts, as we have seen. At Silchester cat bones have been found indistinguishable from those of the present domestic variety, and this evidence is borne out by the prints of a cat's paws on some of the tiles. These must have been made while the tiles were drying, and it is highly improbable that a wild cat would have strayed into such a situation.[3] No cat bones were found at Cranbourne Chase. Palladius kept cats to catch moles on his farm, but did not think highly of their prowess with regard to rats and mice. Thick oil dregs in a pan, he says, catch more rats and mice than the cats.[4]

*Fowls and geese*, as we have already seen, were introduced shortly before the coming of the Romans, but were not eaten by the natives. The Roman Colonists, however, would not share this prejudice, and the bones of fowls occur in their refuse-heaps. Palladius considered that "to set hens is a woman's business," and liked it to be done when the moon was between ten and fifteen days old, with an odd number of eggs. He considered black hens the best, then yellow, and lastly white. He gave two cruses of half-boiled barley a day to a hen at large, and

FIG. 27.—ROMAN ENAMEL COCK (London)

---

[1] Pitt Rivers, op. cit., ii. 220-2.
[2] *Palladius*, 99. E.E.T.S.
[3] *Archaeologia*, liii. 287; lx. 165.
[4] Palladius, 33 and 109.

instructed the farmer how to treat the "pippe" and sore eyes.[1]

"The goose with grasse and water up is brought," says the monk who translates him. The Roman writer preferred white geese to skewbald and brown, and hatched out his goslings under a hen, with nettles for her to sit on, feeding them indoors for ten days. He allowed three geese to a gander.[2]

*Bee-keeping* was thoroughly understood by the Romans, and it is probable that they were not domesticated in this country before the Roman occupation.[3] It seems very unlikely that the Roman settlers should not have introduced an art of such great importance at a time when honey was the only form of sweetening. Palladius kept his hives in a "bee-yard gladsum and secrete," a square walled enclosure, full of flowers and herbs and sheltered by trees on the north. The brook or well was to have branches in it for the bees to alight on and drink. His hives were made preferably of thin rind, canes or willow twigs, sometimes of boards made cup-shaped, or from a hollow tree, but never of pottery. The hives were cleansed about April 1st and on a warm day in November, the spare honey being taken in October.[4] There were to be two or three openings the size of a bee, facing south.

Perforated vessels of pottery for straining the honey from the comb are found in Romano-British excavations.

It is considered that the Romans introduced pheasants and fallow deer into this country, but neither of them can have been of much benefit to the farmer, even in those days before the time of game laws.

The life of the farmer was far more secure from raids of both man and beast under the firm and disciplined rule of the Romans than it had been hitherto, however irksome their ways may have been to the freedom-loving Celt. In upland and open districts such as the South Downs and Cranbourne Chase the herds of sheep and cattle were still kept in entrenched enclosures at night, for wolves were plentiful, but on villas and farms in more civilized parts they were housed in the outbuildings of the farm-yard much as at present.

Roman methods of agriculture were necessarily different

[1] Ibid., 22.

[2] Ibid., 26.

[3] J. Thrupp considers that they were still undomesticated in the sixth and seventh centuries, and that the Saxons domesticated them. It is possible, of course, that the Roman art of bee-tending was lost at the Saxon Conquest and had to be rediscovered. *Trans. Ethnological Soc.*, N.S., iv. 169. Greenwell and Rolleston: *British Barrows*, 725.

[4] Palladius, 37-9, 147, 196.

F

from those of the North. They were accustomed to small individual holdings and light soil; hence a light and simple plough, drawn by a single yoke of oxen, was most suitable for their requirements. But two oxen naturally could not plough so long a furrow without stopping to take breath as a team of eight, and consequently the Roman furrow had a length of only 120 feet, and the acre (jugerum), was a short, broad piece of land, of the shape of two squares placed side by side, or in other words twice as broad as it was long, 120 feet × 240 feet.[1] But this light plough (the *aratrum*) was used mainly on small individual holdings and in the more backward villages. On the large estates — the villas, with their wide stretches of open fields—the heavy plough (the *caruca*) was used, and with its team of eight oxen, yoked either four abreast or in pairs, a much longer furrow was possible without a pause, and the acre there became a long narrow strip.

The Roman jugerum, not quite two-thirds of the statute English acre,[2] was considered a suitable day's work for a single yoke of oxen, when ploughed for the first time. For second ploughing they were expected to do a jugerum and a half. The Romans also had a beneficial system of cross-ploughing, i.e. ploughing the field again at right angles to the first furrows, which had probably been practised by the Celts. Pliny considered that a yoke of oxen should plough 40 jugera a year in light soil or 30 jugera in heavy. According to Palladius fat land was ploughed three times and lean land twice. In September manure was laid and at once ploughed in; marl was put on sandy ground and sand on clay. Larger ploughs were used to rig up the land to drain.[3] The Romans usually sowed five modii of wheat seed to a jugerum.[4]

This mention of "rigging up" may plunge us into a fascinating speculation. We know that in the fourth century A.D. Britain had become a peaceful agricultural country, producing an astonishing amount of wheat in excess of its needs, for in the time of Julian the Apostate eight hundred wheat ships were sent from these shores to supply the garrisons in Gaul. Where was the land in our forest-grown and marshy isle that yielded such crops? The Romans we know drained and reclaimed much land. In Northamptonshire, in the fen-lands near Peterborough, in Yorkshire and other parts lie great tracts

[1] F. Seebohm: *English Village Community*, 384.
[2] 3,200 square yards as compared with 4,840.
[3] Palladius, 42, 179, 180.
[4] F. Seebohm, op. cit., 275.

of grass-lands now uncultivated but still showing clearly and indelibly the long low ridges of ancient ploughing when the land was "rigged up" as a means of drainage. Were these the Roman wheat-fields? On some of the ridges stand great trees many centuries old, and these owing to their position must have germinated at some period after the lands had gone out of cultivation. After what great crisis of depopulation did these great tracts of arable land cease to be used?[1] In the opinion of many it was the Black Death of 1348, but others consider that it may have been the withdrawal of the Romans and the Saxon Conquest. We cannot tell.

We have already seen how Roman and British methods of cultivation were suffered to continue side by side, and we can therefore picture the villager, in one part, ploughing his little double square of land with his pair of oxen, slacking their yoke at the end of the furrow to cool their necks, and cleaning

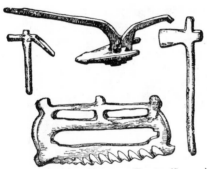

FIG. 28.—SET OF MODEL TOOLS (Sussex)

his share with a scraper fitted to a stick; and in other parts the dwellers on large villa estates ploughing their long strips with heavy, wheeled ploughs drawn by eight little oxen, commonly yoked four abreast, one man guiding the plough and another urging on the team with his long rod and his song. A little bronze model of a Roman plough has been found in Yorkshire, with two beaks (the upper one of which may be meant to represent a coulter), but which is otherwise a simple "over-treading" plough. The ploughman holds the stilt in his left hand and has a short implement, either goad or scraper, in his right.[2]

Rham has an illustration of a Roman plough with a sort of arrow-shaped wooden "sole" in which are set, upstanding

[1] H. Rider Haggard: *Rural England*, ii. 122-5.
[2] Wright: *Celt, Roman and Saxon*, 256.

on each side, stout iron teeth, to turn the soil cut by the share to one side or the other, a function which was later performed by the "mould board" or turn furrow. This corresponds with Virgil's description of a plough in which "the teeth are fitted to the double back."[1]

Spades at this time were made of wood, with a crescent-shaped keel of iron at the bottom, into which the wood was inserted. One found at Cranbourne Chase measures 7 inches in width and 1 inch in depth. This form of spade, owing to scarcity of iron, continued in use until the sixteenth century. Iron points were also fitted on to wooden picks, and to ox-goads. The work of the Romans in developing iron, tin, lead and salt mines must have ultimately benefited the farmer and his equipment to a considerable extent.

Reaping continued to be two processes, the ears being cut first, high up on the stalk, the straw cut afterwards if worth

FIG. 29.—TRIBULUM

it. Small iron sickles were used for the first process, but straw cutting and mowing of hay were much facilitated by the introduction of scythes. Some of these were nearly seven feet long, with a sharp bend about 17 inches from the butt.[2] The inhabitants of the Cranbourne Chase settlements reaped with very small sickles and stored their corn, apparently, in little barns mounted on posts. In more backward regions a few underground store-pits still remained in use. In Roman villas corn was stored in the barns and granaries of the outer courtyard. Palladius liked his barn high, windy, cold and dry, with a hard floor two feet thick, having a place for every sort of corn. He carted his corn in a square two-wheeled cart, boarded, and drawn by a "meke ox."[3]

[1] Rham, *Dictionary of the Farm.*
[2] F. Ward, *Roman Era in Britain,* 201.
[3] Palladius, 18 and 159.

Threshing, which the Romans were accustomed to do in the open air on a hard, beaten threshing floor, in Britain had to be done under cover, and the Celts, as we have seen, were accustomed to dry the corn beforehand. The Romans apparently brought with them a threshing implement called a "tribulum," which was a flat sledge of wood, five to six feet long and two to three feet broad, with the underside furnished with holes into which were inserted flakes or splinters of stone, or sometimes pieces of iron. A heavy weight was placed upon the sledge, or was supplied by the person of the driver standing upon it, and it was drawn over the corn by a yoke of oxen, so that the grain was scattered from the ear.[1] But this seems to have been a primitive survival rather than a new invention, and the usual method of threshing employed was probably the flail.

For drying the corn the Romans introduced a true kiln, having an underground furnace with flues which carried the heat under the floor of the kiln-chamber. These are found in Roman villas and in the Romano-British villages of Cranbourne Chase. This would no doubt be the "piped-kiln" of the Welsh Laws.

Their most common means of grinding the corn was the revolving stone quern (mola versatilis), which was known in Britain in the Bronze Age, as we have seen, but did not become general until this period. In early querns the lower stone was almost conical in shape, and the upper one concave to fit over it, the idea being that the downward slope would enable the flour to fall more easily from between the stones. But as this was found to be unnecessary the shape became more and more flattened, until the faces of both stones became horizontal. The upper stone had a hole in the centre which acted as a hopper into which the corn was poured, and a small hole at the side into which the handle was fixed to turn the stone. A spindle or pin was fixed into the centre of the lower stone, passing through the hopper above and also through a bridge of wood which crossed the bottom of the hopper, leaving a space on each side for the grain to percolate. This bridge was called the mill-rynd, and was later made of iron and often shaped like a cross instead of a single bridge. Various improvements were made in the working as time went on: the faces of the grinding stones were grooved, and the fineness of the flour was regulated by fixing a piece of wood

[1] Sir John Evans: *Ancient Stone Implements*, 284-5. Evans identifies it with the "sharp threshing instrument having teeth" mentioned in Isaiah xli. 15.

into the socket of the rynd, and thus bringing the stones nearer together or farther apart as desired. Another variation in shape was the pot-quern, in which the lower stone was like a shallow pan, with a spout for the meal to escape, the upper stone which fitted into it being somewhat pot-shaped.

The typical Romano-British quern was larger and flatter than the earlier type, the average size of those found being about 15 inches in diameter.[1] Larger ones often had two handles and were worked by two women seated opposite each other, to get up speed.

Until the fourth century, when the establishment of Christianity as the State religion led to the abolition of slavery, the Romans used large slave mills of the type found at Pompeii, consisting of a conical lower stone (meta), surmounted by an upper one (catillus), hollow underneath and crowned by a large hopper, so that the shape was not unlike a dice-box in solid stone. This weighty contrivance was pushed round by projecting handles like a ship's capstan. A small mill of this sort was found at Ham near Poole, each stone being one foot high and the grinding edge two feet in diameter. The stone hopper was, however, absent, being probably replaced by a wooden one, which would be a great saving in weight, the whole machine weighing nearly 3 cwt. as it was. These mills would doubtless be used on Roman villas where there was a staff of slaves. The Roman horse- or ass-mill was worked in the same way, the animal being harnessed to the bar in place of the slaves. This type of cattle mill remained in use until the close of the fourth century, when it was superseded by a newer form.

Alongside of these newer inventions the old saddle-stone called by the Romans "mola trusatilis," or the "thrusting mill," to distinguish it from the quern, continued in use in outlying parts; and the mortar seems to have remained still longer, for some very elegant Romano-British examples have been found.

But in the fourth century A.D. a form of mill came into general use among the Romans which was destined to supersede all these. This was the *watermill*, a contrivance known in the days of Pliny, but not adopted to any great extent until the abolition of slavery under Constantine and the increased demand for free meals in Rome brought about an urgent need for the use of water power. The watermill was described by Vitruvius (20-11 B.C.) and was very probably invented by him,

[1] F. Ward: *Roman Era in Britain*, 192.

although the edict protecting it and thereby establishing it in general use was not issued until A.D. 398. The principle of its action was a vertical waterwheel connected with a cog, which impelled a horizontal wheel turning the millstone. Whether this form of mill was introduced into Britain during the half-century between its general adoption in Rome and the evacuation of this country is not known with certainty, but if it was the case it would only be as the "latest improvement," on large and flourishing villas, or perhaps in enterprising townships. The number of quern stones found among Roman remains testify that that was the implement commonly used for grinding. Already the famous Andernach volcanic rock was imported for making millstones.

It was not until the coming of the watermill that the process of grinding the corn was removed from the precincts of the farm-yard. Until then the mill and the bakehouse were inseparable; the corn was ground in the cottage or at the villa on the spot where it was to be formed into loaves and baked, and on Roman estates the "pistrium" was a combined bakehouse and mill, and the "pistor" (lit. the pounder) was responsible for preparing both the meal and the loaves.[1]

The principal crop grown by the Romans in Britain seems to have been wheat, and this judging by the specimens found at Cranbourne Chase was of very good quality,[2] not inferior to what is grown now at a similar level. Oats, peas, beans and barley were also grown,[3] but evidently in much smaller quantities. A quite new importation at this time was the vine, permission to plant vines and make wine in Britain being granted by the Emperor Probus (276-82). It is difficult to realize that vineyards ever really flourished in this country, or that wine worth drinking could be made from the grapes, but such was evidently the case, judging from the regular occurrence of pictures of vine-culture in Saxon and mediaeval calendars. Once established in this country they seem to have remained far on into mediaeval times, especially on monastic lands.

To the Romans also is usually attributed the introduction of the orchard and herb garden, two very pleasant and healthful

---

[1] For all this section see R. Bennett and J. Elton: *History of Corn Milling*, i. 128 onwards; ii. 31 onwards.

[2] General Pitt Rivers found, on taking a specimen from the excavations and from wheat just grown on a farm half a mile away, that exactly the same number of grains went to the cubic inch.

[3] Roman crops included millet, lentils, vetches, clover, lucerne, trefoil, lupin and hemp, but there is no evidence that these were grown in Britain.

additions to the farmer's homestead.  There seems no reason
why apple orchards should not have existed before, though
possibly the idea of congregating the trees together in a suitable
place near the farm-house may not have occurred to the Celt.
To the apple-tree was now added the cherry,[1] and probably
the pear, quince, medlar and peach; possibly more.[2]

Among trees it is thought that the chestnut and walnut
were now introduced, and the elm naturalized.  Cæsar states
that the beech and fir were not found in Britain, which, if true,
implies that the Romans introduced them also.[3]  The Romans
used elm leaves for cattle fodder and trained vine leaves up
the trunks.  The woods found at Cranbourne Chase were oak,
ash, willow, hazel, beech, elm, walnut, and edible chestnut.
Of these the last four were not found in the pre-Roman village
of Glastonbury.

What exactly was grown in the herb or kitchen garden we
do not know either.  All manner of herbs and vegetables
were known to the Romans, but it does not follow that these
were all brought over, or remained in cultivation subsequently.
Many of them, if grown here by the Romans, must have been
lost again, for they had to be reintroduced in later centuries.
We may safely assume that the cabbage, regarded by Pliny
as the most esteemed of all vegetables, was cultivated, as it
is indigenous in England.  Palladius gravely remarks that
stolen cabbage-seed is said to thrive the best.  Probably leeks
also, afterwards so popular both with Celts and Saxons, were
introduced now, and perhaps onions, turnips, lettuce and
parsley.[4]  Flax no doubt continued as indispensable as before.

What the life of the typical farmer's homestead was like
at this time it would be difficult exactly to say; probably at
no time were there greater contrasts.  At the top of the
scale, as we have seen, was the rich farmer living in luxury
in his great country villa and farming his estates with gangs
of slaves; burning coal, devouring oysters, eating his eggs with
little bone spoons in a thoroughly civilized manner, and using
the pointed handle to pick his edible snails out of their shells.

[1] Pliny: *Nat. Hist.*, xv. 30.

[2] The last four occur in the Anglo-Saxon vocabulary as words derived
from the Latin.  The Romans cultivated them, and also the fig, plum, mul-
berry, almond and medlar, but we do not know if they gained a place in
British orchards at this time.

[3] Caesar: *De Bello Gallico*, lib. v. c. 12.

[4] These last four occur in the Anglo-Saxon vocabulary, derived from the
Latin.  Palladius grew chiefly besides these, radish, coriander, beet, parsnip,
asparagus, and many small herbs.

At the other end of the scale men were living in remote spots in round wattle huts, hunting and guarding their little flocks and herds from wolves, growing a little corn and storing it carefully in pits, boiling their humble pots by dropping in red-hot flints because their pottery was not well enough made to be put on the fire and metal pots had not reached them. But between these extremes must have been a numerous class of fairly comfortable little homesteads. There would be the Celtic tribal homesteads, working towards the stage of development described in the last chapter, and absorbing compulsorily or otherwise a certain number of Roman "improvements"; and there would also be the many small households of "coloni" planted by the Romans, cultivating their separate holdings, and doubtless living with a humble measure of Roman comfort, each with little shrines for the Lares, guardians of the household, and the Penates, protectors of the storehouses. Roman rustics wore shoes made of a single piece of leather with the soles studded with iron nails, an early ancestor of the hob-nailed boot.

Under the peace and prosperity of Roman rule conditions of life would no doubt improve. The Celts, restrained from the perils and delights of tribal warfare, settled down into peaceful agricultural communities, and lived more on their domestic animals (including horses) and less on products of the chase than hitherto. In the Cranbourne Chase settlements sheep and oxen were most usually eaten, and after them horses and pigs, occasionally deer and fowls and perhaps even dogs. Cheese, whether it was much in use formerly or not, now became a universal article of food, and doubtless added to the labours of the farmer's wife. Palladius made his cheese in May, curdling fresh milk either with rennet from a kid, lamb or calf, or with teasel flower, or fig-tree milk, or with "the pellet that closes a chicken's or pigeon's crop." The curd was wrung and pressed, wrapped in salt and pressed again and again, laid on crates, and finally stored in a close place out of the wind.[1] It was sometimes flavoured with stamped pine-nuts, the juice of bruised thyme, pepper, or other flavouring, but these refinements would probably be beyond the capacity of the Romano-Britons.

Lamps now became a common form of lighting, and candle-sticks both of pottery and iron were used, the latter being made pipe-shaped with a spike to stick into the wall.

That very useful article the spoon now came into general

[1] Palladius, 154.

use upon the table, and was used in many shapes and sizes. The farmer's wife, if tender-fingered, was even able to obtain a small bronze thimble to assist her to cope with her bone, bronze or iron needles. The kitchen also would be stocked with bronze or iron pans, or even some of pewter, and her crockery was much improved, besides being supplemented by pots and dishes of Samian ware and cups of glass.[2]

Early in the fifth century the bringers of this foreign and somewhat exotic civilization we have been considering withdrew from Britain, leaving it to fall into confusion. But for four hundred years their rule had continued, and cannot therefore have been without some lasting effect upon the history of the race, perhaps endowing it in some degree with their own ideal of a solid, respectable and orderly life, unaesthetic but practical, which is characteristic of so many Englishmen to-day. Some of their enduring contributions to the English farm-yard we have noted. So strong however did the Celtic element remain that the fifth century became the era of a Celtic revival of considerable importance. The spirit of the Celtic race arose afresh, and emerging from its strongholds in the West, prepared for its last struggle with the invading Saxon.

[1] Wooden ladles were used at pre-Roman Glastonbury, and bronze spoon-shaped objects of uncertain use have been found in various places. Clay and horn spoon-shaped objects often occur in Neolithic interments, but probably were used only in this connection.

[2] Haverfield and Macdonald: *Roman Occupation of Britain*, 236.

# CHAPTER V

# THE SAXON PERIOD

## REFERENCES

| | |
|---|---|
| A.l..I. | Thorpe: *Ancient Laws and Institutions,* vol. i. |
| Att. | F. L. Attenborough: *Laws of the Earliest English Kings.* |
| Gerefa. | Treatise on Gerefa (Reeve) in Liebermann: *Die Gesetze der Angelsachsen,* vol. i. |
| R.S.P. | Thorpe: *Rectitudines Singularum Personarum,* vol. i. |
| Coll. Ælfric. | *Colloquium of Ælfric.* Thorpe: *Analecta Anglo-Saxonica.* |

# CHAPTER V

## THE SAXON PERIOD

FROM the fifth to the seventh centuries this country was subjected to a continuous series of invasions by bodies of Angles, Saxons, Jutes and Frisians, all of whom are broadly referred to in general speech as Saxons, or Anglo-Saxons. They were all of the tall, blond, Teutonic race, fair-haired and blue-eyed, a change in hue from the dark colouring of the Iberians and the Romans, and from the red or sandy-haired Celts. Although originally of tribal organization like most of the "barbarians," they had become to a certain extent disintegrated by centuries of warfare and migration, and seem to have arrived in England as families or groups of followers rather than as tribes, and when conditions became sufficiently peaceful, congregated in villages, "hams" and "tuns," under their chiefs, with a communal organization for defence and agriculture. During the five or six hundred years of their ascendancy many changes and developments necessarily took place, until by the tenth or eleventh century we find the "manorial system" so typical of the Middle Ages already very widely established.

The tendency of the Saxons was to form larger settlements than the little tribal hamlets of the Celts, and having a rough way with them they swept aside the existing inhabitants, and paid but little attention, so it seems, to the civilization left by the Romans, but established their villages, apportioned the land, and organized it for tillage in a manner which we shall presently examine.

Originally a portion of land called a "hide" was assigned to each family for its support, and this consisted of the amount of arable land which would normally be cultivated in a year by a man with a full plough team of eight oxen, together with a share of meadow-land for hay, pasture for his beasts, wood, stream, waste and other common ground of the village. The amount of arable land contained in the hide was probably on an average about one hundred and twenty acres, but varied according to the nature of the soil and the system of agriculture employed. This level distribution of land, a hide to a homestead, could not, however, be kept up as time passed, and men soon came to hold more or less than this amount.[1] Social ranks inevitably developed, and in the typical Saxon society we have the *thane* or lord of the "ham" with his home farm, the *ceorl* or freeman, under the protection of the lord but working his farm quite independently, and various classes of unfree tenants owing service to the thane, the *geneat* or follower, the *gebur* and the *cotsetle,* who had their own share of land in the village to cultivate, but also had to do a fixed amount of work on the farm or "demesne" of the lord. Below these were the *theows, esnes* or serfs, attached to the household of the lord in various capacities, but comparatively few in number.

The sturdy and independent ceorl flourished under Saxon rule, and in Kent particularly formed the typical class. He held from one to five hides of land, which he cultivated with the help of his family and perhaps some serfs, or he could hire another man's yoke of oxen and pay him in fodder or other goods.[2] He was "foldworthy," that is he could fold his sheep on his own lands at night and thus reap the benefit of the manure, instead of having to fold them on the land of the lord as his poorer neighbours had to do. His "flet" or homestead and close was surrounded by a fence or hedge which he was bound to keep up. If he left a gap unrepaired he got nothing for damage done by straying cattle,[3] though he could get compensation for trespass by poultry and geese. The homestead consisted of a group of timber buildings: house, byre, pig-sty, barn, storehouse, stables and sheds for horses and wagons, ranged round a courtyard, with rick-yard and flax plot beyond, and usually an orchard and a garden with its beehives; all enclosed within a fence or wall of mud or stone, according to the standing of the owner. If the ceorl flourished to the extent of owning five hides of land, and could erect a church and kitchen, bellhouse and "burh gate" (gatehouse), he attained the rank of a thane, but if on the other hand he was so misguided as to work on one of the Festivals of the Church he lost his freedom altogether.[4] The holidays appointed for him by Alfred were very liberal; they consisted of twelve days at Christmas, the "day on which Christ overcame the Devil" (February 15th),

[1] By the time of the Domesday Survey the hide had become merely a unit of assessment for taxation without any fixed relationship to the amount of land contained in it. The amount of arable land cultivated by a plough team was then called a " carucate."

[2] A.L.I., 141. Att., 157.

[3] A.L.I., 127. Att., 49.

[4] A.L.I., 173. Att., 107. Laws of Edward the Elder and Guthrum. Ine had allowed him the alternative of a 60s. fine. 105. Att., 37.

St. Gregory's Day (March 12th), seven days before Easter and seven after, SS. Peter and Paul (June 29th), the week before St. Mary's Mass (August 15th), and one day at All Hallows (November 1st).[1] If he were compelled to remain idle on all these days the fortnight at Easter and the first week in August must at times have gravely inconvenienced his farming operations. He could, however, keep his hired labourers and theows at work as they only got "what they deserved" in the way of holidays. He could hunt venison and take game on his own lands, but might not follow the chase into the lands of his neighbours.

All classes of the community, from the thane to the cotsetle, had a separate enclosed homestead such as we have described as belonging to a freeman, i.e. house and outbuildings, garden plots and rick-yard, fenced round in some way, the difference being only one of degree.

The thane had a moat round his wall or stockade, with a gatehouse at the entrance, and a bellhouse, presumably to summon his dependents to work, or assemble them in case of emergency. His hall, timber-built and timber-roofed, varied of course in size, but consisted of one large apartment which was used as a general living- and sleeping-room. A bower for the ladies, a kitchen, storehouse, malthouse and any other apartments that conformed with his means and his ideas of luxury, were built on separately outside, but a bath-house, deemed necessary by the Celts, the Romans and even the Norsemen, does not seem to have been included.[2] The roof was either peaked or rounded.

This, as we may see, was a very complete stage of separation between dwelling-house and farm buildings, and compared with the entirely self-contained establishment of the Continental Saxons and Frisians described in the last chapter seems almost disintegrated. That ancient type may have been that in use when the Saxons arrived, and might linger among the peasantry for a considerable time, but in Northern Europe the tendency seems to have been towards separate buildings.[3] There was, however, an intermediate stage, used a good deal for farmhouses in the eleventh century and long after. In this, although living-rooms, cowhouses and barn were all included under one roof, the English had devised a means of separation by placing

---

[1] A.L.I., 93. Att., 85.
[2] His household gear, however, did include bath-tubs and a soap-box.
[3] G. T. Files: *The Anglo-Saxon House*, 25.

the main floor transversely as a wide walled passage between the house part and the ox-stalls. This passage formed at once the entrance and the threshing floors, doors opening from it to the dwelling on one side and the ox-house on the other, and hence the entrance to a house came to be called the "threshold."

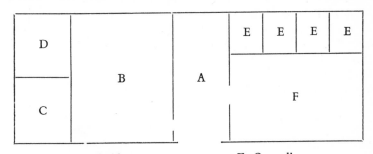

A. Threshold.
B. House-place.
C. and D. Chamber and buttery.

E. Ox-stalls.
F. Barn floor.

The main living-room, the "house-place" or "fire-house," which corresponded to the hall, contained the family hearth in the centre, and usually had a chamber for the women and a buttery or storehouse partitioned off at one end. On the other side of the entrance passage or threshold were the stalls for the plough oxen opening on to the barn floor from which they were foddered. The corn harvest was stored in a loft above and thrown down on to the "threshold" to be threshed.[1] This simple and convenient form of house would doubtless be much used by ceorls and even the better-off of the unfree classes, who would, perhaps, dispense with the chamber and buttery. The dwellings of the humbler geburs and cotsetles were probably merely single-roomed wooden or mud huts with sheds for such beasts as did not share the living-room, particularly for what precious plough oxen they possessed. The Saxons, who had a taste for cheerful colours, painted their houses yellow and blue, yellow ochre being used for the former, and "archil," a colour made from liverwort and mixed with limewash, for the latter.

As regards *stock*, the Saxon farmer was greatly benefited by the introduction of a larger breed of *oxen*. Instead of (or as well as) the little dark *Bos longifrons* he now had good

[1] Addy: *Evolution of the English House*, 60 and 85.

sized white cattle with fairly long horns, a breed from Friesland, Jutland and the Lower Elbe, probably derived from the wild Urus.[1]  A representation of one of these superior animals may be seen on the Bayeux Tapestry, being carried off by the marauding Normans. They were more formidable than the Celtic shorthorn and apparently apt to be troublesome, for the Laws of Alfred dealt very drastically with their behaviour: "If an ox gore a man or a woman so that they die let it be stoned and let not its flesh be eaten."  If the ox had been "wont to push with its horns" for several days before this regrettable incident occurred, and the lord was aware of it and had not shut the animal up, he was liable to be slain himself.  If any beast injured a man the owner had to hand it over to the sufferer and come to terms with him.[2]  Cattle were anciently used as a form of currency, and the value of a man was pretty universally fixed at one hundred cows, but the Saxon kings made an interesting attempt to bring this Celtic custom into line with their more modern coinage by fixing the value of a cow at three ounces or "scores" of silver, one ounce of silver being equivalent to twenty Anglo-Saxon pence.[3]  Curiously enough about the same time that the cow became of less importance in this way, the possession of a certain number of plough oxen came almost to constitute a claim to a corresponding portion of arable land.  Oxen often wore bells, and were used for drawing carts, as is shown in the illustration for June in a Saxon Calendar.[4]  (See Fig. 33.)

*Horses* seem to have undergone some improvement under the Saxons, and Athelstan imported some from Spain to improve the breeds.  A good horse in his reign was valued at half a pound, but this would probably not be a farm animal.  In the laws of Ethelred, however, horses are again valued far above oxen, which seems to imply that they were scarce.[5]  They are said not to have been used by the Saxons for riding until the middle of the seventh century, and even then only in case of necessity.[6]  In the tenth century the geneat had a horse, with

[1] The famous white oxen of Chillingham were thought to be descended from these, but T. McKenny Hughes (*Arch.*, lv. 157) considers the Chillingham ox the result of crossing with Roman cattle and not descended from the Urus.

[2] A.L.I., 49.  Att., 75

[3] F. Seebohm: *Tribal Custom in Anglo-Saxon Law*, 49.

[4] Cott. MSS, Julius A, vi.  Here the men are cutting wood and loading it in a light cart, while a yoke of oxen are standing by ready to draw it.

[5] A.L.I., 357-9.  Att., 161.

[6] J. Thrupp: *Trans. Ethnological Soc.*, N.S., iv. 1681.

which he was bound to "ride and carry and lead loads" for his lord, and similarly the beekeeper and the pigkeeper often had a packhorse which was at the service of the lord for carting or going to market, but while it was thus employed the gebur was free to work on his own land and was not obliged to work for his lord.[1] The Celtic custom of turning mares loose into the forest to breed and catching the colts when required for breaking seems to have continued, at any rate in some parts, for horses untamed or in the woods (equi indomiti or silvestres) occur in the Domesday record for Norfolk, and still more often in Cornwall, as we have seen, where Celtic habits would naturally survive. By the laws of Athelstan no man might "part with a horse oversea unless he wish to give it."[2]

*Asses* were certainly known in England in the days of Ethelred, Alfred's predecessor, but were evidently unusual possessions, as they were valued at twelve shillings.

*Swine* were very important to the Saxon farmer, for their food value, and large herds of them were kept. Mr. Thrupp quotes a will in which two thousand swine are bequeathed to each of the testator's daughters, as their sole legacy. His opinion is that in the seventh century swine were only semi-domesticated, that many were bred on the farms and then turned out into the public forests.[3] They were not usually allowed on the meadow-land or stubble, as their rooting would spoil it for the other animals. As the manorial system developed the lords of the manor seem to have annexed the woods as their property, and in the season of acorns and beechmast, between the feast of St. John (?August 29th) and the New Year, only allowed the villagers to run their pigs there on payment of a certain proportion of the herd. This customary payment (afterwards the Norman "pannage") was known as "denbera." Under Alfred a careful eye was kept on the benefit the pigs derived from their feeding in the woods and the contribution regulated accordingly: of those which had acquired three fingers thickness of fat one third were taken at the end of the season, of two fingers fat one-quarter, and of a thumb thickness one-fifth.[4] For feeding pigs on grass-land a payment known as "gaerswyn" was made, and in Sussex the customary payment came to be one pig in every seven; in parts of Cornwall one in ten.

By the tenth century swine were more thoroughly domesti-

[1] R.S.P., 434-5.
[2] A.L.I., 209.
[3] J. Thrupp, op. cit., 166.
[4] A.L.I., 133. Att., 53.

G

cated, and a class of pig tenants or tributary swineherds had appeared. The tenant was supplied with a herd of swine, of which he surrendered a certain number yearly for killing, usually ten full-grown pigs and five young ones, keeping the rest for himself. It was his duty to have these pigs stuck and to prepare, singe and scrape the carcases ready for the swineherd of the court to make the bacon and lard, and he was then rewarded with the entrails. He kept a horse, which he had to

FIG. 30.—ABEL TENDING SHEEP (from Cædmon's *Paraphrase*)

use in the lord's service at need, and at his death his herd of swine reverted to the manor.[1] They were still high-backed, long-legged creatures akin to the wild boar.

*Sheep* were universally kept, and their milk used for cheese and butter. Judging by the illustration to the Saxon Calendar the ancient, horned breed, long-legged and long-tailed, was more popular than the hornless variety introduced by the Romans. A comparison between the Saxon sheep in the Calendar (Cott. MSS, Julius A, vi) and the drawings of modern St. Kilda sheep

[1] R.S.P., 436.

in Pitt Rivers (*Excavations in Cranbourne Chase*, ii. 222) still shows a remarkable resemblance in length of leg, general build, and curve of horns.

Each gebur gave his lord a lamb at Easter. The usual value of a sheep was a shilling, and its skin might not be used for covering shields,[1] probably because it was needed for parchment.

Herds of *goats* were also kept, and milked, but were much less valued than sheep, being worth only 2d. in the time of Ethelred.

*Fowls* were universally kept, each gebur having his little chicken-yard from which he gave the thane two hens every year at Martinmas.

*Geese* are mentioned in the Laws.

*Bees* are thought by Mr. Thrupp to have been entirely undomesticated at the beginning of the Saxon period. They were not "capable of ownership," being always on the move, and anyone finding them could have the honey and wax. Under the Saxons, however, they finally became domesticated. The first step was to cut down the tree where the wild bees were, and take home the portion containing them, which must have been a precarious undertaking. Next a "rusca" or primitive hive of bark was made to imitate the tree trunk they were accustomed to. Later a sort of box or "bee cist" was evolved, and lastly a "hyfa" or true hive.[2] In the tenth century hives of bees were farmed out to tributary beekeepers, like the herds of swine, the tenant rendering five or more sestars[3] of honey a year, according to the custom of the district. But as his charges did not take so much attention as the swine, the beekeeper had to do other services for his lord, ploughing, reaping and mowing when extra hands were needed. Also, unless his share of land was very small, he kept a horse, which the thane could have for carrying and driving work at need. His hives reverted to the manor at his death.[4]

*Dogs* were still very necessary to guard the flocks from wolves. They wore collars and were expected to behave themselves, for a bite was very expensive to their masters. In the laws of Alfred the fine for a first bite was six shillings, for the second twelve, and for the third thirty, a very large sum in

[1] A.L.I., 209.
[2] J. Thrupp, op. cit., 169.
[3] If this was equivalent to the Latin sextarius it was about a pint.
[4] R.S.P., 435-6.

those days.[1] After that (if not before) the master was doubtless glad to dispose of the animal.

Most animals seem to have been folded or stalled at night, and were driven out to pasture by herdsmen during the day, special pieces of grass-land being sometimes reserved for them. Otherwise they grazed upon the common waste, or upon the arable that was lying fallow that year, or, after harvest, upon the stubble.

In the second half of the seventh century the "Yellow Plague" was raging and even affected the beasts, for about 684 there was "a mortality upon all animals in general throughout the whole world for the space of three years, so that there escaped not one out of the thousand of any kind of animals."[2] We will hope that the chronicler exaggerated.

We have already seen that the owner of a hide of land, or a carucate of arable land, was assumed to be the possessor of a full plough team of eight oxen, and as land came to be held in smaller portions this relation between oxen and plough-land was continued. The usual holding of the gebur, the typical unfree villager, came to be a "yard-land," about 30 acres in extent—a quarter of the average carucate of 120 acres—and the outfit accompanying it was hence two oxen, and also a cow and six sheep, with tools for his work and utensils for his house and, when he was first installed in his holding, seven of his acres ready sown.[3] A man with only one plough ox held a holding half this size, later called a bovate or ox gang.

The cotsetle, or cottar, held as a rule five acres, which he would have to till with his spade and hoe, as he did not possess oxen. Bondsmen were usually apportioned one acre strip.

But all these holdings were not separate and compact farms such as we have nowadays. The homesteads with their little gardens and closes fenced round against stray animals were grouped together in the village, but the acres which each man cultivated were scattered about all over the place according to the "open-field" system of agriculture, which we must now examine. There was no longer such a superfluity of land that fresh and untouched fields could be taken into cultivation each year and the former ones left to return to grass, as in the days of "extensive" culture. Arable land had now become permanently fixed, and some method had to be devised by which it

<hr />

[1] A.L.I., 79.  Att., 75.  Coll. Ælfric, 20.

[2] Adamnan's *Life of St. Columba*.  Ed. Reeves, 183.

[3] F. Seebohm: *English Village Community*, 133.

did not become too exhausted by constant "intensive" cultivation. The arable lands belonging to the village were therefore divided either into two or into three roughly equal portions. If into two portions, one was cultivated and the other left fallow in alternate years, and this was known as the two-field system. If the land was divided into three parts two were cultivated and one left fallow each year in rotation. This three-field system seems to have become rather the more usual, probably because more land was thus cultivated in a limited area, but both systems were used, the two-field being no doubt more suitable for poor soils. These great village fields being marked out, the whole of their area was divided up into long, narrow, acre strips an entire "furrow-long" (furlong) of 40 poles in length and only four poles in breadth.[1] These acres were divided from each other by narrow strips of turf usually called "baulks," but when they ran horizontally along a hillside, terracewise, the little grass banks between were called "linches." The odd triangular pieces of ground which were left over in cutting up a field were called "gores" or "gored acres," and were sometimes cultivated and sometimes left as "no-man's land." Along the top of each group of acre strips ran another strip at right angles called the "headland"; on this the ploughs were turned, and it was therefore ploughed last of all.

These acres were dealt out to the villagers according to the number of oxen they contributed to the plough team, and constituted a permanent claim to so much arable land. Thus the gebur with two oxen and the normal holding of a "yard-land" of 30 acres would get thirty strips scattered far and wide over the parish, so that he shared equally with his neighbours the bad lands and the good, our ancestors being more concerned with fair dealing than with the saving of any man's legs. Under the three-field system ten of his strips would be in each of the three fields, so that each year 20 acres would be under cultivation and ten lying fallow. He was not at liberty, however, to sow just what he liked, for the rotation of crops was fixed by the community. Thus in one year field A (which had been lying fallow) would have all its strips sown with wheat or rye in the autumn; field B (with the wheat stubble of last year) would be ploughed up and sown with barley, oats or beans in the spring; field C would be left fallow. Next year A would have the spring sowing, B would lie fallow, and C would be

---

[1] They were sometimes half-acre strips, of the same length but half the width; but this standardization of size was not fixed by statute till the reign of Edward I.

sown in the autumn with wheat, and so on in rotation, wheat always coming after fallow.

The two fields under cultivation were carefully fenced round each year to keep out the animals, each gebur doing his share, and the third was left open for the cattle of the village to feed on. The best meadow-land, kept for hay and divided among the tenants like the arable, was also fenced, the rest being left as common pasture. To this and to the common waste on the outskirts of the village the cattle were driven daily by the village herdsman, and in the autumn the swine-herd drove the pigs to the woods, receiving six loaves from each gebur for his services.[1] The villagers also had certain rights

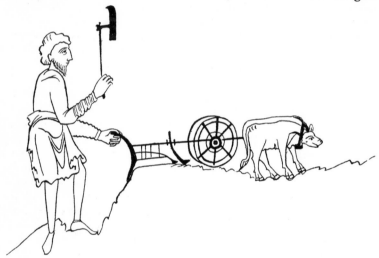

FIG. 31.—TUBAL CAIN PLOUGHING (from Cædmon's *Paraphrase*)

of wood-cutting on the waste lands to supply themselves with the necessary fuel and building material.

The regulation plough was a cumbrous affair, needing, as we have seen, eight oxen to pull it, but judging by illustrations of the period lighter ploughs requiring fewer oxen were also used. In the illustrations to Cædmon's *Paraphrase* of the Bible both Noah and Tubal Cain are seen ploughing with light ploughs, drawn by a single yoke of oxen but provided with wheels. Tubal Cain carries an axe-shaped instrument, probably a cleansing spud, and Noah, who has a ragged staff in his hand, is assisted by a driver with a long goad quite in the orthodox Saxon fashion. The plough in the Utrecht Psalter

[1] Att., 49. R.S.P., 435.

is a most primitive implement, simply a pole ending in the share, and a cross-beam, without coulter or wheels, the plough-man holding the single stilt with his left hand and goading on his pair of oxen with the right. Even in the Anglo-Saxon Calendar only four are shown in the team, but it may be that the artist felt that eight oxen would occupy too much space in the picture. Here we have a heavy plough with large wooden spoked wheels and two handles, and a very solid wooden beam behind the share which was evidently an elemen-tary form of the mould-board or turn-furrow, which directed the sods to right or left as required. The driver walks in front of the oxen with the usual long rod or goad. The Saxons yoked their oxen in pairs, not four abreast like the Celts. Some consider that it was the great eight-ox plough with its unwieldiness in turning which gave that strange twist to the furrows, like a reversed and much elongated S, which has often been observed on anciently cultivated lands.[1]

The old tribal custom of feasting the chief or king on circuit lingered among the Saxons, but in time took the form of a fixed rent in kind rendered at intervals. In the time of King Ine the food rent from 10 hides was 10 vats of honey, 300 loaves, 12 ambers of Welsh ale, 30 ambers of clear ale, 2 full-grown cows or 10 wethers, 10 geese, 10 hens, 10 cheeses, a full amber of butter, 5 salmon, 20 pounds of fodder and 100 eels.[2] Collecting centres were formed for corn, provisions, cattle and dairy produce, known respectively as bartons, berewicks and herdwicks; but as the manorial system developed this arrangement became for the most part restricted to Royal and Monastic lands, for which it was obviously a convenient form of organization.

Except for a few theows attached to the household and probably living in little huts round it, the work of the farm was done by the unfree tenants, the geburs and cotsetles, and even at times the geneats, who were all bound to contribute certain services in return for their holdings. Thus each gebur from January to November had to plough an acre a week for the thane, and in addition to this three acres as "boon-work",[3]

---

[1] A recent writer, however, suggests that, especially on sloping ground, the curve was to prevent the washing away of the soil by drainage which might result from a straight cut (W. M. Tod, *Farming*, 56).

[2] Att., 59. An amber in the thirteenth century was four bushels, but it may not have been the same under the Saxons.

[3] "Precariae" or boon-works were originally supposed to be not compulsory services but those rendered at the lord's request. Consequently it was usual for the workers to have an allowance of food and drink; but they came to be as fixed and unavoidable as the other services.

two acres in payment for his grazing rights (or more if he needed more grass), and three acres ploughed and sown with his own seed. The thane would generally have besides one or two plough teams of his own, and these were driven by his own serfs. A description of the labours of these men is given very graphically in the Colloquium of Archbishop Ælfric, a very attractive little work of the period, designed to teach Latin to Anglo-Saxon boys. The ploughman is asked to describe his work and replies, "O my lord, I labour hard: I go out at daybreak in order to drive the oxen to the field, and I yoke them to the plough. There is not so stark a winter that I dare stay at home, for fear of my lord, but having yoked the oxen and fastened the share and coulter to the plough every day I have to plough a full acre or more . . . I have a certain boy driving the oxen with a goad who also is hoarse from cold and shouting . . . I have to fill the stalls of the oxen with hay and water them and carry out their litter . . . Indeed it is great toil, because I am not free."[1]    The seed was then scattered by the sower, who if he sowed all the seed well for the space of a year was rewarded with a measure of each kind of seed.[2]

Harrowing is not mentioned among the gebur's services, but was done by someone, as it had to be superintended by the reeve. They would be light harrows, used to cover over the seed, not to break up the ploughed clods. This was done with a mattock. Probably the geburs contented themselves with rakes on their own ground, or perhaps drew boughs over it. But besides their ploughing duties the geburs had to do at least two days' "week-work" for the thane, and three days a week, from Candlemas (February 2nd) to Easter, at any kind of farm work that was required. The cotsetles had to work for the thane every Monday.[3]    At harvest-time not only did the geburs have to work three days a week for the thane, but even the geneats were called in to reap and mow. The other duties of the geneats were to ride errands and carry loads with their horses and assist in the making of hedges and roads. If a gebur owned a horse he also had to lend it for carrying duty. The corn being cut and carried it was entrusted to the granger or barn-keeper ("berebrytt"), who was rewarded with the corn-droppings at the barn-door, if the reeve considered he had

[1] Coll. Ælfric, 19.

[2] R.S.P., 439.

[3] They did not have to plough for him as they did not possess plough oxen.    The services of all classes varied according to local customs and needs, but those specified in R.S.P. were presumably average ones.

worthily earned them.[1] The corn was threshed and winnowed (with a winnowing fan) during the winter, a kiln being erected in the yard to dry it before threshing[2] as the Celts did—a testimony to the continued dampness of the climate.

Corn was still ground by the women in hand querns to a large extent, especially among the poorer classes. In the Code of Ethelbert, about 565, the King's bond-maid who grinds at the mill is mentioned[3] (she being worth only half as much as the other maid-servants), but later on watermills seem to have been pretty generally adopted. Besides the Roman type of watermill, with a vertical wheel, described in the last chapter, there was an earlier form, known as the "Norse mill," but originally Greek, remains of which have been found in Scotland, Ireland and the Isle of Man, but not in England. It is thought, however, that it may have been used by the Saxons, but early superseded by the Roman type. This mill had a horizontal wheel which lay flat upon the water; a spout of water was directed upon its sloping blades, and the wheel revolved on a central shaft which was fixed above to the nether millstone. This form of mill needed less water power than the Roman wheel, but was very slow in its action. It was still in use in Shetland and Lewis in the nineteenth century, and is found in Norway, Roumania and parts of China.[4] The earliest allusion to a Saxon mill occurs in 762, at Cert near Dover, but by the time of the Domesday Survey there were five thousand watermills in England. Oddly enough the miller or "millward" does not appear either in the Rectitudines or the Colloquium, but he is mentioned in the Gerefa, one of the duties of the reeve being to supply him with the necessary implements and to see to the construction of watermills.

The thane manured his fields by the simple method of compelling his tenants to fold their sheep on his land at night. The village field which was lying fallow for the year had the benefit of the cattle droppings, part of the duty of the herdsman being to drive the cattle evenly over it. The cultivated fields were also manured in this way after the harvest was gathered in. Manure, presumably litter from the cattle stalls, seems to have been scattered over the corn-fields after harrowing. Marling was occasionally resorted to, but does not seem to have been a general practice.

[1] R.S.P., 441.
[2] Gerefa, 454.
[3] Att., 5.
[4] R. Bennett and J. Elton: *History of Corn Milling*, ii. 9-31, 82-101.

There were also on the estate a number of men with definite work, who were at first village officials elected by the house-holders and later became manorial officials. Some of them were serfs, but all worked for their keep and not for wages.

The *Shepherd* in the Colloquium describes his duties as follows: "First thing in the morning I drive my sheep to pasture and stand over them in heat and cold with dogs lest the wolves devour them, and I lead them back to their sheds and milk them twice a day and move their folds besides, and I make cheese and butter and am faithful to my lord."[1] According to the Rectitudines he was allowed to fold the flock on his own land, for the benefit of the manure, for twelve nights at midwinter; he was given one lamb of the year's increase, and a bell-wether's fleece, also the milk of his flock for seven nights after the equinox, and a bowl of whey or buttermilk every night during summer. The shepherd in the Calendar wears a cloak and has a long staff, curved over at the top, but not so much as a modern crook.[2] From Martinmas to Easter it was the duty of the geburs to take turns in "lying at the lord's fold" at night to guard it from thieves.[3] These would be the winter sheep-sheds near the manor.

The *Swineherd* was either a pig tenant of the sort we have already described, or a theow attached to the lord's household. The latter received the ordinary allowance of a bondsman and also a young pig from the sty each year, and the entrails of a slaughtered pig when he had prepared the hams, bacon and lard for the household.[4] This would be done in the larder of the manor, a part of the herd being kept in a sty in the yard.

The *Oxherd* guarded the plough oxen during the night, and during the day presumably had charge of those which were not being used at the plough. "O my lord," he says, "I labour much: when the ploughman unyokes the oxen I lead them to the pasture, and all night I stay by them to watch against thieves, and again first thing in the morning I commit them to the ploughman well fed and watered."[5] He sometimes had to climb trees and throw down leaves to the cattle, a duty not unattended by accidents.[6] This watch-

[1] Coll. Ælfric, 20.
[2] Cott. MS. Tiber, B. v.
[3] R.S.P., 434, 440.
[4] Ibid., 438.
[5] Coll. Ælfric, 20.
[6] Ælfric's Hom., ii. 150.

ing of the cattle sounds like a summer arrangement. In winter he could no doubt have them stalled at night and could snatch a little sleep; it is difficult to see how he got any at other times. Cædmon was such an oxherd, and it was when he had composed himself to sleep in the ox-stall, according to custom, that his vision came to him. He had a house of his own, but on this occasion had been attending an entertainment at the house of a neighbour.

The oxherd was allowed to pasture two or more oxen and a cow with his lord's herd, and received shoes and gloves in addition to his keep.

The *Cowherd* drove the cows out to pasture and milked them, in return for which he was allowed the milk of an old cow for seven days after she had recently calved, and the "beastings" of a young cow for fourteen days.[1]

The *Goatherd* received all the milk of his herd after Martinmas (November 11th), and before that a share of whey, and one kid of the year's increase, if he tended his herd well.

The *Beekeeper* was usually a bee tenant as we described earlier, but if a theow he received the usual allowance of one.

The *Hayward* was responsible for the upkeep of the hedges round the arable fields and the meadows kept for hay. His strip of arable land was given him next to the pasture where the cattle grazed, so that if the hedges were neglected and the cattle strayed through, his own crops would be the first to suffer.[2]

The *Woodward* had the care of the woods, supervised the cutting of timber and underwood, supplied what was needed for building, fencing, fuel and brushwood, and superintended the wood pasture for the animals. He had the right to all wind-fallen trees.

The *Cheese-maker* was a woman attached to the household, and was in fact a kind of dairymaid. It was her duty to provide the lord's buttery with a hundred cheeses yearly, and she was given for her own use all the buttermilk except that which the shepherd had. She also made butter for the lord's table of the wring whey. As this was the residue left after the curds had been strained off for cheese it has a thin sound, but a good deal of cream may be raised from whey by gentle heating, and this form of butter was considered in the eighteenth century little inferior to ordinary butter, and it

[1] R.S.P., 439.
[2] Ibid., 440.
[3] R.S.P., 441.

was still made late in the nineteenth century in some parts. It may have been preferable to that made by the shepherd from the ewes' milk.[1]

All female serfs in the household were allowed 8 lb. of corn for food, a sheep for winter food, a sester of beans for Lent food, and whey in summer. Male serfs, such as the herdsmen and others described above, received 12 lb. of corn for food, and the carcases of two sheep, and a good meat cow, with wood according to custom, a "harvest handful," and feasts at midwinter and Easter.[2]

Besides all these persons there were the beadle or general messenger, fishers, fowlers, hunters, salters, bakers, smiths, wrights and greaves (to look after moors and dykes), and other craftsmen, according to local requirements.

All the workers on the thane's estate were supervised by a wonderful person called the *Reeve*, whose duties were so multitudinous that to fill the post adequately he must have been extraordinarily gifted. He had to attend to everything "in tune and dune" (i.e. in the village and all the surrounding country belonging to the manor), in wood and water, on ploughed land and fold, within and without. He was usually chosen from the villagers, and was in the difficult position of not allowing himself to be too proud or them to be too familiar, a situation calling for both tact and firmness. He had to see that everyone in the village performed all those duties we have already enumerated well and truly and at the right time, he also provided all the necessary implements for every sort of occupation, and saw that they were in good order: carpenter's tools, implements for ploughing, harrowing, weeding, haymaking and reaping; forks, flails, sieves, winnowing fan, and and baskets for the barn; wagon covers, troughs[3] and wheelbarrow, weighing machine and measures, horse-comb and shears, cauldrons and branding iron, beehives and honey baskets, implements for the miller, shoemaker, smith and other trades as required, all had to be found. For the thane's household spinning and weaving apparatus[4] down to needles and sleekstone (the predecessor of the flat-iron); bath-tubs,

---

[1] Ibid., 440. Low, *Domesticated Animals of the British Isles*, 273, says that butter made from sheep's milk is soft, "leaving a fatty impression like tallow in the mouth," and the cheese has a strong stimulating flavour. For whey butter see below, Ch. XI.

[2] R.S.P., 438. The pound of that time was much larger than at present.

[3] "Trogas": trough-shaped garden baskets made of thin slips of wood are still called "trugs" in Sussex.

[4] The loom had no treadles and the spinning apparatus no wheel as yet.

soap-box and comb; candlesticks and lamps,[1] salt-cellars, horn pepper-box and spoon-rack, chests of various sorts, hatches and stools, beer-casks, buckets and bowls, fire-tongs, an eel container and an oven fork; he was even responsible for the mouse-traps, and "what is still more trifling, the pin for the door-latch." Also he had to see that no misfortunes overtook the granary, the stores of salt meat and bacon, the cheese and other provisions, and no doubt saw that the tenants duly paid their food tributes. The geburs in many places provided 23 sesters of barley and two hens at Martinmas and a young sheep or 2d. at Easter, besides a small money payment, and in some places a fixed amount of honey, meat or ale.

In the farm-yard the reeve had to keep the outhouses in good repair, clean them out, build new ones when needed, make what improvements he could, look after hedges, ditches and paths, and "always do whatever is useful."[2]

From the catalogue of his duties, combined with the illustrations to the Saxon Calendar, we may construct pretty fully the Saxon farmer's programme for the year.

January in the Calendar shows ploughing and sowing, which seems full early for the latter; the reeve seems to have begun ploughing in February, Candlemas being the most usual date for the geburs. The tenants were given a "drinking feast" for ploughing. This was, perhaps, the origin of "Plough Monday," afterwards fixed as the first Monday after Twelfth Night, when ploughing was supposed to begin. On this day, up to the Reformation, money was collected for the upkeep of lights kept burning before the images in the churches to obtain a blessing on the work in the fields.

During the three months of spring (February, March and April) he was employed in ploughing and grafting, sowing beans (or presumably any other spring crop that was appointed, e.g. pulse, barley or oats), planting vineyards, digging gardens and closes, and cutting down thickets. Vineyards therefore must have been a Roman survival fairly widespread among the Saxons. The February picture in the Calendar shows men with billhooks pruning very curly trees not unlike vines, though again it seems too early in the year for such large shoots to have grown. A little later in the spring flax was sown, that all-important plant the fibres of which provided linen, the tow wicks for lamps and the seeds oil; madder plants were

[1] These were new-fangled articles adopted from the Romans and called by the Saxons "light-vats" for want of a better word.

[2] Gerefa, 453, etc.

set, the roots being used for red dye, and woad seed sown.   The leaves of the woad plant were fermented, hence the use of the special rounded iron "woad spade" mentioned among the farm tools to shovel them up, and were afterwards formed into balls and used for dyeing blue.   At this time the gardens were also planted, both with flowers and vegetables.   Leeks were so popular with the Saxons that the kitchen garden was often called the leek garden, and the gardener the "leekward," but it also contained cabbages. peas, turnips, lettuce, parsley, parsnips, onions, mint, garlic, and perhaps globe artichokes and beets.[1]

The March illustration shows men engaged in field-work, one with a spade with a crescent-shaped iron tip (as described in the last chapter), another breaking clods with a mattock,

Fig. 32.—Cain and Adam Digging (from Cædmon's *Paraphrase*)

and a third sowing seed from his lap.   The spade in this picture and in the illustration to Cædmon's *Paraphrase* do not have the handle set in the centre of the blade, as in modern spades, but at one end, so that all the projection is to one side, for the foot to be placed on for pressure.

In April they are represented feasting, which would be the Easter feast in which everyone took part, down to the bonds-men, the freeman getting a fortnight's holiday.   This would

[1] A. Amherst: *History of Gardening in England*, 3.   Wright: *Homes of Other Days*, 306-7.

FIG. 33.—FELLING TREES (Cott. MSS. Julius, A, vi)

doubtless take the place of the ancient heathen feast of Beltane on May 1st.

In May cows and sheep were milked three times a day, and it was hence called the month of three milkings. In the Calendar shepherds are seen tending their flocks, and the reeve is directed to put up sheepfolds and later to see to the washing and shearing which took place after Midsummer.[1] During May, June and July the fallow fields were ploughed, manure scattered, weeding done among the crops, and fish-ponds and watermills constructed. Wood was now cut for all sorts of building, repairing and fencing, and in the Calendar men are shown felling trees with large axes, and loading the wood in a rough cart with sides of stakes, a yoke of oxen standing by ready to draw it. In July five men are mowing hay with scythes, and one lifting it with a pitchfork. The scythes have long straight handles with one peg for the left hand, the blades being at right angles to the handles and almost straight. One man is sharpening his scythe with a whetstone and a bucket of water. A feast, as in ancient days, was held at Midsummer, with straw fires and flowers.

The reeve, as usual a little behind the Calendar, did not mow till August, but in the next three months, August, September and October, he was busy gathering in crops of all sorts. According to the picture in the Calendar the harvesters were called together by the blowing of a horn, a practice which occurs in the Welsh Laws and continued for many centuries. Barley and pulse were mown, and corn was still cut high up on the stalk, with sickles like modern ones in shape, the straw being cut afterwards, when worth it, for thatching or litter. The corn is seen being loaded with a pitchfork into

[1] Att., 59. If sheared earlier 2d. had to be paid on each fleece.

a cart with neatly wattled sides.    The tenants were given a
"boon-feast" at harvest-time, a mowing reward, a "rick-
meat" on the completion of the hayricks, a "stack-cup" at
corn-carrying, and a "wainstick" at wood-carrying.[1]  The
woad plants were also taken up at this time, and the fruit
gathered in.    Apples were still the principal fruit crop,[2] for
making cider, but they would also have cherries, pears, plums,
chestnuts and hazelnuts, perhaps medlars, peaches and mulberries.[3]  After the harvest the fences were removed from the
stubble fields, so that the cattle could wander over them, and
later the swine were driven to the woods to fatten on acorns
and beechmast, a picture of this appearing in the Calendar
for September.[4]  One man has a horn and one or two dogs,
and both he and his companion carry spear-like weapons, which

FIG. 34.—HAY-CUTTING (Cott. MSS. Julius, A, vi)

were probably used for defence against wolves, or perhaps as
swine goads.
    The reeve must then prepare for the winter, see that all
roofs were sound and the necessary thatching done, clean out
the sheepfolds, and erect stalls and sheds for the cattle which
wintered in the farm-yard.    Lastly he had to start the autumn
ploughing of the field which was to be sown with wheat or
rye.    Meanwhile his superiors occupied themselves, according
to the Calendar, in hawking; heron and wild duck being the
game portrayed.

[1] This last was, perhaps, the same as the later " wenbote," the provision
of a log or tree trunk for the repair of each tenant's wain before the carting
of the hay began.

[2] An orchard was an " aeppel-tun."

[3] A. Amherst: *History of Gardening in England,* 3

[4] This illustration is sometimes called "Boar Hunting," but it seems more
probable that it refers to the usual practice of driving pigs to the woods in
autumn.  Besides the appearance of the animals is fairly domesticated.

November opened with the feast of All Souls, the Christian equivalent of Samhain, the custom of bonfires being preserved.[1] In the Calendar one man is tending the fire with tongs, while another is bringing logs from a neat wood pile, three others are warming themselves at the blaze with some satisfaction. At Martinmas (November 11th) great stores of food were laid in for the winter. Cattle and sheep were slain and salted, that they might not have to be fed during the winter months, and later the swine fattened in the woods were killed for bacon and salt pork, hence perhaps the Saxon name of "blood-month." The late ploughing was completed, the winter wheat crop sown and the gardens dug over. Heavy frosts were taken advantage of to split wood for building purposes, As work in the fields ceased everyone became busy indoors during the short

FIG. 35.—DRIVING PIGS TO MAST (Cott. MSS. Julius, A, vi)

days. The reeve looked over his implements and fittings and had fresh ones made where required, even to the hen-roosts. A kiln was prepared in the yard to dry the corn ready for threshing, and a drying oven for malt.[2] In the barn the corn was threshed with very long flails, as shown in the Calendar for December, winnowed with a round fan, and carried to the granary in a deep wicker basket with a pole passed through the handles and resting on the shoulders of two men supporting themselves with ragged staves. A man (perhaps the reeve himself) stood by with a notched stick, ticking off the measures of grain. Peas and beans were threshed as well as corn.

[1] Some say that these were bone-fires, in which were consumed the bones of the animals killed for winter stores, but November 2nd seems a little too early for that.

[2] The corn-kiln is described as "cylne" and the drying oven or malt-kiln as "aste." Malt is barley soaked until it germinates, and then dried again before grinding.

H

Mid-winter, or Christmas, was enlivened by a feast in which all, from the highest to the lowest, joined, and the celebrations were kept up until Twelfth Night. Ale, mead and cider would be made, and wine in the places where there was a vineyard. Thus passed the winter until ploughing-time came again and the year's round began afresh.

We have now seen how the typical manorial farm was worked, and how the geburs and others had for the most part a co-operative system of farming; but although this was the normal method throughout the country, it must not be supposed that single, compact farms belonging to a single, free owner did not exist also. Even at the time of the Domesday Survey a number of these survived; as many as thirty-six are recorded in Essex, but in other parts they were rarer. The large village estate, however, was the most prevalent.

The farmer's bread was made in round flat cakes, of various grains, wheat for the well-to-do, barley for most people, and a mixture of rye, oats and beans for the poor. Yeast was now used, for a yeast chest occurs among the household gear. Salt meat and bacon were much eaten, but horse-flesh was forbidden after 787 on account of its association with heathen rites. Every man might take game on his own property to eke out his larder. The boy in the Colloquium gives as his diet flesh food, vegetables and eggs, fish and cheese, butter (salt) and beans, with ale and water to drink.[1] Eels and broth were also common articles of food. Ale and mead were everywhere brewed and drunk, but wine much less often. That made from local vineyards cannot have been very choice, and imported wine was only for the wealthy.

The Saxon farmer after the establishment of Christianity paid tithe of all his produce to the Church. This comprised the tenth foal, calf, lamb and young pig, every tenth sheaf of his harvest, tenth cheese or tenth day's milk, the tenth measure of butter, tenth fleece and a tenth of the yearly yield of his bees, groves and meadows, mills and waters, parks, stews, fisheries, brushwood and orchards. One-third of these payments went to support the parish priest, one-third to the upkeep of the village church, and one-third to the poor.

The farmer's wife was as busy as ever, supervising the poultry-yard and dairy, spinning and weaving her flax and wool, dyeing them with madder and woad from her garden, and doubtless many wild plants as well, and making clothes for the household, although if illustrations of the time can be

[1] Coll. Ælfric, 34 and 35.

trusted she did not trouble to clothe her children while they were young. If she were very poor, or there was no mill on the manor, she still ground her own corn with a quern. According to the Laws of Cnut it was her right and duty to keep the keys of the storeroom, chest or cupboard in the cottage, but she might not forbid her husband to "put into his cot what he will."[1]

With all this careful organization of agriculture which we have described, no export of corn took place as under Roman rule, which argues that either very much less was grown or that the population was larger and could only provide enough for its own needs. Each Saxon village aimed at being a self-sufficing agricultural and pastoral community, with little need or desire for intercourse with its neighbours beyond occasional marketing at the nearest town,[2] and the necessity of obtaining salt for the preservation of food.

We have already examined the intricate tribal arrangements of the Celts, and the highly organized Roman villa-estate worked by slave labour, but the Saxons evolved still a different method of dealing with agriculture. They had a co-operative system of work and a sharing of land among the members of the community not unlike that of the Celts, but combined with it an overlordship and an obligation to cultivate the lord's farm not unlike that of the Roman villa, but without its accompanying element of complete slavery. This new manorial system was established in many places by the end of the period, but was not universal. Single farms, independently worked and owned, still remained in considerable numbers, especially in the Danish districts, and in a few places free communities survived without an overlord at the head. It remained for the Normans fully to carry out and crystallize this organization into the "feudal system."

A few words must be inserted here with regard to the *Danelaga*, that part of the north and east of England which from the eighth to the tenth century received so many Scandinavian immigrants. As a race the Norsemen were not greatly different from the Saxons in either speech or custom, and amalgamation was therefore not difficult. They did not arrive in any tribal formation, but were for the most part free men who settled down as small, hard-working farmers, either in villages or on independent farms.

---

[1] A.L.I., 419.
[2] Buying and selling beyond 20d. worth might only take place in a town Att., 115, 135.

They seem to have had much the same classes as the Saxons, the thrall corresponding to the theow, the leysing or newly freed man and the bondi or ordinary freeman to the ceorl or twyhynde man, and the odalman, the free landholder of inherited land, similar to the Saxon twelfhynde man or thane.[1] Vinogradoff considers that the typical free Norse peasant (bondi or socman) in England had a holding of only 12 acres, i.e. the Swedish "mantal" or "man's share."[2] It is noticeable that in Domesday the Danish counties of Lincoln, Notts and Derby were supplied with teams in excess of the arable land, which may have been the result of many small independent holdings, each with their own teams, instead of the more economical co-operation of teams in the typical village.[3]

It would not, perhaps, be too far from the truth to take the Icelandic Sagas as an approximate picture of the life of these men, as they were settlers of the same race and time although in another country. The Norse house, like that of other early peoples, consisted originally of one great hall or "skali," other apartments being added separately as required. The Norse hall was a long wooden building constructed internally like a nave with low side aisles. There was a door with a porch at each end of the nave, one for the men and the other for the women. Little bedrooms for the different members of the family were shut off between the pillars of the nave, with a passage round behind where others could sleep. Sometimes there was a loft formed by planks laid on the cross-beams and reached by a ladder, which also could be used as a sleeping-room.[4] Long hearths for fires ran down the centre of the hall, with movable tables running parallel, and benches along the walls, with a "high seat" for the master of the house in the middle of the side. The floor was of rammed earth.

Here the family lived and the cooking was done, unless there was a separate kitchen or "seething house." The windows were not large, and were shuttered or often covered with the caul of a newly born calf stretched on a frame, and the smoke escaped through apertures in the roof. Sometimes the house was roofed with turf and the active cattle were able to spring on to the low, sweeping eaves to feed.[5]

Occasionally these halls were built so that a stream ran

[1] F. Seebohm: *Tribal Custom in Anglo-Saxon Law*, 271, 500 on.
[2] Vinogradoff: *English Society in the Eleventh Century*, 281.
[3] Maitland: *Domesday Book and Beyond*, 428.
[4] Njala, lxxvi., and see Dasent's Introduction to the *Story of Burnt Njal*.
[5] Njala, lxxviii.

through, for the convenience of a water supply, or across a road to ensure visits from passers-by.[1] A storeroom was added nearby, guarded by the house-dog; and there was often a bower or apartment for the women, where they did their spinning and other work and sometimes slept. Some establishments had a separate guest-house, a kitchen, and frequently a bath-house, as the Norsemen were a cleanly people, combing their hair every day and bathing every Saturday in a manner which seemed very frivolous to their less particular Saxon neighbours.[2]

The cow-byre was placed close to the hall, often with a communicating door for the convenience of foddering in winter. At the homestead of Thorgrim at Saebol there were thirty cows in the byre next the hall, tethered in rows back to back. Their beds were made up by a serving woman, and a boy drove them in.[3] Other buildings in the yard were the sheepcote, stable, barn and piggery, and the whole was enclosed with a turf or stone wall, within which the pigs were sometimes allowed to run loose.[4] This homestead and the fields immediately adjoining were known as the "town."[5]

The stack-yard seems to have been at some distance from the homestead and was surrounded by a high wall of turf, the hay and other crops being stacked along the inside of the walls, and carried from there to the homestead barn when needed during the winter on sledges.[6] The sheephouse, if not in the home yard, was often by the stack-yard, but would only be used for milking in summer or possibly at lambing time, for it was customary to leave the sheep out on the hills to fend for themselves all the year round. They were collected for shearing and branding, and at harvest time all of them were driven into muster-folds and sorted out by their respective owners or their shepherds, after which many would be killed and salted for the winter stores. Sometimes there was a reeve responsible for the "sheep-homing." With these wild mountain sheep the shepherd seems to have had his hands full between looking after the sheep he had collected and searching for the missing ones.[7] The milch sheep were

[1] Ibid., xlviii.
[2] Wright: *Homes of Other Days*, 73. Eyrbyggia, xxviii.
[3] Gisla, xxxix.-lii.
[4] Eyrbyggia, xx.
[5] Cf. Saxon "tun" and Scottish "town-place" of the present day. Njala, xlv.
[6] Eyrbyggia, xxxvii.
[7] Ibid., xxiii. Cormac, iii. Havard, 2, v. (i).

fortunately less wild and were evidently very important for the making of cheese and "clotted milk": the "milking post" was conveniently placed by a stream or ford.[1]

*Horses* were kept in considerable numbers, the Norsemen being much more expert riders than the Saxons, but like the sheep they were branded and allowed to run loose on the hill pastures and rarely seem to have been put in stables, for in winter storms they took shelter under the walls of the rickyard. The stud horses were sometimes kept in special pasture between two brooks. They were used continually for riding by all classes, as packhorses for carrying food, hay and fuel, and for the sport of horse fighting.[2] Probably they were eaten also.

*Oxen* were used for draught, the most usual vehicle in Iceland being a sledge to which a pair of oxen were yoked.[3] The Norsemen doubtless brought their own cattle with them to England, and these would be of the larger type customary on the Continent, and akin to those brought by the Saxons.[4] As time went on and the Danish immigrants settled down into peaceful village groups the township fold became a great feature. This "forta" was a large green in the centre of the township where the cattle were collected at night and from which they started for the feeding grounds in the morning.[5]

The Norsemen in England no doubt grew the same crops as the neighbouring Saxons, although in Iceland oats and vetches are those most mentioned. The hayfields were carefully manured.[6] It was characteristic of the greatest chiefs in that country that they personally sowed and reaped their fields, drove sheep, and were expert carpenters and smiths; no honest work came amiss to them, though they frequently worked with their weapons ready to hand.[7] Their barns had wooden floors on which the corn was threshed, and this "kornplada" is sometimes referred to as the drying house or "kylna,"[8] from which we may suppose that they dried their corn before threshing in the ancient way we have met with.

[1] Njala, liv. Cormac, xvi.
[2] Eyrbyggia, xvii., xxxv. Vatzdaela, xxxiv. Njala, xlv., xlvii.
[3] Eyrbyggia, xxxvii.
[4] W. Boyd Dawkins: *Early Man in Britain*, 493.
[5] Vinogradoff: *English Society in the Eleventh Century*, 389-90.
[6] Njala, xliv., cxxiii. Eyrbyggia, xxx.
[7] Njala, liii.
[8] R. Keyser: *Private Life of the Old Northmen*, 70.

They are said to have specially valued angelica in their herb gardens.[1]

At harvest there was a great feast at the larger houses, to which all the neighbours were bidden, and much fresh ale was brewed.

They had two meals a day, the "day meal" in the morning and the "night meal," the most important, and accompanied with the drinking of much home-brewed ale and mead, before retiring to bed. Porridge was a very general food, served in troughs and eaten with spoons of wood, horn or bone; also cheese and clotted milk, but apparently not a great deal of fresh meat. Large stores of meal, "stockfish" and salt meat were laid in for the winter. Boiled suet sausage is also mentioned.[2]

In the early days it was customary for the settler near the coast to sow his crops, "go in Viking" (i.e. on a predatory voyage) during the summer, and return to harvest his crops, sometimes setting out again for an "autumn Viking" before the winter.[3] But by degrees they settled down to a more peaceful agricultural life, either as free socmen working small independent farms, or united in villages worked on the usual open-field system. At the time of the Norman Conquest a considerable number of socmen remained and a few free villages, but the greater part of the Danelaga was coming into line with the widespread manorial type of organization.[4]

---

[1] Ibid,, 71.
[2] Njala, xi., xlvii. Cormac, iv.-xvi.
[3] R. Keyser, op. cit., 105.
[4] Vinogradoff: *English Society in the Eleventh Century*, 477 9 F. Seebohm: *Tribal Custom in Anglo-Saxon Law*, 522.

# CHAPTER VI

## THE NORMAN CONQUEST TO THE BLACK DEATH

### 1066-1348

REFERENCES

D.S.P.  *Domesday of S. Paul's.*  (Camden Society.)

L.N.P.  *Liber Niger Petroburgense,* contained in *Chronicon Petro-burgense.*  (Camden Society.)

R.C.G.  *Rentalia et Custumalia Abbatiae Glastoniae.*  (Somerset Record Society.)

C.B.A.  *Custumals of Battle Abbey*  (Camden Society.)

W.H.  *Walter of Henley.*  Ed.  E. Lamond.  (Royal Historical Society.)

# CHAPTER VI

## THE NORMAN CONQUEST TO THE BLACK DEATH

THE coming of the Normans—that dark and ruthless people who combined the stern qualities of their Norse origin with a superficial polish of French civilization—did not bring any radical change in the organization of the English farm. The manorial system, as we have just seen, had already gained a considerable footing in the land, and it was the work of the Conqueror not to upset this but to see that it was firmly and systematically established in every corner of the realm. The idea of lordship was carried out to its fullest extent; every man must hold his land from some master—from the King himself if there were no great landowner between—and the thoroughness with which this was established is shown in that remarkable document known variously as the Great Survey of 1086, the "Liber Winton," or Domesday Book. There every estate, large or small, over the whole country as far north as Yorkshire (and excepting Wales) was written down, with its owner, its overlord, its assessment for taxation and other details: the number of plough oxen was always recorded, the amount of pasture for the cattle, the number of pigs that could be fed in the woods and, in East Anglia, how many sheep could be kept on the grazing grounds.

It shows the people for the most part grouped in well-organized villages, scattered sparsely over the countryside in favourable spots; surrounded by wide tracts of the waste land so necessary for the feeding of the herds, and often separated by wide tracts of forest, marsh or heath.

The classes of people revealed in this Survey remained very much the same. The Saxon gebur became the Norman villein, and the cotsetle the cottar or bordar, without very much change in their status, though their condition possibly altered slightly for the worse. The Saxon free class of small, independent farmers diminished at the beginning of the period, especially in the south, as would only be natural under a system of consolidation imposed by a conquering race, though their numbers increased again later on. Many of them were reduced at the Conquest to the status of villeins, and those who remained as freemen or socmen had their services increased. On the other hand, the landless serfs decreased in numbers and disappeared altogether in the thirteenth century. If they were household servants they had small wages in addition to their keep, or otherwise had a little land for their support.

The villein, like the gebur, had his own little homestead with its stock, surrounded by a stout fence, to keep the cattle in and thieves out at night. Theoretically he held a virgate of about 30 acres, with its corresponding rights of pasture and meadow, and contributed a yoke or more of plough oxen to the village team; but as time went on actual practice varied more and more from this standard and the number of oxen he owned tended to increase. He owed a variety of services to his lord, which were presently augmented by certain payments or fines, such as a fee on the marriage of his daughter, or on the sale of a horse or ox, but for the rest he was free to cultivate his own strips and tend his cattle or poultry, with the assistance of his family, and, if flourishing, could sometimes hire the services of a poorer neighbour. His position, however, was not too secure; a run of bad seasons, a murrain among his beasts, or a disastrous raid might deprive him of his precious plough oxen and cause him to sink to the condition of a cottar.

The cottars or bordars, with their little holdings of five acres or less, sometimes owned a few oxen, but eked out their living chiefly by gardening and raising vegetables,[1] and were employed as herdsmen, craftsmen or labourers of different sorts on the lord's farm. Their little houses had yards, tofts or gardens, a piece of enclosed land called a croft, and sometimes a few strips in the common field. Other men of this class, holding five to ten acres, were in some places called "lundinaria" or Mondaymen, because they worked on Mondays for their lord. Like everyone else they had their little poultry-yards.

Below them sometimes appears a class known as "coterelli" and "cotteriae," who held one or two acres or none at all except the plot the cottage stood on. These no doubt would be manorial servants, the relics of the former serfs.

The houses of the villeins and cottars were one-roomed huts, often with one end partitioned off for the cattle. The more prosperous either interposed a space between themselves and the ox-house, as described in the last chapter, or accom-

---

[1] In the *Inquisitio Eliensis* there are frequent mentions of cottars and bordars "with their gardens." The difference between a cottar and a bordar has never been discovered, but the terms were often used interchangeably in the twelfth century, and the term bordar disappeared soon after.

modated them in sheds in the yard. These little houses were often built of layers of mud or clay and straw, each layer being left to dry before the next was put on, thus forming a very hard and durable material. The roof was thatched or covered with turf and the windows were either of lattice-work or of "harden," a kind of coarse sackcloth, the door being sometimes formed by a curtain of the same.[1] Horn was also used, which would be less draughty, but glass was only found in great houses.

In manor-houses the hall was the principal and central apartment; it was used as a general living-room by day, manorial courts were held there, straw might be stacked in it for want of a better place, and even threshing occasionally took place there, regardless of the dust which must have been raised. The walls were hung with agricultural implements, such as scythes, corn measures and sacks, with weapons and other adornments, and two "perches," one for birds and the other for clothing. The men and servants slept in it at night, and the women in the chamber. As ideas of architectural planning advanced the other rooms were included under the same roof instead of being built on irregularly in a semi-detached fashion. In addition to the chamber where the women worked and slept, which sometimes had a "tresantia" or ante-chamber screened off, it became customary to construct a cellar or storehouse at the upper end of the hall with a "solar" or retiring-room over it, to which the lord could withdraw from the general life and bustle of the hall, and in which he and his lady slept.[2]

At the opposite end of the hall were the kitchen and other apartments pertaining to it. In large establishments these consisted of the pantry (or spence) for food, and the buttery for drink, presided over respectively by the spencer and butler, as well as the more ordinary bakehouse (sometimes with an oven for melting lead) brewhouse, and malthouse, dairy, washhouse (where the clothes were trodden in vats), larder (where the bacon, lard and pork were prepared), and sometimes a "salser" or salting-house in addition.[3]

In the hall the hearth was usually in the centre of the floor, the smoke escaping through louvres in the roof, but in stone-

---

[1] Addy: *Evolution of the English House*, 39, 40.

[2] D.S.P., 132, xcix. At Thorpe about 1150 there was a hall and "Great Chamber" with ante-chamber; the solar (with fireplace and garderobe) not appearing until the Survey of 1304.

[3] D.S.P., 132, 136. F. C. Davenport: *Economic Development of a Norfolk Manor*, 21.

building districts hearths against the wall with chimneys were occasionally found even in this period, being copied from those in the stone-built keeps of the Norman and Early English castles. But in the ordinary dwelling of timber filled in with wattle and daub or lath and plaster, and often roofed with shingles or thatch, the fireplace was kept well away from the walls.

Another improvement of this period was the cutting off of the lower end of the hall, towards the main entrance, by the erection of screens across it, an arrangement which not only prevented draughts and separated the general living-room from the kitchen regions, but allowed the erection of a gallery for minstrels over the passage or ante-room thus formed. It became customary also to erect a dais at the upper end of the hall for the accommodation of the lord and his more honoured guests.

In the court there were quarters for such farm servants as lived on the spot. In the thirteenth century those thus housed at Forncett in Norfolk were four ploughmen, the carter, swineherd, cowherd and dairymaid, and a harrower three or four months in the year; later were added a granger, a warrener, and a maid to prepare the servants' pottage.[1] Next came the ox-house, sheep-shed, lambcote and cowhouse; and these were no mere sheds, for at Kensworth, on the estates of St. Paul's, in 1152 the ox-house was 33 feet × 12 feet and 13 feet high, the sheep-house 39 feet × 12 feet and 22 feet high, and the lambcote 24 feet × 12 feet and 12 feet high. These palatial outhouses did not compare badly with the dwelling-house, for on that estate the hall was 35 feet × 30 feet, the ante-chamber 12 feet × 17 feet and the chamber 22 feet × 16 feet, the heights being respectively 22 feet, 17 feet, and 18 feet.[2]

In the court were also the stables, the pig-sty, sometimes a goose-house, and as many as three little henhouses, but the poultry were often kept in a separate inner court. Beyond these were large barns, a granary and a hay-house; and early in the fourteenth century a "small house to hold the carts" was often added, and generally a dovecote. The twelfth-century barns were enormous timber structures. The "great barn" of Walton on the St. Paul's estates was 168 feet × 53 feet, and 33½ feet high, but this seems to have been unusually

---

[1] F. C. Davenport: *Economic Development of a Norfolk Manor*, 24.
[2] D.S.P., 129.

large,[1] the others on their estates being smaller, ranging from 53 feet to 117 feet in length. Often there were three or four to one manor. Many of the outhouses would be of clay, with thatched roofs; and the whole manorial homestead was surrounded by a clay wall with a thatched top.

The animals kept in this goodly array of outbuildings and in the humbler sheds of the villeins did not undergo much change, although some improvement in stock-breeding took place towards the end of this period, especially on the lands of the great monasteries.[2] In 1086 the *plough ox* was still the all-important animal, and the team of eight was still intimately related to the measurement of arable land. To take an example: "In Chemere" (Keymer), so the Domesday formula runs, "there is land for 25 plough teams; and in the demesne (i.e. on the lord's home farm) there are two teams, and there are 36 villeins and 11 bordars with 17 teams (between them)." From which we may see that the people of Keymer did not cultivate all the land that they might. The neighbouring village, on the other hand, was much overstocked, being large and wealthy. "In Dicelinges" (Ditchling), we read, "there is land for 60 plough teams; on the demesne there are 8 teams, and there are 108 villeins and 40 bordars with 81 teams."[3] Even if each bordar in this village owned a yoke of oxen, which was not unknown, though very unusual, there would be at least five oxen for each of the villein households, instead of the pair which theoretically went with the virgate of land, so that it is obvious that this village was rich in cattle far above the average. Even in Keymer the villeins seem to have had three or four oxen each, allowing a team or two for the bordars.

A tendency seems to have grown up later to restrict villeins from increasing their plough teams, for about 1290 the Abbey of St. Albans forbade its villeins to rear plough oxen or horses except at risk of forfeiting them, though they might have pigsties and sheepfolds.[4] But Domesday is the last occasion on which the ox appears as the sole draught animal for the plough: early in the twelfth century we find horses being used along with oxen. From this time onward there was much

[1] Addy quotes an instance of a barn at Cholsey 303 feet long and 51 feet high (page 132).
[2] The stock on the demesne farms in 1086 is recorded in Domesday for Norfolk, Suffolk and Essex; in the Exon Domesday for Devon, Cornwall, and parts of Dorset and Somerset; and in the *Inquisitio Eliensis* for the lands of the Abbey of Ely. The totals and averages for Norfolk, Essex, Cornwall and Ely are given in an appendix.
[3] Domesday Book (Rec. Com.), 26a, 27a.
[4] M. Bateson: *Mediaeval England*, 390.

argument as to which animal was the better, but oxen continued to hold their own for many centuries longer. The use of oxen for drawing carts lingered in some counties for centuries, particularly in Sussex and in the West, but horses came to be more usual for this purpose.

An ox soon after the Conquest cost from 2s. 6d. to 3s., but his price rose towards the end of this period, and during the first half of the fourteenth century he cost 5s. in Wales and about 12s. elsewhere.[1] Alexander Neckham, writing at the end of the twelfth century, is loud in praise of the ox: alive it cultivated the land and dead it was important for the table; its skin was put to many uses, and its horns furnished handles, combs, spoons, knife cases (? sheaths), lanterns and other things. Even its dung was useful as fuel when firewood was scarce.

An acre of meadow was considered enough to feed an ox during the summer, but according to Walter of Henley the

FIG. 36.—MEDIAEVAL BULL

plough beasts were stall fed, on oats, for more than half the year, from St. Luke (October 18th) to Holy Cross (Invention of the Cross, May 3rd.) He did not, however, believe in putting them in houses in wet weather as that was liable to cause inflammation. They were fed little and often, bathed and curried twice a day with a wisp of straw "that they might lick themselves more effectively." On the manors of St. Paul's it was frequently one of the duties of the villeins to thresh enough corn for the food of the teams during the winter and spring. Others gave "fodder corn," i.e. a quarter of oats, at Martinmas. When at pasture it was the duty of the oxherd who drove them there to bind them when required with "langhaldes and spanells," i.e. ropes connecting the fore and hind legs, to prevent the animals from leaping ditches and fences, and wandering where they were not desired.

Cows, owing doubtless to the lack of winter fodder, were for the most part only milked from May to Michaelmas, and during that period each cow was reckoned to give sufficient

---

[1] Cunningham: *Growth of English Industry and Commerce*, 1890, 163 n. F. Seebohm: *Tribal System in Wales*, 23.    Thorold Rogers: *History of Agriculture and Prices*. It must be remembered that the purchasing value of money was then ten or twelve times greater than it was at the beginning of the twentieth century.

milk for seven stone of cheese and one stone of butter. But some were kept on for milk where there was sufficient fodder, as we shall see later. At Martinmas the herds were inspected and only a number of the best retained, according to the supplies of fodder available to keep them alive during the winter; the rest were either sold, or killed and salted down for the use of the household. The effect of this poor winter feeding upon the food properties of the animals is curiously described at a much later date (eighteenth century) by Martin in his *Description of the Western Isles of Scotland.* The cows in Skye, he says, become mere skeletons during the winter, many of them not being able to rise from the ground without help, but recover as the season becomes more favourable and the grass grows again, and "then they acquire new beef which is both sweet and tender; the fat and lean is not so much separated in them as in other cows, but as it were larded, which renders it very agreeable to the taste. A cow in this isle may be twelve years old when at the same time its beef is not above

FIG. 37.—SHEEPFOLD (Luttrell Psalter)

four, five or six months old." The unfortunate cows cannot have felt the advantages of the situation, and many must have found the winter a hardship in mediaeval times when the uses of root crops were unknown. Moreover, wolves were still so common that we read of many cows being strangled by them in Lancashire,[1] and traps were set to catch their feet.

*Sheep* became more important as the value of their wool increased, and the English woollen industry began to grow up: indeed the sale of wool and wool-fells became an important source of income to the manor. The price of sheep varied from 4d. to 6d. in the early part of this period, but later was more than doubled. In an illustration to the Luttrell Psalter (early fourteenth century) the ewes are hornless, from which we may gather that this breed had at length supplanted the

[1] M. Bateson: *Mediaeval England,* 378.

earlier horned one, but they were small and their fleeces light. They were still valued for their milk as well as for their wool, and it was reckoned that ten sheep supplied as much butter and cheese as one cow.  Butter in Walter of Henley's day cost 6d. a gallon.  The sheep were milked through the spring and summer but not after the Nativity of Blessed Virgin Mary (September 8th), because it was important that they should be strong for the winter and its lean fare.  The old and feeble were sorted out between Easter and Whitsun, shorn early to distinguish them and fattened for sale about St. John's Day (June 24th).  It was recommended that this should be done not later than August, as mutton fed on the fresh spring grass was best.  During the winter, from Martinmas to Easter, they were kept in houses, but were allowed to lie out in the folds when the weather was dry and serene.[1]

At Alciston in Sussex the tenants of each half-hide had to erect four gables (cheveruns) with roofing for the lord's sheepcote.  In some places it was the duty of the tenants to "carry the lord's fold" (i.e. to move the hurdles) three times a year; at Appledram the wealthier cottars carried two hurdles each, with the stakes, and the lesser cottars one hurdle.[2]  The sheep-houses were cleaned out, marled and well littered once a fortnight.  The sheep were fed during the winter on hay mixed with a little straw, or on pea-haulms.  For grazing purposes an acre of fallow was supposed to support at least two sheep a year.  The sheep were marked by their respective owners either at shearing time or when they were brought in for the winter, and they often wore bells.  The question of where the sheep should be folded for the night for the sake of the manure was still a burning one, and most villeins owed "fold-soke" to the lord of the manor.  On the manors of St. Paul's many tenants had to fold their sheep on the demesne from Hokday (the second Tuesday after Easter) till August 1st. The tenants therefore could fold them on their own land after their harvest was gathered in. until Martinmas, when they would be housed.

The anonymous writer of the thirteenth-century *Seneschaucie* recommends his lord to have three good folds, one for the wethers, another for the ewes, and a third for the hogs,[3] with respectively 400, 300 and 200 in each, if he had sufficient pasture.  This shows an enormous increase compared

---

[1] W.H., 27, 29, 97.  Fleta, ii., c. 79.

[2] C.B.A., 29, 56.

[3] A young sheep up to its first shearing.

with the numbers kept on demesne farms in 1086. The average number kept in Essex, a particularly good county for pasture, was then about 100, although a flock of 810 occurs once. On the lands of the Abbey of Ely the average was rather higher, and on one farm, Walton in Norfolk, there were 1,300 sheep, but that was most exceptional. The monastic orders came to be great sheep farmers, especially the Cistercians, who paid much attention to sheep breeding on their lands in Yorkshire.

The diseases of sheep are first mentioned late in the thirteenth century. In 1283 there was a great "sheep rot," and scab first appeared about 1288. The early remedy for this was verdigris, copperas and quicksilver, or an ointment made of quicksilver and lard, but about 1295 a tar dressing, mixed with butter or lard, was introduced and became general after 1307, although the earlier treatment lingered on conservative farms for a quarter of a century longer. Grease was also used to protect the sheep's heads from fly.[1]

*Swine* continued of the utmost importance for the larder. Oxen were needed for work and cows for the dairy, sheep both for the dairy and for wool, but the sole duty of the pig was to supply food, and pork and bacon were always needed. Consequently every household, large and small, had its pigs if the land was at all capable of supporting them.[2] The village swineherd daily collected the pigs and drove them out either to the waste, to grass-land or to the woods according to the season, and the villeins paid certain dues for their feeding. A very usual payment for "pannage," i.e. for feeding in the woods, was one pig in ten, and at Belchamp on the St. Paul's estates all the pigs that had been fed in the woods were driven to the hall of the manor to be taxed. The period during which the pigs were fed in the woods seems to have steadily decreased. At this time it was about six or eight weeks, in October and November, the pigs being killed in December. Under the Saxons the period stretched from August 29th to the New Year, and under the Celts extended to January 15th. The shortening of the pannage period may of course have been due to decrease of woodland. Domesday Book bears evidence to the importance of the custom by always recording the woods

---

[1] Thorold Rogers: *History of Agriculture and Prices*, 7, 460.

[2] In Domesday Book, out of the 465 manors on which the demesne stock is recorded in Essex, 413 kept pigs, the average number being 31. In Norfolk, 460 manors out of 515, the average being 17. On the lands of Ely, 104 out of 115, the average number being 25. In Cornwall they were much scarcer, only 87 manors out of 243 kept them, and the average was 5. See Appendix.

I

in their relation to swine, sometimes stating how many pigs they were capable of feeding, but generally how many they yielded to the lord in pannage dues.  On the lands of Battle Abbey in the late thirteenth century this was commuted for a payment of 2d. a pig, and 1d. for a small pig, or in some places 1d. for all sizes.  Herbage dues, later called "stubble silver," or sometimes "garsavese" (thought to be a corruption of the Saxon gaerswyn), are mentioned in Domesday in the south-east.  In Sussex one pig in seven was paid, and in Battersea and Streatham one pig in ten.  Later (1222) 1d. a pig was paid for "garsavese" at Walton on the lands of St. Paul's

This feeding would doubtless be on rough pasture or stubble, not on good land needed for the cattle.  Walter of Henley says that pigs should be able to dig, in order to get roots for food, but in February, March and April needed extra fodder.  The grain left from the brewhouse, called "drasch," was often used for this, but sometimes they were fattened with barley and peas, as at Forncett.[1]  The number of pigs kept on the demesne varied according to the nature of the land.  On an estate where there was plenty of wood, waste or marsh they could easily support themselves and many could be kept.  Pigsties were built in the woods or on the feeding grounds during hard weather, and the pigs kept there under the charge of the swineherd, only the sows that had farrowed and other weaklings being taken to the farm-yard of the manor.  On less well-provided estates where the pigs had to be fed from the grange (i.e. on corn or peas), the anonymous *Seneschaucie* recommends that only as many should be kept as could be fattened on the stubble and leavings, and that these should then be sold, as winter feeding would be too expensive.  At Forncett the accounts from 1270 to 1307 show that pigs were kept on that manor for only seven out of the thirteen years.  Presumably the supply of bacon in between whiles was derived from the tenants.  The author of the anonymous *Husbandrie* (thirteenth century) states that a sow should farrow twice a year, or five times in two years, having at least seven piglings in a litter.  Walter of Henley gave her thirteen pigs at three farrowings and observed that it was good for swine to lie long in the mornings and to lie dry, a feeling that was doubtless shared by the swineherd.

We may see from the illustrations to an early thirteenth-

[1] F. G. Davenport: *Economic Development of a Norfolk Manor*, 29, 35. W.H., 29.

century Psalter[1] that the swine continued high-backed, long-legged creatures, with hairy bodies, a well-defined ridge of bristles along the spine, prick ears and long sharp snouts; active and lively-looking animals. The illustration for November shows them being fed in the woods with enormous acorns, and in December they are killed, by a blow from the back of a hatchet, and cut up.

Pigs in the twelfth century were worth from 4d. to 10d., sows 8d. to 10d., and boars 1s.; towards the end of the period about 2s. 6d., 2s. 8d., and 4s. 3d. respectively.[2]

*Goats* were much less popular than they had been, but they were still kept in some places, probably where the feeding was inferior and therefore less suited to sheep, but mainly because the kids were much valued for food. Neckham observed that goats gave more milk than sheep, and that it was "efficacious against many diseases." Rough cloaks and coverings were made of their hair; "but," he says, "the goat is a melancholy animal."[3] In 1086 goats occur on 127 demesne farms in Essex, on 102 in Norfolk, and 77 in Cornwall, but on only 13 on the lands of Ely. At Kettering (about 1125-8) the cottars paid 1d. for each goat, for feeding dues, and ½d. for each she-goat. On the Berkeley estates in the time of Edward III herds of goats were kept on those of their manors which adjoined the Forest of Dean and Micklewood Chase. They were under the charge of a master goatherd, and at least 300 kids a year were provided for the lord's table. In the twelfth century they were valued at 4d. or 6d.[4]

*Horses* for farm use were still very scarce in 1086, but were increasingly kept later. At that time it was usual to have one or two of the little work horses called rounceys (Lat. *runcinus*) on the demesne farm, but rarely more. In the twelfth century they are usually referred to as stotts or affers (i.e. an animal that bears burdens), and they were used for harrowing, pack-carrying and often (combined with oxen) for ploughing. Their value at that time was 3s., later it rose to about 11s. 8d. Cart horses seem to have been a stouter breed, for they were valued in the twelfth century at 6s. and in the early thirteenth at about 18s. They would doubtless need to be strong animals to pull a cart or sledge over the deplorable mediaeval roads.

---

[1] Royal MS, 2B, vii. Generally known as Queen Mary's Psalter.

[2] The classes of pigs, according to Thorold Rogers, were as follows; porci, pigs fit for eating; porculi, lean pigs for fattening; porcelli, little pigs; hoggets, young boars of the second year. D.S.P., 131-6.

[3] A. Neckham: *De Naturis Rerum*, 161.

[4] J. Smyth: *Lives of the Berkeleys*, 302. D.S.P., 122-3.

Mares were apparently much used for farm work, for Fitz-stephen in describing the market outside the walls of London in 1174 notes the mares standing there "fit for the plough, the sledge and the cart." "Horses," when referred to as such, seem to have been superior animals, probably for riding. They sometimes occur on demesne farms in 1086, but mostly on those farms of Cornwall and Norfolk, where they were reared in the woods. A "horse" at that time was worth about 10s.

The farm horses shown in the Luttrell Psalter (before 1340) are thick bodied animals, brown or dappled, shod with pointed shoes, and wearing rope harness. On Forncett manor the stotts were fed on barley mixed with oats. Walter of Henley considered that plough horses needed one-sixth bushel of oats each every night, and fed them also at midday. Their provender might be mixed with oats or wheat straw but not barley. The bailiff was expected to see that no servants of the manor rode on, lent, or ill-treated the farm horses. They were kept in stables under the care of the wagoner, who sometimes washed and combed them down. Almost every villein would probably have a stott or cart horse which he had to use for carrying

FIG. 38.—HORSE CARRYING WATER-SKINS

services for his lord, and for harrowing. According to Neckham he also had "if fortune smiles an ass or mule," but this seems very improbable, considering how rarely these animals are mentioned.[1]

The *ass* was considered useful for carrying burdens, and in the opinion of Bartholomew Anglicus "is fair of shape and disposition while young and tender . . . but the elder he is the fouler he waxeth from day to day . . . and is a melancholy beast." Neither his flesh nor his skin was any use.[2]

The *mule*, says Neckham, is "a cunning animal, obstinate,

[1] Cuningham: *Growth of English Industry and Commerce*, 163 n.
[2] Bartholomew Anglicus, ed. R. Steele, 119.

disobedient and revengeful."[1]  Neither of these animals was popular in 1086.  Twenty-six asses are recorded in Essex, but only two in Norfolk; there was one mule in Norfolk and two in Essex, and none of either in Cornwall or the lands of Ely.

*Dogs* are not much heard of during this period, but they were, of course, used to guard the flocks and herds and assist the shepherd, as in former times.  In the Luttrell Psalter two dogs are depicted in the illustrations of husbandry; one is going in front of the sower, chasing and barking at an immense crow, and the other is sitting cheerfully on the projecting beam that turns a windmill to catch the wind.  Both are middling-sized brown dogs, with long bushy tails and large prick ears, and both are wearing collars.

The farm-yard *cat* does not seem to have been very firmly established, and Neckham states that weasels were not infrequently domesticated to take her place, as they were skilled hunters of rats and mice.[2]  He also mentions that the domestic weasel was accustomed to lay its spoil before the feet of the lord or lady whose house it inhabited—a distressing habit not unknown among cats.  Arsenic was already used to keep down

FIG. 39.—SOWER WITH DOG (Luttrell Psalter)

rats, which were of the old English black variety, not the present grey, which are a modern importation.  But cats also did their share; at Thorpe, on the lands of St. Paul's, there was recorded among the stock in 1150 an old tom-cat (solemnly referred to as a "cattus senex") and two young cats.  Neckham includes a cat among the necessary equipment of a villein; this is no mere "cattus" but "muscipula," a mouse-catcher.[3]  The tom-cat evidently behaved in much the same way then as he does now, for, says Bartholomew Anglicus, he "maketh a

[1] A Neckham: *De Naturis Rerum*, ii. 159.
[2] Cf. Palladius, who in the fourth century recommended tame weasels to catch moles, 109.
[3] A. Neckham: *De Utensilibus, iii.* D.S.P., 132.

ruthful noise and a ghastful when one proffereth to fight with another."

*Rabbits* were much valued as a form of fresh meat by the lord of the manor, particularly from the thirteenth century onwards, and towards the middle of that century it became customary to obtain a grant of "free warren" from the king, and to keep large tracts of ground for rabbits, hares and pheasants, under the care of a warrener. Good land was wasted in this way, and, what was more serious, in the absence of wire netting the crops of the unfortunate villagers suffered severely from their depredations. In 1340 it was reported from West Wittering in Sussex that year after year the wheat in the village fields was devoured by the Bishop of Chichester's rabbits, and at Ovingdean that year 100 acres of arable were "lying annihilated" by the destruction wrought by the rabbits of the Earl de Warrenne.[1]

*Poultry* were universally kept by high and low. Every tenant, whether villein or cottar, must have had his little poultry-yard, for it was part of his duty to present his lord with one or two hens at Christmas, and a consignment of Easter eggs, varying from five to several hundred, according to custom or the size of his holding.[2] Sometimes he had to give a goose also. At Wye in Kent a servant of the manor went round and collected the hens and eggs, but if they were not ready the tenants had to bring them to the manor-house themselves within twelve days, on pain of a fine of 21 pence.[3] Neckham states that at the manor the poultry were usually kept in an inner courtyard, and that it was a good thing to have a magpie building in the poultry-yard, owing to the great clamour it made on the approach of a thief or intruder.[4] The fowls were under the care of the dairymaid, and were sometimes farmed out to her at the rate of 12d. for a goose, and

---

[1] *Sussex Arch. Soc.*, i. 62. Nonae relating to Sussex.

[2] In the Boldon Book (1183) the custom seems to have been usually one hen from a bovate, the ordinary villein therefore giving two, with 10 or 20 eggs. The cottar with 12 acres gave one hen and five eggs, and with less land either the hen *or* the eggs. In the Black Book of Peterborough (1125-8) the numbers are quite irregular, a round number often being required from the village as a whole. The Saxon gebur gave two hens, as we have seen, at Martinmas, not Christmas, but the eggs seem to have been a Norman addition. On the other hand, the lamb given by the Saxon gebur at Easter was discontinued.

[3] On this manor one tenant owed 1½ hens and another 2½ eggs, which it must have puzzled the messenger to collect, unless he took one hen and two eggs one year, and two hens and three eggs the next. C.B.A., 127.

[4] A. Neckham: *De Naturis Rerum*, i. 69; ii. 165.

3d. or 4d. for a hen.[1] Walter of Henley, always unduly optimistic, considered that a hen should lay 180 eggs a year, which would be quite a good yield for a hen at the present day. The anonymous *Husbandrie* lets her off with 115 eggs and seven chicks, from which one would suppose that they either put fewer eggs under a hen than we do now, or were less successful in hatching and rearing the chicks. The same author allows five hens to a cock and five geese to a gander. A goose, he says, will hatch once a year if she is good, but will not do it every year, and should produce five goslings at a time.

Capons were much reared, and were fattened in coops for the table. Hens with chickens were not always indulged with coops, for in the Luttrell Psalter a hen appears tethered by one leg to a peg driven into the ground. On this occasion she is endowed with a full family of thirteen, one of which is seated on her back after the manner of small chicks at the present day. She has a little square water trough in front of her, and a woman with a distaff tucked under her arm is scattering seed to the chickens from a platter.

FIG. 40.—POULTRY
(Luttrell Psalter)

Pullets cost 1d. practically throughout this period, and eggs were 1d. a score.

Geese in many places were driven out to feed by a gooseherd. He appears in the Luttrell Psalter scaring a crow from his five goslings, waving a hood in his right hand by the long

FIG. 41.—GOOSEHERD (Luttrell Psalter)

tail so pleasantly called a liripipe, and holding a stick in his left.

[1] W.H., 33, 75.

The most useful addition to the poultry-yard, and indeed to the farmer's stock in general, during this period was the domestic *duck*. It is not clear exactly when this bird was introduced from the Continent, but it does not seem to have been at all generally kept before the latter half of the thirteenth century. Alexander Neckham, writing about 1190, includes ducks in his list of domesticated fowls, but as he spent part of his life in Paris it is possible that he may have been describing what was usual in other countries, and which may only have penetrated in a few isolated cases to England. Walter of Henley and his anonymous contemporaries, writing with great detail of farm affairs during the first half of the following century, make no mention whatever of ducks, which could hardly have been the case if they had been generally kept. In the reign of Edward I, however, they seem to have been plentiful upon the Berkeley estates. Under Thomas, who was Lord Berkeley from 1281 to 1321, they were priced at 1d. each, but in 1321-2 at 1¼d.[1] At that time two of his manors supplied him yearly with 288 ducks. They were included in the schedule of tithes issued by the Council of Merton in 1305, and pigeons also. By 1333 ducks were well established on the estates of Merton College; only one manor out of eleven did not keep them, the others having an average of about nine each.

*Pigeons* seem to have become popular about the same time, or a little earlier. In the second half of the thirteenth century they were bred on a large scale for the larder and became a regular article of diet on the lord's table, to the harrying of the unfortunate peasants who had to scare them off their crops. Bartholomew Anglicus however looked upon them in a more poetical light. "The culvour," he says, "is the messenger of peace, the example of simpleness, clean of kind, plenteous in children, the follower of meekness, the friend of company and the forgetter of wrongs." In any case, they were a toothsome dish when fresh meat was scarce. and large dovecotes were added to the manorial buildings. The Berkeleys had at least one upon each of their manors and almost upon each farmstead, on some as many as three, and from these they derived thousands of pigeons yearly. They priced them at 3d. a dozen throughout the reigns of the three Edwards.[2] We read in the estate book of Henry de Bray that

[1] J. Smyth: *Lives of the Berkeleys*, i. 155, 162, 166. The average price given by Thorold Rogers between 1261 and 1350 is 1¾d.

[2] J. Smyth, op. cit., i. 302.

in 1303 he built him a dovehouse at the corner of his herb garden, and added another in 1305. By 1326 houses of 600 holes existed.

In addition to the ordinary inhabitants of the poultry-yard many manors kept *peacocks,* both as a dish for festivals and for the value of their feathers, for, says Walter of Henley, "a peacock gives as much for his feathers as a sheep for her wool." This bird did not appeal to Bartholomew Anglicus, for "as one saith he hath the voice of a fiend, the head of a serpent, and the pace of a thief. For he hath an horrible voice." At Forncett a small number of peacocks and peahens were kept, and the Berkeley manors were stocked with them. Humble persons seem to have kept them also, for in 1247 at Ogbourne in Wilts one Roger Pleade was called upon to prove at the manor court that he did not kill Nicholas Croke's peacock.[1]

*Swans* were fattened in coops on oats and peas, and it was part of the bailiff's duty to see that a good stock of them was kept.[2] Neckham adds herons, cranes, pheasants and various water birds to the ideal "court," but we need not assume from him that their presence was the general rule. He also says that the poultry sheds should not be thatched with straw, perhaps because the birds were liable to pick the straws out.

*Bees,* he says, excel all the domestic animals in worthiness and utility, for they provide light and sweet food. Great stores of wax and honey were obtained yearly from the Berkeley estates. In Domesday beehives appear on comparatively few demesne farms; in Essex about a quarter of the estates where the stock is recorded kept them, and in Norfolk less than a fifth, but as honey was universally required we must suppose that on the majority of estates either the beehives were farmed out in the Saxon fashion to a beekeeper, or the peasants paid dues in honey to the lord. In the Boldon Book (1183), Ralph the beekeeper of Wolsyngham held 6 acres for his service of keeping the bees, and very probably this was a common arrangement. Mr. Wright gives an interesting illustration of peasants of the Norman period tending bees: the hives are conical, tapering gracefully to a fine point.[3] The bees depicted are not unlike small chickens. Two gallons of honey was considered the average yearly yield from a hive in the thirteenth century.[4]

[1] Maitland: *Select Pleas,* 9.
[2] Fleta, ii. c. 76.
[3] Wright: *Homes of Other Days,* 103.
[4] Anon., *Husbandrie* (W.H.), 81.

The organization of the manorial estate under the Normans continued on very much the same lines as under the Saxons, but somewhat elaborated.  In some places, notably on Royal and Monastic lands, the ancient provender rents survived; for instance, each of the manors of an abbey would be charged with the food supply for the establishment for periods varying from one night to a fortnight, the amount being provided by the villeins as a whole.  Or again, the manor would be leased to a "firmarius," who supplied a fixed amount of provender as rent, and returned the demesne farm at the end of his lease with precisely the same amount of stock, implements and buildings on it as when he received it, and even the same amount of corn of different sorts stored in the barns, the hay stacked, and the fallow ploughed and manured.[1]  This delivery of food was a cumbrous system unsuited to very scattered manors, but was a step further in development than the custom of feasting the abbot or lord for one day and night when on tour.

On the ordinary manor farming was conducted by the lord and his officials much in the same way as in former times.  The all-important Saxon reeve, with his multitudinous duties, was relieved of some of them by the appointment of two other officers above him, the seneschal and the bailiff, but of course lost proportionately in standing, and became merely the representative of the tenants assisting the bailiff in his work.

The *Seneschal* or steward represented the lord in his absence, or where he owned a number of manors, a circumstance which became much more common after the Conquest.  In this capacity he held the manorial courts, overlooked the expenditure and accounts, was acquainted with everything concerning the extent, cultivation and stock of the manor, arranged for the fields to be divided and ploughed on the three-field or two-field system, as the case might be, and saw that the bailiff carried out his duties.  Among other things he was to suffer no beast to be skinned until the bailiff and reeve had ascertained the cause of its death.[2]

The *Bailiff* had to rise early and supervise the agricultural work of the demesne.  It was his business to calculate the amount of ploughing, reaping and other work that was provided by the customary daywork and boonwork of the peasants, and to see how much more would have to be hired for payment.  He overlooked all the operations of the farm,

[1] D.S.P., 130, etc.
[2] Fleta, ii. c. 72.

saw that the animals were well treated, that the tenants folded their sheep on the demesne land at night when they should, that the land was properly dressed, attended to many other details, and saw that the farm was well stocked, especially with horses, swans, bees and fresh-water fish. He saw that all the work was well done, and was exhorted to be just but not oppressive.[1]

The *Reeve* or provost (praepositus), annually elected by the tenants from among themselves, was the bailiff's assistant in all his duties, being especially responsible for the collection and scattering of manure. He also had charge of the granary, unless there was a special granger for this purpose, and gave out the corn by tally for baking, brewing and fodder. He saw that the horses and oxen were properly fed and rubbed down, and with the bailiff inspected the dairy to see when the cows and ewes were at full yield. He was answerable for the deaths of any animals in the courtyard, and for lack of proper increase among them, and also had to collect the horse-hairs carefully for making ropes.[2] In the Boldon Book the reeve's holding varies from 12 acres to two bovates (a virgate).

These two officials together examined all the stock between Easter and Whitsun, and separated the weak and sickly from the sound. The bailiff and reeve were present on every manor, but the seneschal only where the lord was habitually absent, or of very high standing. Their assistants were also more numerous than before.

The *Shepherd* was expected to cover, enclose and mend the sheepcote (this would be the permanent winter house) and sleep in it with his dog. In summer he had a little hut beside the movable fold, for himself and his dog, which was to be a good barker. The sign of a shepherd's benignity was taken to be that the flock was not dispersed but feeding quietly around him. He marked the sheep once a year, and milked them and delivered the milk to the dairymaid, but did not make the butter and cheese himself, as the Saxon shepherd did. He looked after the sheep at lambing time and was instructed to cut the wool round their udders, lest the lambs should suck it and die of the wool remaining in their stomachs.[3] He was provided with candles for his lantern, which were quite an item of extra expense at that season, and we even read of two pipkins being bought to warm milk for the lambs, at a

[1] Ibid., ii. c. 73. Anon, *Seneschaucie* (W.H.), 91.
[2] Fleta, ii. c. 76. Anon, *Husbandrie* (W.H.), 63 on.
[3] Fleta, ii. c. 79. W.H., 113.

cost of ¼d. each.[1]  He usually had the right to fold the lord's sheep on his own land for twelve or fifteen nights from Christmas to Epiphany, and was given a lamb (or occasionally two) and a fleece yearly: an almost exact survival of the Saxon shepherd's rights.  The shepherd of Longbridge in the Glastonbury Custumals (about 1250) was more generously treated than he of the Rectitudines.  He had the milk of twenty-nine of the sheep which had no lambs living, and (shared with the dairymaid and assistant) all the milk on Sundays between Hocktide and St. Peter's Chains (the second Tuesday after Easter to August 1st); also a portion of food daily for himself and another for his dog.  He was given two lambs "of the best" and one skin each year, an acre of land for himself, and a "small parcel of land" for the feed of his dog, also a sheaf on every reaping day except one.  He could feed thirty sheep free of charge in his lord's fold and had fifteen days' "foldage" of his lord's flock.[2]  This, however, seems to have been more generous treatment than most.  At Appledram, Sussex, the shepherd was merely quit of all his customary agricultural works except in autumn, when he received a quarter of barley, and the lord had to plough for him two acres in winter and two acres in Lent.[3]  At Bromham, Wilts, he received sixty sheaves of medium corn if he watched the lord's grain at night.

On some manors there was a "keeper of the wethers" or "muttons," and a "keeper of the ewes" besides.  At Bright Walton, Berks, the latter was not a customary tenant in 1272, but a permanent farm-servant, probably living in the court and superintending the lambing and milking there.  In 1293, however, one John Lord was presented by the vill as "a good man needful to the lord for the keeping of the ewes," with John Atgreen for keeping the "muttons."[4]  Occasionally there was also a "keeper of the lambs," who officiated from the Purification (February 2nd) to the Nativity of St. John the Baptist (June 24th).[5]

The mediaeval shepherd was a cheerful soul who played on bagpipe and horn to enliven the monotony of his task, and who often provided music for the benefit of his companions on festive occasions.

The duties of the *Swineherd* have already been explained

[1] Thorold Rogers: *History of Agriculture and Prices*, 568.
[2] R.C.G., 138.  The "parcel of land" was perhaps to grow corn for the "dog loaves" that were then customary.
[3] This rather suggests that he was a cottar holding six acres.
[4] Maitland: *Select Pleas*, 168.
[5] C.B.A., 57, 67, 82.  R.C.G., 235.

when dealing with the swine. On manors where few animals were kept one man sometimes did the work of both shepherd and swineherd, as at "Castre" on the lands of Peterborough about 1125. There, and at Glinton and Kettering, the swineherd held eight or nine acres of land.[1] On the lands of Glastonbury Abbey the swineherd received at the end of the twelfth century very much the same perquisites as in Saxon times; he was given one sucking pig each year (or in some places two little pigs, one of each sex), the interior parts of the best pig slaughtered, and in addition the tails of all the others killed. The dairymaid had charge of the sucking pigs, but the swineherd watered them after they were weaned, and fed them with the corn supplied to him by the "berebrit."[2] The swineherd's duties seem often to have been combined with those of the woodward or forester, which would be a convenient arrangement on manors where the woods were not large in extent and not a great number of pigs were kept.

The *Oxherd*, as before, had charge of the plough oxen when not at work; he took them to pasture and hobbled them to prevent their wandering, and looked after them and fed them in their stalls. He was helped in the feeding and cleaning by the ploughman. But besides this, if there was not a special "driver" on the manor, he often had the work of the boy who assisted the Saxon ploughman. For he yoked the oxen to the plough, "and pricketh the slow with a goad and maketh them draw even, and pleaseth them with whistling and with song to make them bear the yoke with the better will for liking of the melody of the voice."[3] This is more reminiscent of the Celtic driver with his "rude song" than the Saxon boy, who merely made himself hoarse with shouting, and seemed altogether a rather depressed person. The driver was exhorted to love his oxen, sleep with them and groom them well. The oxherd yoked and led the oxen in whatever tasks they were employed for besides ploughing, such as carting or even threshing.[4] On estates rich in oxen there were several oxherds. some manors on the lands of Peterborough Abbey (about 1125)

[1] L.N.P., 163, 164, 157.

[2] R.C.G., 95. This official occurs frequently in the Glastonbury Custumals, and seems to be a direct survival of the Saxon berebrytt of the Rectitudines, who was the barn keeper or granger.

[3] Bartholomew Anglicus, ed. R. Steele, 120, and cf. Fleta, ii. c. 78, who says drivers must not be melancholy or bad-tempered, but gladsome, singing and joyful.

[4] According to C. M. Andrews (*Old English Manor*) the oxherd proper was "bubulcus," and the attendant upon oxen at work "bovarius."

having as many as eight. There they usually held from five to ten acres for their services. If there were few working oxen, on the other hand, they would sometimes be put under the charge of the cowherd, while on large manors the ploughmen were often assisted by several drivers (tentores or tinctores) who relieved the oxherd of the duty of assisting with the plough.

The *Ploughman,* or his assistant driver, slept with the oxen at night, a duty formerly pertaining to the oxherd, and was therefore exhorted, like the other herds, never to carry a naked light into the byres, but only to use a lantern. He was either a permanent servant living in the court, or a tenant (called akermannus or carucarius) who held his land by the service of following the lord's plough. It was his duty to guide the plough, help yoke the oxen, and drive them without beating or hurting them. He sometimes carried a "clotting-beetle" in his right hand to clean the ploughshare, but after the plough had acquired a second handle this was obviously inconvenient. The ploughman had to be a "man of intelligence," able to repair or even make the woodwork of ploughs and harrows, if necessary; and he and the drivers when not otherwise employed had to sow corn, thresh, make fences and help in draining the land. He generally had the use of the lord's plough on his own land every second or third Saturday, and at Bromham, Wilts, he received one meal a day from Michaelmas to sowing time.[1]

The *Cowherd,* as before, looked after the cows and calves, milked them and drove them to pasture, saw them well housed

FIG. 42.—PLOUGHING (Luttrell Psalter)

at night during the winter and well provided with litter or fern, and slept with them. At Sturminster Newton he had from each cow after it had calved the milk the calf did not take, for two weeks, and from each heifer the same for four weeks. He had to prepare the stalls of the cows in the cow-

[1] C.B.A., 81. R.C.G., 102. Fleta, ii. c. 77. W.H. (Anon., *Sen.*), 111.

house and lie there with them in winter, and in the fold when they were there in summer. His wife helped with the cheese-making.[1]

The horses were cared for by the *Wagoner*, or *Carter*, who now first appears, owing no doubt to the increased importance of his charges. He fed and curried the farm horses and slept beside them, and it was his business to mend the harness and gear of the wagons, and cart marl, manure, hay, corn, timber and firewood when required. He was expected to be steady, experienced, modest, and not wrathful, skilfully directing his horses and not overburdening them.[2]

No "keepers of beasts" were allowed to go to fairs, markets, wrestling matches or taverns without leave, and all were exhorted against the danger of taking naked lights into the sheds.

The *Dairymaid* (or daye, if masculine) was bidden to "keep herself clean and know her business," which was con-siderable. She made the milk of the cows and sheep into butter, cheese and "salt cheese" from May until Martinmas, and sometimes up till Christmas, but from then until May it was considered more profitable to sell what little milk there was, as it was three times as valuable as in summer. Cheeses were made in three sizes, great, middle and small, and were sold at 3d., 2d., and 1d., their weights being probably about 2 to 6 lb. Rennet for curdling the milk was usually made at home also. On some estates the cows were let out to the "daye" at 5s. to 6s. 8d. a cow, and ewes at about 1s., the lessee to restore the animals in similar numbers and condition at the end of the term.

Besides doing this ordinary dairywork and looking after the vessels carefully so that they did not need to be renewed every year, the dairymaid had charge of the poultry-yard with its cocks, hens, capons, geese, ducks and peacocks, with all their various offspring, and the collection of the eggs, and also tended the sucking-pigs, which she kept at the dairy or near by. She was expected to winnow the corn, with other women to help her, to sweep, and "to cover the fire that no harm arise from lack of guard."[3]

The *Punder* or *Pindar* was a new official who appears at this time. His duties seem to have been an offshoot from the many

[1] R.C.G., 94. Fleta, ii. c. 86. W.H., 113.

[2] Fleta, ii. c. 85. W.H. (Anon, *Sen.*), 111.

[3] Thorold Rogers: *History of Agriculture and Prices*, 14, 397. W.H., 75, 117. Fleta, ii. c. 87.

tasks of the hayward, and they were still carried out by the latter on manors where there was no separate pindar. It was his business to collect any animals found straying, and confine them in the village pound or pinfold until claimed by the owner. If they were not claimed within a year and a day they became the property of the lord of the manor. As chickens and even swarms of bees were treated in this way the pindar must have had a varied career, but since his duties were not as a rule very onerous he seems to have often filled up his time with chicken-farming. In the Boldon Book the pindars render extraordinary numbers of hens and eggs to the Bishop, 40 hens and 500 eggs being quite a usual amount, while as many as 120 hens were occasionally given, and 1,050 eggs once recorded. These pindars generally held about 12 acres of land for their services, and received one thrave of corn (24 sheaves) from each plough team.

We now come to a little group of officials whose duties seem to have overlapped at certain points, or to have been to some degree interchangeable; these are the Woodward, Forester, Hayward and Messor. A *Woodward* was appointed on manors where there was a fair amount of wood-land to be cared for, and was responsible for seeing that valuable trees were not wantonly cut down or burned. He received the dead brushwood, and when timber was cut for ploughs he had the branches. If his work were light he sometimes, as we have seen, performed the duties of the swineherd as well, and had the care of the lord's pigs, an employment which in autumn particularly would combine very well with his other tasks. At Bright Walton, Berks, where these two offices were combined, he received for the care of the pigs four bushels of barley and three meals, and at killing time the intestines of one pig, except for the fat, which was needed for the larder. He also impounded any animals found straying in the woods.[1]

The *Forester* would have charge of larger tracts of woodland, probably suitable for the chase, and would have something of the duties of a modern gamekeeper. He also sometimes kept the pigs, as at Bromham, Wilts, where he received one meal a day from Michaelmas to sowing time, and had to take his part in threshing and carrying duties.[2]

The *Hayward* was still another person having charge of the woods, according to the anonymous *Seneschaucie*. This might

[1] C.B.A., 67. Maitland, *Select Pleas*, 171. R.C.G., 57.
[2] C.B.A., 81, 82. R.C.G., 37.

be the case on manors where there was but little wood-land—
what is recorded in Domesday as "nemus ad sepes," sufficient
wood for the fences; for the chief duty of the Saxon hayward,
as we have seen, was to keep up the hedges or fences round
the crops in the common fields. The thirteenth-century
hayward however had a good deal more to do, for he had
general charge of the "woods, corn and meadows," and also
superintended the sowing or did it himself, overlooked the
harrowing, haymaking and harvest, and tallied the corn with
the provost, so that his interest in the fields was considerably
widened. In this last connection he seems to have been
identical with the *Messor,* an official appointed to take charge
of the harvest, but it seems improbable that the titles were interchangeable. In Suffolk the official overlooking the haymaking
was known as the *Lurard.*[1]

In some places the hayward acted as pindar also, which
would be a natural outcome of his ancient duty of keeping the
cattle from the corn-fields. At Ashbury he had charge of the
pinfold, sowed the lord's seed, and there and in other places
on the Glastonbury Estates rendered 100 eggs at Easter, which
recalls the pindars of the Boldon Book. He sometimes had the
right to all the wayside grass, and at Winterbourne he had to
carry litter for the sheep to the lord's fold and visit them twice
a day. It is quite possible, therefore, that a large manor might
have a hayward to look after hedges and fences, and separate
officials for the care of the woods, the haymaking and harvest
and the care of strays, whereas on a smaller manor the hayward
attended to them all. At Longbridge on the Glastonbury
lands the hayward did the sowing and also superintended the
gathering in of the sheaves. The size of the sheaves made by
the tenants was tested in a curious way: "If any sheaf appear
less than is right, it ought to be put in the mud, and the hayward should take hold of his hair above his ear, and the sheaf
should be drawn through his arm, which if it can be done without the soiling of his clothes or hair it should be considered less
than is right, but otherwise it shall be adjudged sufficient."[2]
Such a process would certainly not encourage the hayward to
find fault lightly. At Bright Walton the messor received three
meals for watching the lord's grain at night in autumn until
it was cut—an important part of his duties—and a sheaf daily
as long as the reaping continued. The size of the sheaf there
was fixed by binding it with a cord $1\frac{1}{2}$ ells long. At Bromham

[1] Vinogradoff: *Villeinage in England,* 319.
[2] R.C.G., 53, 57, 64, 95, 136.

K

he received 60 sheaves of medium corn, and could pasture a mare and foal on the lord's pasture in summer. Fleta says the messor shall be endowed with the qualities of courage, health, harshness and fidelity.[1]

Men suitable for all these offices were chosen by the villagers and presented at the manorial court for their appointment to be confirmed by the steward.[2]

Besides these officials there were a number of craftsmen in the village, varying of course according to its size, of which the most essential were the carpenter and smith. The carpenter made and repaired the woodwork of the ploughs and harrows, and the smith the all-important plough-irons, horse and ox shoes and nails. Skilled ploughmen and carters were sometimes able to do the carpenter's work, but the smith could not be dispensed with. He had a small holding for his services, and was often given a money allowance to find the necessary iron and steel for his work. He usually received three tree trunks a year from the lord's woods to make charcoal for his furnace, often the hide of an "affer" or farm horse when dead, presumably to make bellows, and in one place a "dish of butter" to grease them. It was often his duty to bleed the horses according to mediaeval custom on St. Stephen's day. He had to grind the reaping hooks before haymaking and reaping, and sometimes to mend the cheese vessels with iron hoops.[3] Iron throughout this period was scarce and high-priced, which must have hampered the evolution of efficient implements.

There were also on many estates the miller, baker, weaver, leather worker, brewer, parchment maker, and many persons engaged in other small activities, sometimes including a wontner or mole-catcher. At Worcester Priory a white loaf was given for each mole caught in the garden; in other places the mole-catcher got 1s. a hundred for old moles, and 1s. 3d. to 1s. 10d. for young ones.[4]

---

[1] C.B.A., 67, 81.  W.H., 103.  Fleta, ii. c. 84.
[2] Maitland: *Select Pleas*, 36, 168, 170, 171.
[3] R.C.G., 59, 63, 92, 101.
[4] Reg. Worc. Priory (Camden Soc.), 121a.  Thorold Rogers, op. cit., 282.

## APPENDIX TO CHAPTER VI

The second volume of Domesday Book, dealing with East Anglia, records the stock on the demesne farms of many of the manors, and a certain idea can therefore be gained of the average numbers of animals kept on such establishments about 1086. These figures can be supplemented by those given in the *Inquisitio Eliensis* for the lands of the Abbey of Ely, and those of the *Exon Domesday* for the West of England. In these tables (see pages 152 and 153) Norfolk and Essex have been taken to represent East Anglia, and Cornwall for the West.

#### A. Total Numbers of Stock Recorded.

| | Ox Teams (of Eight) | Sheep | Pigs | "Animals"[1] | Cows (separately mentioned) | Runcini | Horses[2] | Goats | Beehives | Number of Farms |
|---|---|---|---|---|---|---|---|---|---|---|
| Norfolk .. | 931¾ | 43,848 | 7,824 | 2,125 | 26 | 584 | 242 | 3,016 | 446 | 516 |
| Essex .. | 1,004½ | 46,533 | 12,870 | 3,834 | 166 (& 369 calves) | 783 | 34 (& 117 foals) | 3,405 | 612 | 465 |
| Cornwall .. | 332½ | 13,104[3] | 513 | 1,094 | 53 | 21 | 414 | 935 | — | 243 |
| Ely .. .. | 270 | 13,443 | 2,694 | 932 | 4 | 159 | 28 | 355 | 50 | 115 |

#### B. Average of Stock only on those Farms where Each Sort Occurs.

| | Ox Teams (of Eight) | Sheep | Pigs | "Animals"[1] | Cows (separately mentioned) | Runcini | Horses[2] | Goats | Beehives | Number of Farms |
|---|---|---|---|---|---|---|---|---|---|---|
| Norfolk .. | 1¾ | 105.45 | 17 | 6.45 | 2.6 | 1.92 | 7.33 | 29.56 | 4.95 | — |
| Essex .. | 2⅖ | 104.8 | 31.16 | 10.8 | 3.8 | 2.55 | 8.5 | 26.8 | 4.6 | — |
| Cornwall .. | 1½ | 56.96 | 5.89 | 6.55 | 1.7 | 1.5 | 10.35 | 12.14 | — | — |
| Ely .. .. | 2¼ | 130.51 | 25.9 | 12.1 | 2 | 2.4 | 5.6 | 27.3 | 3.84 | — |

## C. Average of Stock on Whole Number of Farms.

| | | | | | | | | | |
|---|---|---|---|---|---|---|---|---|---|
| Norfolk .. | 1¾ | 84.97 | 15.16 | 4.11 | — | 1.13 | 0.46 | 5.84 | 0.86 | — |
| Essex .. | 2⅛ | 100.07 | 27.67 | 8.24 | — | 1.68 | 0.07 | 7.32 | 1.31 | — |
| Cornwall .. | 1¼ | 53.51 | 2.1 | 4.5 | — | 0.08 | 1.7 | 3.84 | — | — |
| Ely .. | 2¼ | 116.89 | 23.42 | 8.1 | — | 1.38 | 0.24 | 5.0 | 0.43 | — |

1 "Animals," probably includes cows, calves, young bullocks (not yet used for the plough), and bulls, although on a few occasions cows and calves are mentioned separately.

2 The horses in Cornwall, and often in Norfolk, are "indomiti" or "sylvestres."

3 A flock of 240 are entered as "berbices," instead of the usual "oves."

## D. Number of Demesne Farms on which Each Sort of Stock Occurs.

| | Ox Teams | Sheep | Pigs | "Animals" | Cows (mentioned separately) | Runcini | Horses | Goats | Beehives | Total Number of Farms |
|---|---|---|---|---|---|---|---|---|---|---|
| Norfolk .. .. | 504 | 416 | 460 | 329 | 10 | 304 | 43 | 102 | 90 | 516 |
| Essex .. .. | 451 | 444 | 419 | 353 | 43 | 307 | 4 | 127 | 132 | 465 |
| Cornwall .. .. | 228 | 230 | 87 | 167 | 31 | 14 | 40 | 77 | — | 243 |
| Ely .. .. | 115 | 103 | 104 | 77 | 2 | 66 | 5 | 13 | 13 | 115 |

Two asses are mentioned on the lands of Ely, 2 (on 2 farms) in Norfolk, and 26 (on 7 farms) in Essex, making a total of 30. Two mules are mentioned in Essex (on 2 farms) and 1 in Norfolk. Total, 3.

# CHAPTER VII

## THE NORMAN CONQUEST TO THE BLACK DEATH

### (Continued)

THE routine of the farmer's year during the Norman and Early English period did not differ in essentials from that of the Saxon, but was slightly more elaborate in its processes as farming continued to develop from the rude simplicity of early days. The villeins continued to cultivate their scattered strips in the common fields in the same co-operative manner as before, on the two- or three-field system, according to the custom of the place. Each manor being a self-contained unit, practically the same range of crops was grown everywhere without regard to suitable locality; wheat, for instance, was grown in places where it would not be considered profitable now, although rye probably predominated in the North, and was indeed grown everywhere to mix with it. Vines, however, had perforce to be confined to the South, though they existed as far north as Norfolk. Thirty-eight vineyards are recorded in Domesday Book, and they long continued to be cultivated, especially by the monasteries, the somewhat inferior wine they yielded being often sweetened with honey and flavoured with blackberries and spices. In the favoured vale of Gloucester, however, this does not seem to have been necessary, if we may believe William of Malmesbury. Writing in the first half of the twelfth century he says: "This vale is planted thicker with vineyards than any other province in England, and they produce grapes in the greatest abundance and of the sweetest taste. Wine made in them hath no disagreeable tartness in the mouth, and is little inferior to the wines of France."[1] The writer may have been prejudiced in favour of his own district.

The hay meadows of the village, as before, were fenced off from February or March to Midsummer, and were allotted in portions of a half-acre or less to the villagers. Sometimes the right to a certain meadow strip went with an arable holding, so that the tenant received the same one each year; but sometimes lots were cast afresh for them every season, each plot being known by some sign such as a cross or crane's foot, and the corresponding signs being put in

[1] Sir John Fortescue: *Works*, i. 58.

a bag and drawn out.[1] There were also specially reserved pasture-lands known as "stinted pasture," because each tenant might only send a fixed number of animals to feed on it.

The arable land of the lord of the manor was sometimes scattered in strips among the rest, and sometimes separately enclosed as a whole. The main work of the manorial farm was still supplied by the customary tasks of the villeins and cottars, which they rendered in exchange for their holdings, but these tasks had become more varied and numerous. A "full villein" on the lands of Peterborough and Durham continued to do two or three days' "week work" throughout the year, or sometimes two days from Pentecost to Martinmas and one day a week for the rest of the year. In some places as much as four days a week was exacted between August and Michaelmas;[2] in other parts, notably at Glastonbury and Battle, there seems to have been no regular week-work for the rest of the year, but the villeins commonly worked for the lord every day from August 1st to Michaelmas, which must have been an inconvenient arrangement for their own farms unless they had grown-up sons in the family. Besides this there were the "precariae" or boonworks, originally a matter of "request," but now obligatory, consisting of an acre or two of ploughing and harrowing in Lent and late autumn, usually an acre of ploughing of fallow in summer, and extra work at haymaking and harvest. It is little wonder that friction arose over the existence of these multitudinous duties, and we often read of the villeins being fined at the manorial court for not performing them to the satisfaction of the bailiff.[3]

There were also carrying services, varying very much according to the locality, but particularly heavy on monastic lands and other large estates where provisions had to be conveyed to a distant centre. Lastly there were "occasional works" of the most varied description, from gathering nuts, gardening, building and repairing, to heating the water for the lord's bath.

In January there was not much outdoor work, though ploughing theoretically began on "Plough Monday," the first Monday after Twelfth Night. Oats were threshed in the barn

[1] Lord Ernle: *English Farming*, 3rd edition, 26.

[2] *Rot. Hund.*, ii. 539 (Hatley, Cambs.). The duties here seem to have been exceptionally heavy; and the tenant had to give 4 hens, 1 cock, 5 geese and 80 eggs a year.

[3] Maitland: *Select Pleas*, 10, 12, 19, 30, 91.

ready for the spring sowing.  As a Welsh poem of the early fourteenth century has it:

> The cow is lean and the kiln is cold,
> The horse is slender, and the bird silent,

but in the next month, "Busy are the spade and the wheel."[1] The year's work really began with the spring ploughing, early in February, of the village field which had been sown the year before with wheat.[2]  The plough, which Neckham refers to enthusiastically as "a divine work for the invention of which we ought to thank Heaven," did not undergo much alteration. The Bayeux Tapestry shows a simple form (much elongated) with ploughshare, coulter, wheels and two handles, but there were not always two handles, and wheels were by no means always used.  Neckham does not mention them in describing the parts of the plough[3] (about 1190) and the plough depicted in the Luttrell Psalter (early fourteenth century) is without them, but they occur fairly often in early accounts.  The mould-board, or pair of mould-boards, to guide the fall of the turned sods, now became general and is shown very clearly in the latter illustration, but was a straight piece of wood which merely pushed the sod aside and did not turn it over.  "Plough clouts," or plates of iron, were sometimes nailed to it for extra strength.  Ploughshares were, in some places, given as rent. The plough team theoretically consisted of eight oxen, but varied in practice, both in numbers and composition, and on their own strips the villeins probably often used light ploughs with four oxen.  The teams on the lands of St. Paul's were often of ten heads, of oxen and horses or stotts mixed, and sometimes apparently even more.[4]  The numbers of oxen and horses in the teams were sometimes equal, but usually the oxen preponderated.  Mares were often used for the plough, and occasionally cows.[5]  Walter of Henley goes thoroughly into the relative advantages of horses and oxen for this

---

[1] This and following quotations from the same poem, T. Stephens: *Literature of the Kymry*, 287 et seq.

[2] See above, p. 108.

[3] Alex. Neckham: *De Naturis Rerum*. ii. 169.  *De Utensilibus* (ed. Mayer), 112.

[4] The wording is ambiguous.  "Two teams of 16 heads" (Wickham, D.S.P., p. 33) by comparison with other entries could mean that *each* team was composed of 16 animals.  Similarly "Two teams of 20 heads," and two good teams "habentibus 20 capita in jugo" (Nastok, 75); but considering the unwieldy numbers it might with greater probability be taken to mean that the two ploughs *together* had 20 animals, i.e. 10 each.

[5] C.B.A., 60.

purpose and is in favour of oxen. He admits that two horses with a team of oxen increase the speed, but the "malice of the ploughmen" will not let the horses go faster than the pace they choose, so that little is gained. Horses also cost more to feed and only their skins were of value when old, whereas an ox past work could be fattened for the larder on 10d. worth of grass. Oxen with their slow, steady pull would go forward on a hard piece of ground when horses would come to a standstill. A modern writer points out that besides its advantages over the horse on steep slopes, the ox was particularly suitable for the old ill-drained fields because his hoofs spread and gave him a better foothold.[1] Some but not all oxen were shod in the Middle Ages. The amount of ploughing that a team was expected to do in a day varied from 3 roods to an acre. Walter of Henley considered 3½ roods enough at ordinary seed-time ploughing or first fallowing, but an acre at the lighter summer fallowing. At Bocking in 1309 an acre a day was considered an easy day's work, with a team of four horses and two oxen.[2] Walter of Henley reckoned that each of his teams could therefore keep in cultivation 180 acres on the three-field system, or 160 acres on the two-field, but this is usually considered too idealistic. The statute acre was fixed by Edward I as a long narrow strip 40 rods or poles (1 furlong) in length and 4 poles wide, and the village fields were therefore divided into these strips, or ones half the size, as nearly as their boundaries would allow. The old tradition seems to have been a furrow to each foot, and if the land were "rigged up," three or four ridges to the acre strip.

When the early ploughing was done, in small furrows that the seed might fall evenly, the hayward saw to the sowing of the spring crops, viz. barley and oats, peas (grey and white), vetches and beans. Beans were planted by hand, "dibbled," and this is often mentioned as a special service. Vetches were a valuable addition to the farmer's fodder-crops[3] and lentils were occasionally grown.[4] In the Luttrell Psalter the sower is seen scattering his seed from a little triangular wicker basket, his dog busily scaring away the crows. Barley and oats were sown 4 bushels of seed to the acre, and beans and peas 1½

---

[1] W. Johnson: *Byways in British Archaeology*, 466-8.
[2] Vinogradoff: *Villeinage in England*, 315.
[3] Possibly these were introduced earlier by the Norsemen, as they are mentioned in the Icelandic Sagas. See above, p. 121.
[4] *Documents of the English lands of the Abbey of Bec*, Camden Soc. lxxiii, pp. 135, 136, 143, 144.

quarters to 5 acres (i.e. about 2¼ bushels to the acre) according
to the anonymous *Husbandrie*; but at "Wnfrod" on the lands
of Glastonbury they allowed 3½ bushels of barley, 5 bushels
of oats and 3 bushels of peas and vetches to the acre.[1]

Sowing was quickly followed up by harrowing, for "when
the ground is sown then the harrow will come and pull the
corn into the hollow which is between the two ridges."[2]
Harrowing now appears as a regular duty owed by the tenants,
and an addition to their tasks since Saxon days. It was always
done with horses, one horse to a harrow except on very heavy
soil, and a boy followed behind with a sling from which he shot
stones at the birds to scare them off the seed. The harrows
shown in the Bayeux Tapestry and the Luttrell Psalter are of
stout crossed timbers with teeth inserted on the under side, but
Neckham describes one made of thorns, which would be for
humbler farmers. He also mentions a kind of stone roller,
and as this implement also occurs on the Berkeley Estates we

Fig. 43.—Harrowing (Luttrell Psalter)

may presume that it was used on up-to-date farms of the period.[3]
In some places food was given for harrowing service, the men
of Appledram, Sussex, receiving a meal of bread, broth and a
dish of meat, and their horses "as much fodder as can be taken
between two hands."[4]

Early in the year also, before the droughts of March, the
litter and manure was gathered up from the court and the
sheep-houses, and carried out to the fields to be sprinkled, under
the superintendence of the reeve, the work being done by the
villeins with their carts.

In March "Bitter blows the cold blast o'er the furrows."

[1] W.H., 67. R.C.G., 235.

[2] W.H., 15.

[3] A. Neckham: *De Utensilibus*, 113. J. Smythe: *Lives of the Berkeleys*,
i. 303.

[4] C.B.A., 53.

According to the illustration in an early fourteenth-century psalter, the trees were drastically pruned, and during this and the following month the gardens would be dug over and planted, and the vineyards attended to, where they existed. Gardening sometimes occurs among the occasional duties of the tenants; at Marley, Sussex, certain men were required to work in the garden thirty days in the year. At Damerham one of the tenants filled the post of gardener and was given the crop from one apple-tree and two gallons of cider daily while they were making cider.[1] These gardens were mainly "herb gardens," for vegetables and cooking herbs, or orchards, supplying chiefly apples and pears for cider and perry; but "noble" gardens were stocked also with cherries, plums, quinces, gooseberries, medlars and peaches.[2] Neckham, who is always ambitious, adorned his garden with roses, lilies, heliotrope, violets, poppies and daffodils, mixed in with the herbs,[3] but these were probably unusual, except for the first two. A great advance in gardening was made between 1250 and 1350, and under Edward I the little wild strawberries and raspberries were sometimes transplanted and cultivated.[4] The famous Warden pear also appeared at this time, being cultivated by the Cistercians of Warden Abbey, and baked in pasties called "Warden pies."

Flax plots were cultivated as before, both the blue- and yellow-flowered varieties, for, says Bartholomew Anglicus, "none herb is so needful to so many divers uses to mankind"; and occasionally the preparing of land for flax was a villein service.[5] Hemp was also widely grown, which seems to have been a new departure. These two were sown about April. The spades used were still wooden implements with crescent-shaped iron tips, and the gardener was sometimes provided with a wheelbarrow.

April, after the barley sowing, was considered the best month for "first fallowing" (warectum), i.e. the first ploughing up of the field which was lying fallow, and which had to be ploughed

---

[1] R.C.G., 14. C.B.A., 10.

[2] T. Hudson Turner: "State of Horticulture in Early Times," *Arch. Journal*, v. 303-5.

[3] A. Neckham: *De Naturis Rerum*, ii. 166 (about 1190).

[4] G. Henslow: *Uses of British Plants*, 62, 64. The large garden strawberry was a late importation from America. Other sorts of pears known in the thirteenth century, but probably confined to distinguished gardens, were the Caillou, St. Règle, Pessepucelle, Martin, Dreye, Sorell, Gold Knob. Cheysill and Janettar. The only apples named are costards and pearmains. T. Hudson Turner, op. cit., 300-2.

[5] D.S.P., lxxiv.

three times before being sown with wheat in the autumn. Most villeins owed a day's work at this ploughing, and by the end of the month it could be well said,

> The oxen are fatigued and the land naked.

May was the beginning of the dairy season, and animals which had been stall-fed during the winter and early spring were then turned out to grass. As the Welsh poem puts it:

> The caller of the oxen is relieved from care,
> And the hedge affords comfort to the friendless.

Hedges and fences round the growing corn were attended to so that the cattle should not stray on to the crops, and field work such as breaking clods and ditching was carried on. For the first time an elementary sort of draining of the fields was attempted, and it was the duty of some villeins to let off the water accumulated in the ploughed fields by opening the furrows between the ridges.[1] For those who owed ditching service it was considered that two men should make in a day 1 perch of new ditch 5 feet wide, or repair 2 perches.[2] Others, both men and women, were occupied in breaking clods of

FIG. 44.—CLOD-BREAKING (Luttrell Psalter)

earth in the fields with mallets, as is shown in the Luttrell Psalter.

In June "The genial day is long and women full of activity." Haymaking began, and was a "boonwork" to which all tenants were called for a varying period according to their holding. At the beginning of the hay season in some places each villein was given a beam or tree trunk from the lord's wood for the repair of his hay wain, and a holiday in which to effect the repairs. This allowance of wood was known as "wenbote."[3]

[1] D.S.P., lxxix.
[2] C.B.A., 28.
[3] R.C.G., 83.

Even the socmen in many places gave a day's help at hay-making, or in other parts rode about with rods to see that their humbler neighbours were attending to their work. The miller also joined in the work, and the wives of the herdsmen were expected to mow half an acre a week. On the Durham lands all the villein's household were required to turn out except the housewife, who was considerately left at home. In one instance an old woman carried water to the mowers, holding her cottage and croft by that service.[1] All these people between them, supervised by the hayward, or lurard, cut the hay, tossed it, made it into cocks, and presently lifted and carted it to the manor court and there made it into ricks. Food was generally provided for mowing service, as for other "boonworks." At Longbridge on the Glastonbury lands a she-lamb, four cheeses and two loaves were given to the men, and at Boile on the lands of Canterbury they received three quarters of wheat, a ram, a pat of butter, a piece of cheese of second quality from the lord's dairy, salt, oatmeal for cooking a stew, all the morning milk from all the cows in the dairy, and every worker a load of hay. In many places the villeins had the right to as much grass as could be lifted on a reaping-hook, a custom known as "haveroc" or "over-hook." In others they might take as much

FIG. 45.—WEEDING (Luttrell Psalter)

as could be lifted with the middle finger as high as the knee: this was called "medknicche" or "midknee."[2]

After Midsummer Day there was much to be done. The sheep were washed and shorn, an operation in which the villeins helped, the miller occasionally taking part and rolling up the fleeces. For shearing service at Longbridge the men received a cheese "made in the lord's hall the same day." At this time or a little later the fattened sheep were sold.

In July the growing corn was weeded and thistles uprooted, the tradition being that if thistles were cut before St. John's

[1] Vinogradoff: *Villeinage in England*, 285.
[2] R.C.G., 92, 98. Vinogradoff: *Villeinage in England*, 175.

Day two or three would come up for each one that was there before.[1]  In the Luttrell Psalter two women are depicted weeding the corn, and in Queen Mary's Psalter three men, a woman being employed to carry away bundles of weeds.  In both illustrations the workers hold in the left hand a long stick forked at the end, and in the right another long stick with a small hooked blade.  At Bright Walton the tenants doing weeding service were given barley bread, broth and whey at noon, and bread and whey in the evening.[2]

At this time also, between haymaking and harvest, the field lying fallow was given its "second stirring" with the plough (rebinatium).  Most tenants were bound to do an acre or two of this, which was done more lightly than the other ploughings, just to destroy the weeds.

In August, says the Welsh poem,

> Bees are merry and the hives are full,
> More useful is the reaping hook than the bow,
> And ricks are more numerous than play grounds.

At Lammas (August 1st) the fences round the village hay meadows were taken down and the cattle allowed to roam over them to feed upon the aftermath and to manure the ground for the next crop of hay; hence these fields got the name of 'Lammas lands."[3]

August was the month of harvest, and again practically

FIG. 46.—REAPING (Luttrell Psalter)

all the community except the housewife were called together to deal with it.  Barley, oats, peas and beans were mown with scythes, but wheat was reaped with sickles about halfway up

[1] W.H., 17.

[2] C.B.A., 67.

[3] It must be remembered that until the readjustment of the calendar in the middle of the eighteenth century all the Saints' Days mentioned occurred twelve days later in the season than they do now.  Old Lammas Day was therefore equivalent to our present August 13th.

the stalk, and the best of the straw collected later on for thatching; the rest was either ploughed in (after the cattle had been turned on to it) or sometimes burnt "to help amend the land."[1] Five men or women constituted a band, and were expected to reap two acres a day. In the Luttrell Psalter the sickles are saw-edged, but not apparently in Queen Mary's. This may be merely inaccurate drawing, for the saw-edged

FIG. 47.—COLLECTING SHEAVES (Luttrell Psalter)

shearing hook persisted as long as the corn was gathered in a bunch in the left hand and severed high up on the stalk. The sharp-edged "swap-hook" did not come into use until the corn was cut by slashing it close to the ground. Gloves were given to the reapers to prevent the left hand, holding the corn, from being cut.[2]

In Queen Mary's Psalter the messor is directing the reapers

FIG. 48.—CARTING CORN (Luttrell Psalter)

with a horn slung at his belt, which suggests that they were still called together by that method. The corn was bound into sheaves with two strands of corn knotted together by the

[1] Bartholomew Anglicus, ed. Steele, 89. Thatch-cutting and carting was quite often a villein service.

[2] R. M. Garnier: *Annals of the British Peasantry*, 150.

heads. Women gleaners followed the reapers and sometimes helped to bind the sheaves. There was sometimes a special cart-reeve, besides the messor or reap-reeve, to supervise the carrying of the corn.[1] A spirited picture of the carting is given in the Luttrell Psalter. Three horses in tandem, the first dappled, draw the cart, which has stout projecting nails all round the outside of the wheels. The driver, with a long whip, stands on the shafts astride the horse, boys are pushing behind and helping the wheels round, and another man is keeping the sheaves on the cart with a pitchfork and helping to push. In many places the wains were drawn by oxen. The corn was stacked in the great barns of the manor, each sort by itself, ready to be threshed during the winter months.[2] If men were only called out once for reaping it was usually a "dry request," i.e. food was provided but not drink, but if a second time ale was as a rule also provided. On the lands of St. Paul's the allowance of food, with ale, for two men at a "precaria" was at noon two loaves, one white and the other "meslin,"[3] and a piece of meat; and in the evening a small meslin loaf and two pieces of cheese. At a "dry request" they had two large loaves, "potage," six herrings, a piece of some other fish, and water; but these allowances varied very much. Each villein received an allowance of sheaves when the corn was carted.

When the harvest was in, or during any intervals of the summer work, there was woodcutting to be done. Firewood and brushwood were collected and carted to the manor court, the latter being much used for roof covering. Thorns, rods and pales were cut to repair the fences, timber of various sorts brought in, and turf dug, and rushes and heath collected in the localities where that was one of the villein's duties.

In September and early October there were fruits to be gathered and honey to be collected.

> Men and horses know fatigue,
> Every species of fruit becomes ripe,

and in October,

> The birch leaves turn yellow
> And the summer seat is widowed;[5]
> The milk of cow and goat becomes less and less.

[1] F. G. Davenport: *Economic Development of a Norfolk Manor*, 26.
[2] Descriptions of these barns, with their great heaps of every sort of corn, carefully measured, occur often in the leases of St. Paul's manors.
[3] Wheat or barley mixed with rye.
[4] D.S.P., lxviii.
[5] This is a reference to the Welsh custom of migrating to the summer feeding grounds.

Where there were vineyards the grapes were brought in deep baskets to a great wooden vat, where they were trodden by bare-footed men with the skirts of their tunics tucked up.

In most places the apples and sometimes the pears were made into cider, ten quarters of apples and pears yielding by estimation seven tuns of cider.[1] Crab apples were gathered to make verjuice. Everywhere ale was brewed, the villeins being sometimes expected to help in the malt making. Barley was of course the grain most commonly used for brewing, but oats also were sometimes made into malt, and occasionally wheat, or all three mixed.[2] If a villein wished to brew ale for sale the lord often supplied the necessary implements, receiving in payment nine gallons from each brewing.[3] Being made without hops this ale had to be drunk pretty fresh and brewing was continually going on.

At Michaelmas the tenants again brought their carts and took the manure from the farm-yard to the fields and sprinkled it. Sometimes marl also was spread to enrich the land.

When all these operations were complete, and before the winter frosts set in, the "hyvernagium" or autumn work of cultivating the fields began. The part of the village field which had been lying fallow during the past year, and which had been twice "stirred" during that period, was now ploughed up for the third time for the winter crop of wheat or rye, the villeins bringing their ploughs and harrows as in the spring. Wheat was sown two bushels to the acre or a little more,[4] and rye the same. A picture of October sowing is given in Queen Mary's Psalter, a very small horse bringing a sack of seed-corn on his back for the two sowers, doubt-less the hayward and his assis-tant, who are scattering it with an air of graceful negligence. It was recommended that the winter seed should be bought, or brought from a distant manor, but the spring seed might be taken from the lord's own granary.[5] At the end of the "hyvernagium" the villeins were often given a "solemn repast."

Fig. 49.—Threshing
(Luttrell Psalter)

[1] W.H., 79.
[2] D.S.P., 165. J. Smythe: Lives of the Berkeleys, 303.
[3] R.C.G., 82.
[4] At "Unfrod," on the lands of Glastonbury, three bushels per acre were allowed. R.C.G., 235. W.H., 67.
[5] W.H., 93.

L

Threshing was done throughout the autumn and winter whenever the weather was unfavourable for field work, but oats were not threshed until after Christmas, just before they were needed for sowing. Socmen often owed a day's threshing, and the villeins threshed either for a certain number of days or a fixed amount of corn. The hayward and the reeve tallied the corn as it was threshed, and the yield was calculated either by measuring every eighth sheaf separately and reckoning from that, or by getting "prudent, faithful and capable men" to estimate the yield of a stack of fixed size.[1] At Forncett in Norfolk the average yield of wheat was fivefold, barley fourfold, and oats three or fourfold. Walter of Henley states that all corn should yield more than three times what was sown, but the anonymous *Husbandrie* is more optimistic, putting wheat at fivefold, barley eightfold, oats fourfold, rye sevenfold, peas and beans sixfold, barley and oats mixed (sometimes called dredge) sixfold, and wheat and rye mixed the same. On the estates of Merton College 1333-5 the yield was: wheat 10 bushels per acre, barley 16, dredge 14, rye 11, oats 10, peas 11, beans 10.[2] On the whole 10 bushels, or sometimes less seems to have been the usual yield of wheat in the thirteenth century, decreasing towards the end, and falling to 6 bushels in the fourteenth century.[3]

It was carefully reckoned how much corn would be needed for seed, allowances to servants, bread and ale for the household, almsgiving and fodder, and the surplus (if any) was then sold, or stored against emergencies. Threshing was done in the great barns, usually with long flails, but in a few places there seems to have been a curious survival of the Roman "tribulum," the threshing-sledge with stones or iron ridges fixed into the under side.[4] This instrument is mentioned in a Register of Spalding Priory, in which a villein is bound to work for his lord with cart, cauldron, spade, flail, "tribulum," fork and reaping-hook; and it is also included in the list given by Alexander Neckham of the villein's implements.[5] Occasionally the primitive method

[1] *Rules of St. Robert.* W.H., 127.
[2] A. H. Inman: *Domesday and Feudal Statistics*, 116. W.H., 17, 67. The average yield of crops for the nine years 1940-49 was as follows: Wheat about 34 bushels per acre, barley 35½ bushels, oats 43 bushels, peas 28 bushels, and beans 22½ bushels, an extraordinary increase from six hundred years ago.
[3] H. Bradley, *The Enclosures in England*, 51-4.
[4] See above, p. 91.
[5] Vinogradoff: *Villeinage in England*, 17 on. A Neckham: *De Utensilibus*, ed. J. Mayer, 45, 111. The word "tribulum," however, is often translated "rake" by modern annotators, as if it had undergone a change of meaning.

of treading out the corn by oxen was resorted to.[1] The winnowing was sometimes done by villeins, but often by the dairymaid,
with a woman to help her. On the lands of Peterborough it
was done by the wives of the oxherds.

The business of getting the corn ground was a matter that
caused much dissension during this period. Watermills, as we
have seen, came into use under the Saxons, and in Domesday
Book some 5,000 are recorded, and with their general establishment under the Feudal System arose the claim of the manorial
lord to make all his tenants grind their corn at his mill. To
enforce this and gain the "multure" or toll of corn paid from
each sack for its grinding was naturally very profitable, and led
to strenuous efforts to put down the use of the humble hand
quern with which the women of the household had been
accustomed to grind their corn at home. The right of mill soke
(i.e. the obligation on everyone in the district to grind and pay
toll at the lord's mill) was never enforced by statute, but became
in most places a binding local custom which caused continual
friction. The earliest extant charter, giving the mill of Silsden
to Embsay Priory and forbidding the use of handmills on that
manor, occurs in 1150, and many accounts of the struggles
between lord and tenants are found in monastic and manorial
records.[2] At St. Albans, Cirencester, after half a century of
conflict, all the private quern stones were confiscated and broken,
and "the imperturbable Abbot gravely paved the floor of his
private parlour with them in testimony of the thorough sub-
jugation of his foe,"[3] and there they remained until the time
of the Peasant Revolt. The mill-toll was in early days always
paid in kind, and in the thirteenth century was nominally fixed
at one-twentieth or one-twentyfourth part of the grain ground,
but the average toll from the twelfth century to the fifteenth
was one-sixteenth part. The miller took it in a measure known
as the "toll-fat" or toll-vat, which seems customarily to have held
one-sixteenth of a quarter, i.e. half a bushel, and it had to be
taken razed or striked. When mills were leased out to a miller
for a fixed rent paid to the lord, as was very early done, this was
apt to lead to dishonest practices on the part of the miller in
order to swell his "multure corn." In London a money commutation of 3d. per quarter came into use as early as 1281, but
in the country payments in kind long continued.

[1] Bartholomew Anglicus, ed. R. Steele, 120.

[2] For example, Maitland: *Select Pleas*, 36.

[3] R. Bennett and J. Elton, *History of Corn Milling*, i. 216. For succeeding
pages, i. 162; ii. 235; iii. 153, etc.

It must not be supposed however that the ancient and homely quern disappeared; it continued in use throughout the Middle Ages and long after.  In some places there were no suitable sites for mills, and in others their use was allowed by special licence.  In Domesday the county of Lancashire is practically without record of mills, from which it would appear that hand-mills were in ordinary use in the north.  Large mediaeval querns often had a long iron bar fixed into the upper stone in place of the short handle, and the top end of this being passed loosely through a hole in a roof beam, or if out of doors tied to a branch, one or two persons could swing the bar round with much less fatigue than with the ordinary short handle.

The villein who ground his corn at home was equipped with a sieve to "bolt" his flour and remove any inequalities or frag-ments of grit; otherwise this would be done at the mill.

Tidal mills were sometimes constructed, and even streams which ran dry in summer were utilized in their season, these being known as winter mills.  Horse-mills were also used in places where water power was not obtainable, the animal turning a central shaft working a cog and spindle on the same principle as the watermill.  Such a mill as this was used by the canons of St. Paul's.

At the end of the twelfth century a new feature in milling

Fig. 50.—Windmill (Luttrell Psalter)

appeared in the introduction of the *windmill*.  The first authentic mention of an English windmill occurs in 1191, when a violent dispute arose between Abbot Samson of Bury St. Edmunds and his Rural Dean on the subject of a windmill which had been erected to the prejudice of the Abbot's mill. The poor old Dean was promptly compelled to demolish this very unwelcome novelty.[1] In the thirteenth century, however, windmills became common, not as private ventures, but replacing manorial watermills, and being erected on manors which were without water power, the manorial "soke" then became attached to them. In 1284 the Abbot of Peterborough put down the use of handmills in Oundle to the prejudice of his windmill.[2]

The early windmill was a simple, cabin-like construction containing the millstones and sail beam with its gear, the whole structure being mounted on a post so that it could be turned by means of a projecting beam for the sails to catch the wind. Such a mill, with its ladder, is shown in the Luttrell Psalter. The post upon which the mill turned was supported by a tripod of timber at the foot, which was at first merely fastened to the ground by stout pins, but a little later this foundation was sunk into the ground to give the structure greater stability, the earlier variety being doubtless liable to be overturned by storms.

Thus more and more widespread became the dissolution of the ancient companionship of mill and bakery under the same roof; a partnership so ancient that the Latin word *pistor*, a baker, had been derived from the Greek πτίσσω, to crush or pound, and should have belonged rightly only to the grinder of the corn, but the name persisted long after corn was ground and not pounded at all, and adhered to the bakers long after they had ceased to grind. The Saxon kitchen probably, like the Roman "pistrium," combined both processes, but the feudal bakehouse was for the most part far removed from the mill. Where there were horse-mills the combination was longer kept up, as at Bridlington, and at St. Paul's, where horse-mill, brew-house and bakehouse were all together. Mills, we must remember, had to grind the malt as well as the corn.

But to return after this long digression to the work of the farmer's year.

In late autumn, usually in November, the pigs were driven to the woods for fattening on beechmast and acorns, and the villeins turned out to gather nuts for the lord, usually for one day each. Sometimes they were entitled to keep half of what they collected for themselves, the other half going to the lord. The nuts were used not for food but for the oil which could be

[1] For an account of this controversy see Carlyle, *Past and Present*, Book II, ch. xv., quoting Chronica Jocelini de Brakelonda.

[2] Chron. Pet. (Camden Soc.), 67.

extracted from them, one quarter of nuts yielding four gallons of oil.[1]

In November, says the poem,

> Swine become greatly fat,
> Shepherds go and minstrels come,
> Butchers' blades are bloody
> And the barns full.

At Martinmas, as in former days, the stock was carefully thinned out, the draught oxen and breeding stock being kept through the winter and the rest either sold, or killed and salted for home consumption.   Great activity prevailed at the salting-house (*lardarium*), and in some places the villagers paid regular dues at Martinmas or Christmas, which would probably be for the privilege of having their meat salted there with the rest. Salt was for the most part obtained by evaporation of sea-water in shallow pans on the coast, and its price therefore varied according to the sunniness of the summer, the average price at this time being $5\frac{1}{8}$d. per bushel.[2]   The carcases of the cattle and sheep were dealt with at the salting-house during November, and in December the fattened pigs were killed and bacon and salt pork prepared.

In December,

> The land is heavy and the sun drowsy,
> The cock is happy and the feathered owl.

At Christmas the tenants were in many places allowed a "yule-log" from the woods, and were feasted in the lord's hall.   On the Glastonbury manors very careful details are given about this function; the villein and his wife came, bringing a platter, mug and napkin "if he wished to eat off a cloth," and also a bundle of firewood for the cooking of his food unless he wished to eat it raw. The couple were allowed two white loaves, two dishes of meat, sometimes

Fig. 51.—Pig-killing

cheese and "potage," and "sufficient ale, of good quality and clear."[3]

---

[1] Anon., *Husbandrie*. W.H., 81. C.B.A., 61.
[2] Thorold Rogers: *History of Agriculture and Prices*, 484.
[3] R.C.G., 83, 97.

Throughout the year, but probably chiefly in summer owing to the fearful roads, the villeins owed a good deal of "averagium," i.e. carrying service of various sorts, sometimes with packhorses, and sometimes with carts drawn by horses or oxen. On small, self-contained manors they would not have very much of this beyond the carting of crops, manure and fuel, which we have already noticed, but on monastic and other large estates supplies had to be taken to a fixed centre, and to market. Varying food allowances were made for these expeditions. On the Battle Abbey lands, mostly along the coast, the carrier of one thousand dried herrings or two wagon-loads of hay was awarded two loaves and six herrings (the same sustenance being given for ten rods of fence-making or an acre of ploughing). In other places each horse was given a loaf or an allowance of fodder.[1] Water was carried in skins on horseback to where it was needed, and corn to the mill.

During the long winter evenings, and at other times when agricultural pursuits allowed, there were unending home occupations for the farmer and his family which are now relegated to special workers. Nearly all the things now bought in the shops of the nearest town had to be made at home. Candles were made of the hard fat from oxen and sheep, with peeled rushes or "hards"[2] for wicks; wool and the finer fibres of hemp were spun and woven into rough friezes for clothing; hemp was also made into canvas for covering the windows, making mill sails, winnowing fans, dairy cloths, harness linings, sacks of various sorts (which were fastened by a pin or skewer),

Fig. 52.—Women Spinning and Carding

and for packing fish and covering garden seeds and many other uses; ropes and bindings were also made of it. Flax had to be elaborately prepared,[3] spun, and woven into linen for clothing, sheets, table coverings and bags, and flax was also used for sails,

[1] C.B.A., 10. R.C.G., 82.
[2] Hards—the short and rough bits of hemp or flax.
[3] For a description of this process see Bartholomew Anglicus, ed. R. Steele, 90.

nets, thread, lines and bowstrings. Nettles were often used in much the same way as flax.

Spoons, platters and bowls were carved of beechwood, mugs made of horn and jugs of leather, and all kinds of farm implements, baskets and harness made, the last of rope or tawed[1] horsehide. Brooms were made of birch twigs "to sweep and clean houses of dust and other uncleanness."[2] Hair cloth was made for drying malt and straining milk, and rough cloth was also made of goat's hair. Winter garments were lined with dressed sheepskin, and leather sandals were made at home, while gaskins and gaiters of soft leather were usually provided by the village tanner. Moleskins and the fur of woodland beasts could be made into headgear, and judging from illustrations the villein usually preferred to have his head covered, in marked contrast to the Saxon gebur, who nearly always worked bareheaded.

Holidays were not wanting, for besides the scattered Church Festivals upon which work was not permitted, there were periods of a week or more at Christmas, Easter and Pentecost during which agricultural operations seem to have been suspended. In many places there were entertainments known as "Scot-ales," given by the lord of the manor three times a year, notably on the Glastonbury lands. The custom there was for the married men and youths to come to the hall after dinner on Saturday and be served three times with ale; on Sunday husbands and wives came and paid 1d. and could return next day if they wished. The young men might also come on Sunday if they paid 1d., but could drink for nothing on Monday as long as they remained standing, but if they sat down they had to pay 1d.[3]

The farmer's food during this period, and indeed throughout the Middle Ages, consisted more of salted foods than can have been altogether wholesome; hence the prevalence of such diseases as leprosy and scurvy. He ate salt meat all through the winter and salt fish throughout Lent, but for fresh meat he had his fowls and pigs, the latter, as we have seen, being the most important of animals for food. His supply of bread and beer depended on his harvest; wheat was usually considered "bread-corn," for the better-off, and barley "beer-corn," but the

[1] Dressed and whitened.
[2] We are also told that at spring and harvest men slit the birch bark "and gather the humour that cometh out and drink it instead of wine." Bartholomew Anglicus, 92. Birch-wine was made as late as the eighteenth century.
[3] R.C.G., 82, and notes.

poorer would probably eat bread of rye or "meslin," mixed rye or barley and wheat, which was moister than wheaten bread, and "abode longer in the stomach." Investigation shows that the proportion of rye to wheat was larger in early records than in later ones; (two-thirds rye to one-third wheat, as opposed to the other way round) from which it may be argued that rye alone was the original bread-corn of the poor, as it long continued to be in times of scarcity.[1] Beans were sometimes mixed with wheat flour to make it heavier, but Bartholomew Anglicus was doubtful of their wholesomeness: "Beans," he remarks, "are said by Pythagoras to dull the wits, and cause many dreams. Others say that dead men's souls be therein, therefore Varro saith that the bishop should not eat beans." However, the poor in times of scarcity could not afford such scruples. The villein farmer would have besides his home-cured bacon, plenty of eggs, milk, cheese, and "pot herbs" from his garden, besides wild herbs, so did not as a rule fare badly. Every villein had his cheese vat and press; the children drank the whey, and the cheese was carefully covered with leaves to keep away flies and mice.[2] "Potagium," a sort of porridge or thick broth made of peas, beans or other cereals, was also a common dish.

If a sheep died many farmers did not scruple to soak the flesh in water from daybreak to three o'clock, hang it up to drain thoroughly, and salt and dry it for the labourers: Walter of Henley, however, did not consider this practice desirable. The tables of the better-off were furnished, in the absence of fresh meat, with all kinds of game, deer, rabbits, pigeons, rooks, starlings, magpies, sparrows, hedgehogs and even squirrels, by way of variety, but the poor were rigidly restricted by game laws.

Fires were still situated in the middle of the floor, and the food either cooked in cauldrons or roasted on spits, but when the weather permitted a fire was often built out-of-doors for cooking purposes, so that the chimneyless interior was not filled with smoke.

We have now examined the great manorial farms of the period, scattered here and there over the country, and have also seen how every villein within the manor had a little farm of his own, but alongside of these we find occasionally, as in other periods, little isolated establishments which are hardly the one or the other. Such a one is described at "Melrede"

---

[1] Professor Ashley, in *Economic Journal*, vol. 31, pp. 292-301.
[2] A. Neckham: *De Utensilibus*, 110. *Inquisitio Eliensis* (Rec. Com.), 502.

on the lands of Ely about 1086: "In the same vill a certain knight holds under Harduin (de Scalers) one and a half hides and a church. There is land for two plough teams and the teams are there. There are three cottars and one serf, and a mill rendering 5s. 4d. There is meadow for the two teams and pasture for the herds of the village. [On the home farm are] three cows with calves, sixty-two sheep and eighteen pigs." Here we have a little farm of perhaps 240 acres of arable land besides meadows, cultivated by a knight who evidently did not scorn to work with his hands, for of his four helpers, one would presumably be a shepherd, one a swineherd and the other two would be needed either to plough or to look after the cattle. Very possibly the cottars or the knight himself had stalwart sons who could help, and the miller must have had a good deal of time over from grinding the corn of so tiny a community. The congregation in the little church, doubtless served by a neighbouring priest, cannot have been very large.

We have seen the feudal manorial farm at its most elaborate perfection, a self-sufficing organization combining the interests of the villagers on the one hand and the lord with his home farm on the other, both being mutually dependent, and all their relations regulated by "customary" services and allowances. Money, at any rate in the early days, was a very scarce commodity which hardly entered into the villein's life, except for the few pence he was bound to pay his lord yearly. But when markets were set up within reach where the peasant could dispose of his few surplus goods a little money began to circulate. Hence it became possible for him to offer money to his lord instead of his customary work, and thus a start towards the commutation of services was made.

In the outer world the passing to and fro of bodies of men to the Crusades and the French Wars opened up fresh markets for produce to the larger farmers, brought fresh ideas, new products, and again money, and with it the possibility of change and improvement.

Already towards the end of this period the manorial system was declining from the ordered completeness of its plan, and changes were taking place with which we shall deal in the next chapter. The fourteenth century was a time of distress, the poorly manured and almost exhausted soil was yielding less and less return; the years 1314 to 1321 particularly were years of terrible famine and scarcity, owing to continuous wet seasons, bad harvests, and murrain among the cattle. According to Stow's Annals, in 1314 "no flesh could be had, capons and geese could not be found, egs were hard to come by, sheepe died of the rot, swine were out of the way; a quarter of wheat, beans and pease were sold for 20 shillings, a quarter of malte for a marke, a quarter of salt for 35 shillings." In the following year "Horseflesh was counted great delicates; the poore stole fatte Dogges to eate." But Stow was speaking of London, where things would be at their worst. Thorold Rogers gives the price of corn in 1316 as 16s. a quarter, (the average price from 1261 to 1350 being 5s. 11¼d. a quarter, and in 1287 only 2s. 10½d.) Holinshed merely says that "by reason of the murrain that fell among cattle beefs and muttons were unreasonably priced." But poverty and simmering discontent became prevalent, and would no doubt have brought about in course of time the changes which were precipitated by the crowning disaster of the pestilence of 1348-9.

# CHAPTER VIII

## 1348 TO 1500

In the period now before us we have not to record the coming of any new race with its consequent effects on the arrangements of the farmyard, but, on the contrary, that terrible visitation known as the "Great Pestilence" or the "Black Death," which reduced the existing population by at least a third and perhaps as much as a half. In the words of Henry of Knighton's Chronicle, "the fell mortality came upon them and the sudden and awful cruelty of death winnowed them"; and when the scourge abated the dazed and stunned remainder had perforce to carry on the business of life as best they could, or starve.

It was little wonder that the manorial system, already showing signs of change and disintegration, fell rapidly into decay. For the last half-century or more the commutation of the villeins' services for money payments had been introduced more and more, and during this period became practically universal. This arrangement, by which the villein paid his rent in money instead of in labour, a few pence for each day's service that he owed, had obvious conveniences. His time was now entirely his own and he could work his own little farm undisturbed. The perpetual friction which must have existed in many places between the lord and his tenants would be much reduced when the reeve and the bailiff no longer had to exact the requisite amount of labour from reluctant men and bring them up before the manorial court if they were not considered to have performed their duties adequately. The lord, on his side, would find it much simpler to hire what labour he needed, just when he needed it, than to work his farm on the complicated system of customary services, under which each man had to work for him on certain days and no more. His troubles, however, began afresh when the hired labourers demanded higher wages than could be covered by the sums he received for the commuted services.

It was the regular agricultural services that were most universally commuted; the "irregular incidents," such as the obligation to keep the mill and mill-pond in repair, the sheep-folding, and merchet, heriot, and other fines were continued for some time longer, especially in the North. There on the Borders, where it was always necessary to have an organized

body of men at hand to repel Scottish raiders, the system of
regular holdings of the normal size, with customary services,
survived for centuries longer than in the South. In the North
conditions of life were hard and primitive, and long continued
unaffected by the growing industry and commercialism of the
more southern parts of the country.[1]

The ancient rough uniformity of size in the villeins'
holdings had been changing for some time past. A prosperous
villein would take over part or the whole of the holding of
another which fell vacant through misadventure or death, or
would even be able to purchase or lease odd pieces of land
adjacent to his own. This process, which was noticeable at the
beginning of the fifteenth century, was much accelerated by
the Black Death and consequent events.

But the great underlying cause of the breakdown of the
manorial system was the increasing exhaustion of the soil.
Farm-yard manure was the only form of dressing available for
the fields, and that was very limited in quantity owing to poor
feeding and the diminished numbers of flocks and herds which
could be kept in winter. Even so the lion's share of it was
compulsorily bestowed on the demesne lands, and the village
fields got little but the droppings of the cattle feeding on the
stubble and the fallow fields. With the average yield of wheat
fallen to six bushels, and other crops in proportion, the villein's
holding was no longer enough to support his family, much
less pay his little rent to the lord. The more fortunate were
able to increase their holdings, but others paid fines to be
released, or simply fled. Even before the Black Death many
a lord found it difficult to keep his full number of villeins, in
spite of money commutations advantageous to them, and as
land fell in it was often hardly worth while to keep it under
cultivation and it reverted naturally to pasture.

The disaster of the Black Death brought about still further
changes, for in many places the depopulation was such that
sufficient labour could not be hired to keep the home farm of
the manor under cultivation even if worth it, and the landholder was often driven to adopt one of two expedients. One
of these was to enclose the demesne land for sheep-farming, a
course which was expedient because it gave the land a much
needed rest, and also because so few men were required to look
after the sheep. As long as the lord only enclosed his private
lands for the purpose and did not encroach upon the common
pasture of the village, and while the population was still too

[1] Tawney: *Agrarian Problem in the Sixteenth Century*, 66, 103.

thin to feel the loss of employment, there was no hardship to the community in this practice; it was not until the latter half of the fifteenth century that it gave rise to abuses and complaints. The other expedient was to let out the demesne lands on lease, either as a whole or in portions, on what was known as a "stock and land" lease. The tenant paid a yearly rent and was bound to return the same amount of seed-corn, stock and implements at the end of his lease as he had received with the land. He did not, however, acquire any manorial rights over the land, as did the "firmarius" to whom monastic manors were farmed out in earlier times.[1]  At first these leases were short, but in the fifteenth century they came to be granted for life or a long term of years, and thus a class of sturdy yeoman farmers grew up, often cultivating compact blocks of land, on which they were free to herd sheep or to grow what crops they pleased instead of being subject to the common regulations of the village, and which in appearance were much more like the farms of to-day.

The lord of the manor thus tended to lose his close connection with his lands and in many cases became a mere landlord instead of a working farmer. Whether he leased his demesne or enclosed it for sheep-farming, in either case he was saved the expense of the small army of servants and officials needed to run the mediaeval manorial farm. At Forncett, for example, in 1376 many of the actual buildings of the home farm were let, viz. the sheephouse and fold, the chambers east and west of the gate, the carthouse, granary, and stotts' stable, also the pasture and fruit of the garden, the herbage of the "pound yard," and the kitchen yard and orchard. In 1377-8 the demesne was stocked with two hundred sheep, but by 1406 nearly all the land was converted to leasehold. In 1409 the manor was let to farm, and the buildings decayed, and by 1491 the manor house itself had disappeared and only one stable was left.[2]

Those who did not adopt any of these expedients, but attempted to exact the customary services from their tenants as before, found themselves involved in the troubles which led to the Peasant Revolt of 1381. The monasteries, however, clung to the old system, and for the most part did not let out their lands in the new way until late in the fifteenth century.

It was at this time too that the word "farm" changed its meaning. Hitherto it had meant the rent, in money or

[1] See above, p. 142.
[2] F. G. Davenport: *Economic Development of a Norfolk Manor*, 51-9.

kind, produced by the land, the "farm" paid by the "firmarius"; but now the term was applied to the definite area of land required to support the occupier, i.e. the man's "living"; and it was not until after Tudor times that it meant, as at present, any area of land occupied by one tenant for arable or pasture farming.[1]

Besides the tenant farmers, and small freeholders, the "substantial freeholder" known as a *franklin*,[2] who ploughed his own land and took his own produce to market, seems to have flourished. Immortalized by Chaucer and noticed again by Fortescue about 1470, he often became knight of the shire or even sheriff, and his chief characteristics were apparently his good and abundant fare and his profuse hospitality, both of which testify to the success of his farming.

The houses of the poor remained much the same, and Chaucer, writing early in this period, gives us a picture of two of them. Even the household of the patient Griselda, described as the poorest in the village, had a "litel oxe stalle," which was probably under the same roof with the living-room, as Griselda, when bringing home the household water-pot from the well, set it down "besides the threischfold of this oxestalle."[3]

A little higher in the social scale was the poor widow of the *Nonne Prestes Tale*, who was a manorial servant. Her "narwe cotage" had two rooms, doubtless of humble proportions but still referred to as her hall and bower, according to the custom of the time. Both were "ful sooty" from the fire on the chimneyless hearth, and the poultry which were her pride roosted on a perch in the hall at night. Outside she had a yard "enclosed al aboute with stikkes, and a drye dich withoute." She also had three sows, three kine and a sheep called Malle.[4]

Except in districts where stone was the natural building material, the houses of the better-off yeomen were usually built of timber and lath or wattles, filled in with a plaster of clay and chopped straw. The floor was of earth or split flints,

[1] Lord Ernle: *English Farming*, 3rd edition, 50.

[2] During the fifteenth century the name " yeoman," which in Chaucer's day was used for a household servant, such as a forester, came to be applied to freeholders of small landed estates, but apparently less "substantial" than franklins. Fortescue mentions the two in that order, but later the name franklin was disused, and any farmer of free birth, "well at ease and having honestlie to live, and yet not a gentleman," whether tenant or freeholder, was called a yeoman.

[3] Chaucer: *Clerkes Tale*, 9, 11, 95.

[4] Chaucer: *Nonne Prestes Tale*, 39, 64, etc.

and a ladder led to the sleeping-rooms above, which were mere lofts in the roof. Certain improvements, however, began to appear. As an intermediate stage between the ladder and the staircase triangular blocks of wood were sometimes set one above the other, resting upon one angle, so that the foot could be placed underneath the slope as in a notch. The first few blocks were placed perpendicularly, and after that slightly receding.[1] In more important houses spiral stairs usually of

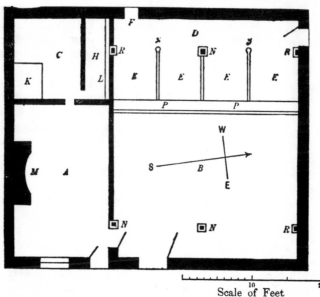

FIG. 53.—PLAN OF COMBINED HOUSE AND OXSTALLS

| | | |
|---|---|---|
| A House-place. | L Shelves. | R Responds |
| C Store-room. | F Manure-hole. | P Hecks. |
| K Hole in floor for ladder giving access to bed-room. | D Cowhouse. | B Main floor. |
| | N Pillars with stone stylobats. | S Skell-boosts or boskins. |
| H Pantry. | E Standings for cows. | M Fire-place. |

stone, were often used. The general plan of the manor house remained the same as formerly, though with more elaboration of apartments. On the one side of the central hall more and more chambers were added for the accommodation of the family, and on the other side an increasing number of servants' quarters and rooms for every conceivable household function. A tendency towards symmetrical arrangement was making itself felt, and the buildings were often ranged round

[1] Addy: *Evolution of the English House*, 57.

a courtyard with some unity of plan. But these mansions were becoming "gentlemen's seats," and were further and further removed from their former position as the dwelling of the principal farmer of the village for whom all worked, so that they need not detain us further.

At the end of the fifteenth century chimneys, not unknown before in stone-building districts, began to appear in some of the more substantial farm-houses, but they were still unusual. Bricks once more were beginning to come into use as a building material, after a long period of oblivion since Roman times. They were used to build on the new-fangled chimneys, and a little later to fill in the spaces between the timberwork of the house walls, and next to build the lower storey of the houses; but it was not until the reign of Henry VIII that brick building was really developed. Wooden houses long continued the most usual, and were known as "reared houses," the rearing of the timber framework being the main work of their construction.

In the lawless times of the Wars of the Roses farms were frequently moated for defence, and fierce dogs kept, while the farm labourers slept in the house. Under Henry VII the farmer was allowed to keep a crossbow for his protection.[1]

No additions to the farmer's *stock* occurred during this somewhat stagnant period, and what he had probably tended to deteriorate rather than to improve in quality. The unfortunate animals, like their masters, suffered severely from the pestilence, which seems to have spared no living creature, dogs, cats, poultry and even bees falling victims as well as the larger quadrupeds. These losses of stock, which must have been a heavy drain upon the farmer, continued till the end of the fourteenth century or longer (under the generic term of "murrain"), as is shown by the records of Heacham Manor in Norfolk. Sheep were particularly hard hit, and during many of these years more than a hundred were lost annually on that farm. In 1392-3 four peacocks and five peahens were lost, but their poultry and geese came off lightly  Ten beehives out of eleven were, however, lost in 1371, and five in 1377.[2] But besides losses by actual disease many poor creatures must have died of sheer neglect and starvation in the confusion of the general mortality of 1348-9. To quote Knighton once more, "sheep and cattle roamed about, wandering in fields and

---

[1] W. Denton: *England in the Fifteenth Century*, 186, 198.  F. Davenport, op. cit. (*Wills of Bondsmen*, App.).

[2] *Archaeologia*, xli. 2.

M

through the growing harvest, and there was no one to drive
them off or collect them, but in ditches and thickets they died
in innumerable quantities in every part for lack of guardians,
for so great a dearth of servants and labourers existed that no
one knew what to do." The losses of the Priory of Christ-
church, Canterbury, about 1350 are recorded at 257 oxen, 511
cows and 4,585 sheep.[1] Truly it must have been a desolate
time in the farm-yard.

During these unhappy years persons with the unsavoury
name of "cadaveratores" were appointed to remove and bury
the carcases of dead animals.

But in spite of all these disasters farming in that self-
supporting age was so vitally essential an industry that it somehow weathered the storm and struggled through the troublous
times of the fifteenth century, with its wars at home and
abroad, to a more prosperous period under the Tudors. All
sorts of stock seem to have been numerous once more by the
end of the fifteenth century, at any rate in the south-east, for
there we are told there was an "immense profusion of every
comestible animal, such as stags, goats, fallow deer, hares,
rabbits, pigs, and an infinity of oxen . . . above all an enormous
number of sheep."[2]

*Oxen* at this time seem to have been very poor, which is
hardly to be wondered at, and according to some writers oxen,
cows and steers in the fifteenth century were not more than a
third of their present average bulk.[3] They are said to have
shown a considerable reversion of type to the original Celtic
shorthorn (*Bos longifrons*), and continued to deteriorate until
improved by the introduction of fresh breeds from Holstein
and the Low Countries late in mediaeval times.[4] The writer
of the *Italian Relation* at the very end of the fifteenth century
observes that the English cattle have much larger horns than
the Italian, which looks as if new blood was already taking
effect.

At Haunchford in Surrey, and doubtless in most places,
the fore-feet of the oxen for ploughing and of the heifers for
harrowing were shod, at 3d. each.[5] At Hawsted they were shod
in winter only.

[1] Gasquet: *The Great Pestilence*, 140-79.
[2] *Italian Relation of England* (Camden Soc.), 10.
[3] W. Denton: *England in the Fifteenth Century*, 171.
[4] T. McKenny Hughes: *Archaeologia*, lv. 158.
[5] *Archaeologia*, xviii. 282.

*Horses* were important still mainly for purposes of carriage. As the horse in Lydgate's poem points out:

> al vitaille
> off the cariage
> hors have the travaille.

It was their duty to carry hay and oats to garner and barn, to bear home the sheaves in August, and the "roweyn" or aftermath in autumn. The poor man took his corn to market and to the mill upon his own little horse. In winter they brought home wood and turf for the fire, and wine and victuals for the table; at other times water from the river or well in "bowges" or leather water-skins, and lead, stone and timber for building. They also helped with the ploughing, though they were not as indispensable to it as Lydgate's horse wished one to believe.[1] Sir John Fastolf had a practice of sending home his horses to work on the farm when they were too outworn for service in London,[2] which does not sound too humane.

Horse-bread, by a statute of Edward III, was made of peas and beans only, and its manufacture was a regular part of the baker's business. Horse dung mixed with clay was used as fuel by the poor.

*Sheep* in these troublous times of agriculture became much more important. The harvest of wool was more reliable than the harvest of corn, and there was always a good market for English wool. But the animals were very small, their fleeces weighing only a pound or two.[3] Ruddle was used to mark them. Even in the fifteenth century, before sheep-farming reached its height, it was common for farmers to own hundreds of sheep, and as many as 14,000 are recorded in one instance in Dorset. Enclosure for this purpose, as we have seen, was not at first a hardship, but during the second half of the century complaints were becoming numerous because unscrupulous lords enclosed the common waste with their demesne, thus depriving the unhappy villagers of the means of feeding their cattle and making it almost impossible for them to carry on the cultivation of their holdings.

Lydgate's ram, with his brazen bell, is very proud of his wool, which he declares to be England's greatest wealth, and he states that "in all the world there is no better woolle."

---

[1] Lydgate's *Horse, Goose and Sheep*. *Political, Religious and Love Poems,* ed. Furnivall. E.E.T.S.

[2] *Paston Letters*, No. 123.

[3] W. Denton, op. cit., 171. Thorold Rogers: *History of Agriculture and Prices*, i. 53. Welsh mountain fleeces may average as little as 2 lb. nowadays but shortwools average 4½ to 5½ lb., and long wools 13 to 15 lb.

He claims to provide clothing for rich and poor, good fur both black and white, cloaks and gloves to keep out the cold, and parchment to write on.   Mutton, he points out, is wholesome both roasted and "sodyn" (boiled), and broth is good after illness.   His horn and bone were used for knife hafts and other implements, his fat for plasters, his guts for harp-strings, and his head boiled down made a "royal ointment."   Black sheep's wool with olive oil was used to staunch bleeding: altogether no part of him was wasted.[1]   It is noticeable that stress is no longer laid on the value of sheep's milk for butter and cheese, so perhaps in the interests of better fleeces milking had been partially abandoned.[2]

*Pigs* continued to be universally valued and were kept even in towns, which sometimes gave rise to curious regulations. In Norwich, for instance, they were kept either in the owner's house or in pigsties, which were cleaned out once a week.   For this purpose the pigs were all turned out on Saturdays, and ran about the streets from noon till evening, when their quarters were once more ready for them.[3]   It must have been a day to avoid taking a walk in the town.

The *poultry* of the end of the fourteenth century is described for us by Chaucer, who paints the "poure wydow's" cock Chauntecleer in glowing colours:

> His comb was redder than the fyn coral
> And bataylld, as it were a castel wal.
> His bile[4] was black, and as the geet[5] it schon;
> Like azure were his legges, and his ton[6];
> His nayles whitter than the lilye flour,
> And lik the burnischt gold was his colour.

His voice moreover was

> merier than the merye organ
> On masse dayes that in the chirche goon,

and his crowing was as punctual as an abbey clock.[7]   This fine yellow cock was accompanied by seven hens of the same colour, of which the favourite was called Pertelote:

> Curteys sche was, discret, and debonaire,
> And compainable, and bar hire self ful faire,

but her egg-laying capacity is not recorded.

[1] Lydgate, op. cit., 27-31.
[2] Tusser, in the sixteenth century, milked his ewes from May 1st to Lammas, the three best months.
[3] A. Abram: *English Life and Manners in the Later Middle Ages*, 194.
[4] Bill.
[5] Jet.
[6] Toes.
[7] Chaucer: *Nonne Prestes Tale*, 31, 38, 44.

A delightful picture is drawn of the cock strutting proudly about his yard, and casting his eye around for food:

> He chukketh, when he hath a corn i-founde
> And to him rennen than his wives alle.

But poor Chauntecleer, like many a humbler fowl, fell a victim to a wily fox, who made his way through the widow's fence of "stikkes" and lay hidden in the vegetable garden.

Lydgate, writing in the next century, deals with the goose, who attempts to compete with the horse and sheep in usefulness. Goose grease was used in medicines for gout and various aches and pains, his quills for making pens, his down for pillows and feather-beds, and his droppings mixed with butter and oil for curing burns; but there his services seemed to end except his proud claim that roast goose was served at the King's table.[1]

Swans were evidently very plentiful, and we are told were "eaten like ducks and geese."

The only hint of any fresh addition to the poultry-yard is the occasional mention of a bird, referred to as "Africanus," which seems to have been the guinea-fowl. In 1407 two of these birds were bought at Banbury for the dairy of Le Brech for 15s. 2d., including the expenses of getting them to Bicester,[2] but evidently this was an unusual experiment, and guinea-fowls were not at all common until much later.

*Dogs,* as we have already seen, needed to be large and fierce in those troublous times to guard the farmer's homestead from undesirable wanderers, but little dogs were also kept, for at "Sheles" (Durham) in 1370 the tenants were warned not to let their little dogs stray into the "Conygarth" or rabbit warren.[3] Lydgate, in his *Proverbys of Howsholdekepyng,* is of the opinion that "smale whelpes" should be left to "ladyse and clerkys."

*Cats* are mentioned twice by Chaucer, and seem to have been comfortably circumstanced. The sick husbandman in the *Sompnoures Tale* had a cat, which the visiting friar somewhat unsympathetically drove from the bench where it was settled.[4] But good living did not cause the mediaeval cat to forget his business:

[1] Lydgate, op. cit., 22-3.

[2] J. Dunkin: *History of Bicester,* 231. The bird is here translated as "turkey," but that bird was a native of America and not discovered by the Spaniards until the following century. The guinea-fowl was of African origin.

[3] *Durham Halmote Rolls* (Surtees Soc.), 92.

[4] Chaucer: *Sompnoures Tale,* 67.

> Let take a cat and foster him wel with mylk
> And tender fleisch, and make his bed of silk,
> And let him see a mous go by the wal,
> Anoon he wayveth mylk and fleisch and al
> And every deynté which is in that hous,
> Such appetit hath he to ete the mous.[1]

However, they ran risks from itinerant pedlars, who coveted their fur: avarice, according to Piers Plowman, has "as moche pite of pore men as pedlere hath of cattes, that wolde kille hem, yf he cacche hem myghte, for coveitise of here skynnes."[2]

The cat's labours were assisted by the recitation of charms: Lydgate gives one against rats, invoking the four Evangelists and various other saints, and adding generously—

> God save this place from alle other wykked wytes, bothe be dayes and be nytes.[3]

The farmer's *implements* do not seem to have undergone any change or improvement during this period. Ropes, halters, traces and sacks continued to be made of home-grown hemp, which earned the somewhat grim name of "neckeweed" from its connection with ropes for the hanging of malefactors, as well as the more innocent halters.[4]  Pairs of "autumnal gloves" were dealt out to labourers, but whether as a protection against the sickle in reaping, or a provision against winter cold is not stated.  Piers Plowman wore cuffs "for colde of his nailes," and "cokeres" or leggings.

As the manorial system fell to pieces, the farmer no longer had all his implements made on the spot by his own servants. The smith, carpenter, wheelwright, etc., instead of being manorial servants who were supplied with the necessary materials and plied their craft in return for their keep, now became independent craftsmen, either doing the required work for wages, or finding their own materials and selling the finished article.  Piers Plowman does not profess to be able to make his own plough, as the early manorial ploughmen could.

No additions were made to the farmer's crops during this period, which in agriculture was one of complete stagnation owing to the troubles of the period and the increasing exhaustion of the soil.  There are some indications, however,

---

[1] Chaucer: *Manciples Tale*, 71.
[2] *Piers Plowman*, passus v. 258 (Skeat).
[3] Lydgate, op. cit., 43.
[4] Will Bulleyn on Neckeweed: *Babee's Boke*, etc., ed. Furnivall, 241.
E.E.T.S.

that crops were beginning to be grown according to the suitability of the district, instead of growing all the crops required
by the community in rotation regardless as to whether they
gave good returns on that land. Thus by about 1500 the best
corn-growing districts were just north of London, from Suffolk
in the East to Gloucester in the West; rye flourished particularly
in the North, the best barley for malt was grown in Hunts,
Cambridgeshire, Northants and Bedfordshire, and especially
good peas and beans in South Leicestershire.[1] The author of
the *Italian Relation* (about 1497) reported that agriculture was
not practised at that time beyond what was required for the
consumption of the people, but that this negligence was atoned
for by the immense profusion of every comestible animal and
bird.

Vines were still grown, but evidently little local wine was
made, probably on account of the cheap and excellent wine
which could be obtained from Gascony. Wine, he says, *might*
be made in the southern parts, but would be harsh;[2] which
seems to imply that it was not customary.

Draining and dressing of the soil were neglected, for the
short leases granted at first, and the precarious times of civil
war between the factions of Lancaster and York, did not
encourage farmers to spend much money or effort on their
lands. Marling, hitherto the favourite dressing, went out of
fashion for the same reason, and farm-yard manure was not
very plentiful. Litter from the floors of houses and churches
was sometimes used for the purpose, the custom being to spread
straw, hay, tree-loppings, fern or rushes on the floor and leave
it there until it became too offensive to be tolerated longer even
by mediaeval noses.

The crops also suffered a good deal from the incursions of
wild animals. Doubtless during time of war the lords allowed
their rabbit warrens to get overstocked and out of hand, a state
of affairs which existed even before the Black Death in some
places, as we have seen; and Piers Plowman complained that
hares, foxes, boars and badgers broke down the hedges of his
croft, and wild-fowl ate his wheat.[3]

Trouble over the grinding of corn continued, and the dispute
between the lord and his tenants concerning the obligation to
grind their corn at the manorial mill was particularly bitter
during the Peasant Revolt of 1381. Handmills seem to have

[1] W. Denton, op. cit., 144-145.
[2] *Italian Relation of England* (Camden Soc.), 10.
[3] *Piers Plowman*, vi. 30.

been used for oatmeal and malt, and to grind pepper and mustard.[1] The primitive windmill turning on an upright post underwent an improvement which gave extra accommodation to the miller, the lower part with its post and tripod being enclosed in a brick or stone turret. This was called a turret post-mill. A later improvement on this was to raise the tripod supporting beams on stout buttresses, thus leaving a clear circular room underneath.

The miller, like the village craftsmen, became in time an independent worker leasing his mill, but up to the time of Elizabeth he was only allowed to sell the corn he took as toll, and no other.[2]

Packhorses were still mainly used for the carriage of goods, owing to the terrible state of the roads, the irregularities being such that persons in carts were not infrequently thrown out and killed by sudden and violent jerks. We read that Margaret Roost, a small child of eleven, when driving her mother's cart, jumped off to put the harness right, fell on her head and broke her neck.[3] Children began farm work young in those days, and it was not unknown to put them to the plough at eight years.

The farmer's life was becoming less circumscribed; he had opportunities of mixing with his fellows when he took his produce for sale to the nearest market town. Clement Paston at the beginning of the fifteenth century, a "good plain husband" [man] with 100 or 120 acres of land, drove his own cart with "divers corns" to Wynterton to sell and had a stall in the market there. He also rode to the mill on bare horseback with his corn under him, bringing the meal home again when ground. At the end of the same century Hugh Latimer, a prosperous yeoman with some 200 acres of rented land, also rode to market on the King's service; he kept one hundred sheep, and his wife and her damsels milked thirty kine. Clement Paston's son became a judge, and Hugh Latimer's a bishop, so farmers' sons were not without opportunity of rising in the world.

The garden also is said to have been neglected at this time, but we have the earliest known treatise on gardening written by one John Gardener about 1440 or 1450.[4] Special importance seems attached by this writer to onions, cabbages and kindred plants (coleworts), parsley and saffron; after which he gives a

---

[1] Thorold Rogers: *History of Agriculture and Prices*, 551.
[2] R. Bennett and J. Elton: *History of Cornmilling*, ii. 280; iii. 157.
[3] A. Abram: *Social England in the Fifteenth Century*, 14, 143-4.
[4] *Archaeologia*, liv. 160.

list of about eighty-five other vegetables and flowers, all of which
are referred to as "herbys to make both sawce and sewe."[1]  The
flowers include periwinkle, violet, cowslip, lily, and white and
red rose.  Herbs for "potage" in another fifteenth-century list
embrace daisies, beets, lettuce, radish, cabbage and spinach, a
rather curious selection to our tastes.  Roots in the kitchen
garden were parsnips, turnips, radishes and carrots (karettes),
and Piers Plowman had parsley, leeks and many cabbages in
his.  Chaucer mentions the new "Perjenete" pear, and Lydgate
gives a list of new apples, the Pomewater, Ricardon, Blaundrelle,
Queening, and Bittersweet, and "Large nuts," perhaps walnuts,
were a good deal cultivated.[2]  Orchards were leased out like
other property; for example in 1374 a "garden of apples and
pears" on the Durham Estates was leased to one William
Harpon, for life, at 24s. a year; oxen and cows were not to be
let in to eat the undergrowth, the roots of old trees were to be
taken up in course of time and new apples and pears planted.[3]
Baked apples and ripe cherries were eaten by the poor in Piers
Plowman's time.[4]

Owing to the falling to pieces of the manorial system the
numerous staff of officials and servants needed for the elaborate
working of the lord's farm became very much reduced.

As less supervision was needed the *bailiff* often became an
itinerant official visiting and inspecting several manors instead
of being the resident supervisor of one only.  The *reeve* consequently gained in power and importance and, if we may believe
Chaucer, had become a most objectionable person, and the
herd and other hinds, and even the bailiff, "were adrad of him
as of the dethe."  Instead of being the representative and
natural protector of the peasants, he evidently oppressed them
and wormed his way into his lord's favour by subtle means.
The lord's sheep, cattle, dairy, swine, horses, poultry and other
stock were wholly in his care, and "wel cowde he kepe a gerner
and a bynne."  He is represented as having a private store of
wealth, and he lived in a fair dwelling upon a heath, "with
grene trees i-shadowed," but was also a good carpenter and
wright.  This particular reeve, who came from Balderswell in
Norfolk, is portrayed as a "sklendre colerik man" with long
thin legs, mounted upon a "ful good stot" of dapple grey.[5]

---

[1] Pottage or stew.
[2] A. Amherst: *History of Gardening in England*, 47, 72, 73.  *Piers Plowman*, vi. 288.  *Arch. Journal*, v. 304-5.
[3] *Durham Halmote Rolls* (Surtees Soc.), 124.
[4] *Piers Plowman*, vi. 295.
[5] Chaucer: *Prologue* (Skeat), 587 on.

If the *ploughmen* of both Chaucer and Will Langland are to be taken as typical of their occupation, they must have been for the most part good and kindly men. Chaucer's ploughman, the worthy priest's brother, was a hard worker and so good-hearted that he would help his neighbours with their threshing and digging without payment when he could. Besides ploughing he was accustomed to do these things and cart manure, while Langland's Piers the Plowman, also an industrious and upright man, in addition could sow corn, drive beasts, weave, wind, and ply the tinker's and tailor's crafts. Both owned a few cattle, Piers having a cow and a calf and a cart mare.[1]

Of the other farm servants mention is made now and then. The pindar or punder occurs at Wearmouth in 1364;[2] and the hayward with his horn and his occupation of guarding the

FIG. 54.—WOMAN CHURNING

standing corn in *Piers Plowman*.[3] Chaucer's "poure wydow" was "as it were a maner deye" or dairymaid and supported herself and her two daughters by her occupation and the proceeds of her few cows, pigs and poultry.[4]

The *swineherd* of course continued, as bacon and pork were such important articles of food; at Basingstoke in 1389 he collected the pigs daily from door to door, receiving ½d. a quarter for each pig or for two very little ones, and a dinner annually from each tenant. He had to make amends for any damage done if they ran wild, and vagrants were impounded and 1½d. exacted for each.[5]

Another servant long retained on many manors was the *warrener*, who loked after the warren or "cony garth," which

[1] *Piers Plowman*, v. 1. 550; vi. 1. 289.  Chaucer: *Prologue*, 1. 529.
[2] *Durham Halmote Rolls* (Surtees Soc.), 27.
[3] *Piers Plowman*, C text, passus vi. 12.
[4] Chaucer: *Nonne Prestes Tale*, 26, etc.
[5] W. W. Capes: *Rural Life in Hants*, 102-3.

contained rabbits, hares and pheasants, and often occupied much good ground. These three were much esteemed for the table in those days of scarcity of fresh meat, and the rabbits were also very useful for their fur. The warrener at Haunchford, Surrey, in 1382-6 received 1d. a day, and at Forncett in 1376-8 he and the shepherd were the only demesne servants remaining.[1]

The *shepherd* was, of course, the most universal of servants at this time of the rapidly increasing importance of sheep, although Sir John Fortescue, writing about 1470, draws a rosy picture of rural life in which sheep pleasantly exist almost without requiring attendance. The feeding lands, he says, are enclosed with hedgerows and ditches, planted with trees which fence the herds and flocks from bleak winds and sultry heats, and so well watered that for the most part they do not need the attendance of a hind. He remarks on the absence of wolves, bears and lions, and states that the sheep lie out at night, penned into folds, without a shepherd, the lands improving at the same time. As a result of these beneficent conditions he considers that the inhabitants, being seldom fatigued with hard labour, are therefore more spiritual and refined, "like the patriarchs of old who chose rather to keep flocks and herds than to disturb their peace of mind with the more laborious employment of tillage."[2] Possibly he lacked personal experience of a shepherd's life. The picture drawn by Barclay in his *Eclogues* a century later is very different and probably nearer the truth. The shepherds, who if they did their duty probably had as busy a life as any other farm servant continued to be the musical members of the farm community and Chaucer in his *House of Fame* speaks of the

> Pipes made of greené corn
> As have those little herde gromes
> That keepen beastes in the bromes.[3]

The hours of work in the fifteenth century were very long, as fixed by the statute of 1495, and there is no reason to suppose that this was any very new departure. From the middle of March to the middle of September men worked from 5 a.m. to between 7 and 9 p.m., with half an hour off for breakfast, and one and a half hours for dinner and the midday sleep: during the winter they worked while there was daylight. By an Act of 1403 men might not work for hire on Holy Days, and only

---

[1] *Arch.*, xviii. 282. F. G. Davenport, op. cit., 50.
[2] Sir John Fortescue: *Works*, i. 413-14.
[3] Cutts: *Scenes and Characters of the Middle Ages*, 301.

for half the day on eves of Holy Days and on Saturdays. As these festivals averaged nearly one day a week it was no doubt necessary to work extra long on the days that were available.

Breakfast was a light meal, dinner was at noon, and supper at five, or probably later in summer, after the day's work was done.

The farmer's food varied much according to his means and the season of the year. The typical food of the poor man was beans and bacon: "his beard beslobbered with bacon like a bondman's," as Piers Plowman observes   The same work gives us a vivid picture of the difference in fare before and after harvest time, so dependent were the humbler farmers still upon their own produce. In the lean months some stayed their hunger with a peas-loaf, or the loaves of beans commonly eaten by horses and dogs, eked out with onions, leeks and other vegetables, and cherries and apples. "I have no penny," quoth Piers, "pullets for to buy,

> Neither geese nor grys[1]; but two green cheeses,
> A few curds and cream, and a haver cake,[2]
> And two loaves of beans and bran for my infants.
> And yet I say, by my soul, I have no salt bacon
> Nor no cockney,[3] by Christ, collops[4] for to make,
> But I have parsley and leeks (porettes) and many kole-plants.[5]
> And eke a cow and a calf and a cart mare
> To drawe a-felde my dung the while the drought lasteth.
> And by this livelihood we must live till Lammas time;
> And by that I hope to have harvest in my croft;
> And then may I dight thy dinner as we well liketh."[6]

After harvest it was a very different story, and the poorest refused any but wheaten bread and the best ale, and labourers expected to eat fresh meat and fish, fried or baked hot.

The franklin or "substantial householder" seems to have fared particularly well, and his hospitable board continually groaned with good things. Chaucer's Frankleyn "loved by the morwe a sop[7] in wine," but

> Without bake mete was nevere his hous,
> Of flessch and fissch, and that so plenteous
> Hit snewede in his hous of mete and drinke.
>
> *       *       *       *
>
> Ful many a fat partich hadde he in viewe,
> And many a brem and many a luce in stewe.[8]

[1] Pigs.
[2] Oatcake.
[3] Egg.
[4] Collops were made of eggs and bacon.
[5] Cabbages and similar plants.
[6] *Piers Plowman*, passus vi. 281-93.
[7] Sop = custard poured on rounds of fancy bread.
[8] Chaucer: *Prologue*, 334-50.

John Russell's description of a "Fest for a Franklen" bears out this account of the luxurious living of these well-to-do farmers. The first course of this feast consisted of beef, mutton, pork, goose and chickens, with a pudding of eggs and cream; the second of more delicate meats—the appetite being doubtless slightly impaired—veal, lamb, kid or young rabbit, chicken or pigeon, "roasted tenderly," with more custards. Finally there was a course of spiced puddings, fruits and cakes, with bread and cheese, in case anyone should yet be hungry, and bragot and mead to drink.[1]

Ale was still a very universal drink, for according to Sir John Fortescue the English never drank water, except for

FIG. 55.—FIFTEENTH-CENTURY LABOURER

penance. The fifteenth-century farmer, if he did not brew at home, bought his ale from the village ale-wife, who was fined if she sold it at the wrong price or refused to sell it outside.[2] There he and his fellows could sit until nine o'clock and sing a song

[1] Bragot was made of ale, honey and spices. *The Babee's Boke*, ed. Furnivall, 170. E.E.T.S.

[2] *Durham Halmote Rolls* (Surtees Soc.), 20, 25, etc.

in praise of ale, in which they complained that brown bread was made of bran, beef had many bones, bacon was passing fat, mutton was often lean, "trypes" were seldom clean, eggs had many shells, butter had many hairs in it, capon's flesh was often dear, ducks slobbered in the mire, but ale was always good.[1]  It was made both thick and thin, and known as pudding ale and penny ale, for ordinary consumption; but there was "best ale" also.  Warmed and spiced with a crab-apple floating in it, it was called "lambswool ale."  An inferior drink made of water, honey and spice was known as "swish-swash."[2]  But in the fifteenth century the modern substitute for ale—beer— made its first appearance, its manufacture being introduced by Dutch settlers in the eastern counties.  The inclusion of hops in this beverage gave it better keeping qualities than ale, which was made of malt only; hence probably its growth in popularity.  The writer of the *Italian Relation* remarked that the common people made two beverages from wheat, barley and oats, one of which was called beer and the other ale. "Both," he says, "are most agreeable to the palate when a person is by some chance rather heated," and were not disliked by foreigners after partaking of them four or six times.  About 1500 ale cost 3d. a gallon and beer 2½d.[3]

The farmer's wife and daughter were busily employed, for besides their household occupations they were often expected to help in dibbling beans, weeding the corn, making hay, shearing and washing the sheep, loading and spreading manure, reaping (especially the stubble), "yelming" or laying the straw ready for the thatcher, guarding the corn from cattle, and scaring the birds with a sling.[4]          ,

The housewife of every fairly well-to-do farm-house had some brass pots and pans in addition to her humbler iron and earthenware, and perhaps some plates and dishes of pewter, besides those of "treen" (wood).  Mention is made of a "Rak of yren for to rost eyren"[5] (eggs), which must have been some sort of contrivance for cooking eggs on the open hearth. Chimneys, as we have seen, were beginning to be built, even in humbler houses, towards the end of the period,[6] but until

[1] *Babee's Boke*, ed. Furnivall, 363-4.  E.E.T.S.

[2] *Piers Plowman*, passus v. 220, 222.  R. M. Garnier: *Annals of the British Peasantry.*

[3] *Italian Relation of England* (Camden Soc.), 9, 10.

[4] W. Denton, op. cit., 220.

[5] *Fifty Earliest English Wills*, 102.  E.E.T.S.

[6] In 1498 a "bondman" bequeathed his chamber at the west end of the hall, with a chimney, to his wife.  F. G. Davenport, op. cit., appendix.

the days of the brick oven in the chimney-corner it must have
been more convenient for the farmer's wife to have her loaves
baked in the common oven, or to buy bread when the baker
became a regular tradesman.  At Acley in 1366 we read of one
Gilbert Randolfe taking over the common oven, formerly held
by his mother, for the term of his life at 10s. a year.  He was
to make for the lord and his neighbours what they required,
and keep the oven in repair.[1]

The abundance of wool no doubt kept the women of the
household increasingly busy at spinning and weaving.  As an
improvement on the primitive distaff and spindle, most households probably had by this time a simple form of spinning
wheel.  In this the spindle was fixed horizontally in a frame,
and a band passed from a wheel round the spindle-whorl, to
give it the necessary motion.  The wheel was turned by hand,
for there was as yet no treadle.  An illustration of this kind
of wheel occurs as early as the fourteenth century, but was
probably not in general use until a little later.[2]  Fortescue
remarks that the people were clothed in good woollens, and
their bedding and other "household furniture" was of wool.
The housewife could buy buttons at 6d. a dozen and thread
at 2d. a pound.[3]  Not too much display in dress was permitted,
however, for by a Sumptuary Law of 1464 "servants to
husbandry" were forbidden to wear broadcloth at above 11s.
a yard or hose above 14d. a pair; nor might their wives have
kerchiefs costing more than 12d. or girdles harnessed with
silver.[4]

We have been considering in this period a time of trouble,
unrest and change, which caused a condition of stagnation in
agriculture and little advance in any farm-yard affairs except
sheep-farming.  But already the firm and competent rule of
the first of the Tudors was laying the foundation for a time
of greater prosperity, and the vigorous activity and enterprise
of the sixteenth century was about to arouse England to fresh
efforts.

---

[1] *Durham Halmote Rolls* (Surtees Soc.), 59.
[2] *Encyclopaedia Britannica*, "Spinning" (see Fig. 52).
[3] Sir John Fortescue: *Works*, i. 421.  Thorold Rogers: *History of Agriculture and Prices*, 581.
[4] Rot. Parl., v. 505.

# CHAPTER IX
## THE SIXTEENTH CENTURY

### REFERENCES

HARRISON = Harrison's *Description of England in Shakespeare's Youth.*
Ed. Furnivall.
TUSSER = *Tusser's Five Hundred Points of Good Husbandry.* 1557.

# CHAPTER IX

## THE SIXTEENTH CENTURY

AFTER the troubled period we have been considering, with its disasters by pestilence, famine and war, we now see the coming, not of a new race, but of a new spirit—that mysterious infusion of re-awakened life known in arts and literature as the Renaissance, but which was in reality a movement of far wider influence, a new tide of vigour which seemed to flow through the whole nation and awake fresh effort in every aspect of its life.

At the beginning of the century the population was still thin, probably a bare three millions, and Starkey, writing in the reign of Henry VIII, notes this "grete lake of pepul and inhabytantys," and even complains of the lack of corn, cattle, victuals and other necessaries, attributing this largely to the bad weather.[1] But although the population does not show any great increase during this period owing to the fearful infant mortality, the people were imbued with a great spirit of enterprise, and this, combined with the comparative security of the times, made possible the greater prosperity of the age of Elizabeth. Yet still the poorer farmers and the small tenants who made up their rent by "a little brede of neate, shepe, swine, gese and hens,"[2] lived on the edge of disaster, ever threatened by flood, pestilence and famine.

A great difference appears in the purpose and aims of the farmer. The object of the elaborately organized village and manorial farms had been to supply food for their own community; they raised what they needed, and if there were any over it was taken to market; but now the mediaeval spirit was passing, towns were growing and markets were developing; the trading instinct so characteristic of the Tudors was coming to the fore, and the aim of the farmer was not merely to supply his own needs, but to raise cattle and crops for profit and to accumulate individual wealth. Many of the old village communal farms continued since they could only be broken up by mutual consent, but it was the sturdy independent yeoman farmer with his russet clothes and tin buttons, who made his

---

[1] Starkey: *England in the Reign of Henry VIII*, lvii, 89-90. But this does not agree with the abundance of cattle described by the writer of the *Italian Relation* a little earlier.

[2] *Discourse of the Common Weal*, E. Lamond, 56-7.

mark during this century.[1]  Yeomen, says Harrison, are freemen worth £6 a year in land, and "are for the most part farmers
to gentlemen, and with grazing, frequenting of markets and
keeping of servants do come to great wealth."[2]  Their dis-
tinctive title of address was "Goodman," but according to
satires of the time it was often their ambition to educate their
sons to become gentlemen and be called "Master."

There is a great difference in outlook, significant of the
changed times, between Walter of Henley, the writer of the
thirteenth century, and his successors in the sixteenth, Fitzherbert and Tusser.  The two latter, although their advice
applies to any sort of farm, are not primarily concerned with
great manorial estates run by a bailiff and an army of minor
officials, with the assistance of the villagers; the farms they
have chiefly in mind are evidently separate estates cultivated
and managed by the individual toil of the farmer and his wife
with the help of a few paid servants.  The farmer himself
overlooks everything, and the prosperity of the farm is dependent on his industry; hence the coining of such maxims as
"The slepe of ye husbande maketh a fatte donghyll, and the
eye of the mayster a fatte hors."[3]  Fitzherbert exhorts the
farmer to have in his purse a pair of tables, on which to take
notes of anything amiss when he is going about, but if he cannot write "let him nycke the defautes upon a sticke," trusting
no doubt to his memory to recall the significance of the nicks.

Side by side with the prosperous independent farming of
the yeomen, the old open-field husbandry continued in many
places, but there was continual controversy as to the respective
merits of this "champion" or open-field farming with its ownership of scattered strips, and the newer fashion of exchanging
strips by mutual arrangement, so that all a man's acres lay
together and could be enclosed in a single compact block over
which the owner had complete and permanent control.  This
was known as farming in "severalty" and both Fitzherbert and
Tusser advocated it as giving better returns and more opportunity for individual enterprise.  It is obvious that much time
and trouble was saved when the farmer had not to visit strips
of land scattered far and wide about the parish; his crops were
less liable to incursions from the village herds, or to weeds

---

[1] The name, originally applied to freeholders of the yearly substance of
40s., came to be used for any well-to-do farmer between the condition of
gentleman and peasant.

[2] Harrison, i. 133.

[3] R. Whitforde: *A Werke for Householders*, 1533.

blown over from a lazy neighbour's strip, or to accidents from
untimely gales at harvest time, which sometimes swept across
the unprotected open fields and mixed everyone's crops in
inextricable confusion against the boundary fence. At the
outset the farmer in severalty had the expense of hedging and
fencing his land, but when once that was done the permanent
hedges gave good shelter to his cattle and he was able to arrange
a series of enclosed pastures which fed them to the best advantage. They did not run the same risk of infection as when
pastured with the common herd, and the farmer got the
advantage of the aftermath or "eddish" of his fields instead of
their being thrown open for public use. Thus, undoubtedly,
from the point of view of farming for profit, the enclosed
arable fields where the farmer had a free hand answered much
the best, but the old common-field system was perhaps better
socially for the needs of the whole village, since by that system
everyone was provided for to a certain extent, whereas under
enclosure it was the larger farmers who prospered: the small
farmers and the landless paid labourers were apt to fall upon
hard times.

Already within the villages themselves a certain amount of
exchange of strips and formation of compact closes had been
going on, and beginning chiefly in Kent and Essex, by the
middle of the century had spread far, particularly in parts
affected by the growing centres of trade and industry.[1]

But the burning question of the period, particularly after
the rise in the price of wool in the second half of the century,
was the rapid growth of enclosure of large estates for sheep-
farming, a movement of which we saw the beginning in the
last chapter. This became a real abuse when the pasture
commons so necessary for the feeding of the village cattle were
enclosed, or worse still, when whole villages were depopulated
to make vast sheep-runs. The period is full of outcry against
the "gredy gentylmen whyche are shepemongers and Grasyars,"
these "catapillers of the commonweale,"[2] but in spite of protests
and Parliamentary regulations the process continued. As in
all widespread changes a certain number of people were bound
to undergo great suffering before time brought readjustment,
but even so in the long run this wholesale conversion of
stretches of country to pasture-land was not without its good
results. As a writer on the fifteenth century wisely points out,
under the old system of agriculture the arable fields were

[1] Tawney: *Agrarian Problem in the Sixteenth Century*, 172, 405.
[2] Starkey: *England in the Reign of Henry VIII*, lxxvi.

becoming exhausted and this enclosure under sheep gave them the necessary rest and manuring and prepared them for a return to cultivation with restored fertility at the end of the century, and still more under improved agricultural methods in the seventeenth century.[1]

The prosperous years of the middle and later sixteenth century and the consequent rise in the standard of living, gave a great impetus to building, and many a farmer pulled down his humble abode and rebuilt it in accordance with the times, as the large number of beautiful Elizabethan farm-houses still remaining bears witness.    The typical farm-house of this period did not differ materially in its ground floor arrangements from the earlier ones.    There was still a hall, used as a living-room, and sometimes entered by a porch, a parlour off it at one end and storehouses or other offices at the other; but with the general adoption of wide, open chimneys set against the wall there was no longer any need for the hall to be continued right up to the roof in order that the smoke from the central hearth might find a way out through an opening.    Conse-quently it was possible to have a series of bedrooms on the upper floor approached by a staircase, and even the cottages came to have little bedrooms with dormer windows.    Harrison notes the "multitude of chimneys lately erected," but this inno-vation seems to have given rise to a certain nervousness of fire, and in the Court Rolls of Scotter, Lincolnshire, the tenants are exhorted to keep their chimneys in good repair, to sweep them four times yearly, not to let hay or straw lie within half a fathom of them, and to refrain from breaking hemp in the chimney-corner.    "To help a chimney that is on fire," says a writer of 1594, get a large, thick blanket and with other persons hold it over the mouth of the chimney; and "if you can come easily to the toppe of the chimney" cover it also with a board or wet woollen cloth, and the fire will stop for want of air.[2]    We hope so.

Tusser liked to "cleanse sooty chimneys" in November, when putting things in order for the winter, but the use of soot on the land was not yet understood.    These wide, open chimneys in small establishments took the place of the smoking houses for bacon, "Martilmas beef" and red herrings.    The oven was built conveniently at the side, and cooking must have been much easier for the housewife than on a central hearth.

[1] W. Denton: *England in the Fifteenth Century*, 154.
[2] H. Platte: *Jewelhouse of Art and Nature*, 59.  *Arch.* xlvi. 381-2, 384-5.

Circular stone stairs continued until the reign of Elizabeth, when they began to give place to straight wooden ones.

The larger manor houses were now frequently built of brick or stone, but except in stone districts the usual material for farm-houses was oak. In early days and in well-wooded counties they were constructed "with not more than 6 to 9 inches between stud and stud, but in open places where wood is scarce few posts and many splints or raddles, cast over with thick clay to keep out the wind." But the same writer, quoting Holinshed, adds pessimistically, "When our houses were built of willow then had we oaken men, but now that our houses are come to be made of oak our men are not only become willow, but a great many . . . altogether of straw"; from which we may see that even in that glowing age someone must needs think that the country was going to the dogs. He notes also that in the northern and fenny parts of the country the ancient custom of including stables and other outhouses all under one roof was still continued owing to scarcity of timber.[1]

As a picture of a still humbler establishment, without even a chimney, we have the description in Bishop Hall's *Satires*:

> Of one bay's breadth, God wot! a silly cote,
> Whose thatched spars are furr'd with sluttish soot
> A whole inch thick, shining like blackmoor's brows,
> Through smoke that down the headless barrel blows.
> At his bed's feet feeden his stalled team,
> His swine beneath, his pullen o'er the beam.[2]

To counteract the lingering mediaeval lack of cleanliness sweet herbs were strewn on the floors, and Tusser includes in his list for the purpose basil, balm, camomile, cowslips, daisies, sweet fennel, germander, hysop, lavender, marjoram, pennyroyal, roses, redmint, sage, tansy and violets. Dried wormwood was recommended to keep away fleas.[3] Glass by Elizabeth's time was plentiful for windows, and lattice and horn were disused.

The Elizabethans were not fond of wind and preferred to place their houses in valleys or sheltered situations, if possible facing north or east, for they considered the south wind sickly and the west boisterous, whereas the north was a "preservative from corruption" and the east brought bright, serene weather. Andrew Boorde recommended building near a wood on account of the bees and the consequent "abundance of hunny."[4]

[1] Harrison, i. 233; ii. 337.
[2] Hall's *Satires*, Bk. V. Sat. 1.
[3] Tusser, 121, 172.
[4] Harrison, ii. 130. W. Denton, *England in the Fifteenth Century*, 252.
A Boorde, *Dyetary*, 283.

The dairy, bakehouse, brewhouse and stable were generally separate from the main building, but "not so farre distant in sunder but that the goodman lieing in his bed may lightlie heare what is doone in each of them with ease, and call quicklie unto his meinie if anie danger should attack him." Similarly in larger establishments it was recommended that the baliff's chamber should have a latticed wall from which he could over-look the courtyard and kitchen regions. Andrew Boorde, who for his time was very particular, after grouping the buttery, pantry, pastryhouse and larder by the kitchen, put only the "pleasure stables" near the house, and placed the bakehouse and brewhouse some way off, and the other stables, slaughter-house, and dairy a quarter of a mile distant.[1] The outhouses did not undergo any marked change, though many of them, like the farm-houses, were doubtless rebuilt. Barns were often partly of brick, the upper part boarded, and the roof thatched. Leonard Mascall recommended that cowstalls should run east and west, closed to the north, and with doors and windows facing south. Tusser housed his old and feeble cattle during the winter, but it was generally considered that most cattle did best in a close, well sheltered by hedges, and foddered in warm and dry spots.[2]

The sheephouse for lambing time was an important matter. Mascall liked his "low like a hogsty," opening towards the sun at noon, the consequent unpleasant stuffiness being combated in the true mediaeval manner by hanging it round with rose-mary or other sweet and strong herbs. Racks for fodder were placed two feet from the ground, and a litter of fern or dry straw provided for the ewes with lambs. The houses were placed in a yard well fenced round, so that the sheep could safely go out for refreshment.[3] For the rest the farm buildings usually included a boar-sty, a "hogscote" with a hard floor, in which to fatten hogs for killing, sties for the sows to farrow in, and elaborate poultry-house, stables, and a "couch for the dog." Tusser liked a "neat round dunghill" in the middle of the yard.

The farm servants seem to have been accommodated in any outhouse that came handy, for in an inventory of 1594 at Gilling, mention is made of a "cubborde bedstead" and

---

[1] Harrison, i. 238. C. Heresbach: *Four Bookes of Husbandry*, ed. B. Googe, 12a. A. Boorde, *Dyetary*, 238-9.

[2] L. Mascall: *The First Book of Cattell*, 53. Tusser, 59. Heresbach, 96.

[3] Mascall, op. cit., 208.

mattress in the stable, a flock-bed over the stable, and a mattress in the maltkiln.[1]

Peas, beans and vetches were stacked in "hovels" raised upon "crotches" to keep them from damp and marauding animals, and covered with poles and straw, or at other times with furze or light firewood. When not in use for these crops they were very handy to shelter the cattle or to fatten porklings.

FIG. 56.—SHEPHERD AND COWHERD (from Spenser's *Shepheardes Calendar*)

The yard hogs also made themselves comfortable in winter under the carts in the cart-shed.

The sixteenth century being a time when more attention was given to pasture-farming than to agriculture, the farmer's *stock* naturally underwent marked improvement, especially during the latter half of the century. During the reign of Henry VIII there are complaints of the dearth of cattle, other than sheep, and the lack of regard to their breeding,[2] but this, even if not exaggerated, was evidently remedied later, for Harrison states that "the country is wonderfully replenished with neat and all kind of cattle" so that only about a quarter of it was under crops. "Where," he asks, "are oxen commonly made more large of bone, horses more decent and pleasant in pace, kine more commodious for the pail, sheep more profitable

[1] *Arch.*, xlviii. 130.

[2] More's *Utopia*, Bk. I. Starkey: *England in the Reign of Henry VIII*, 98. But this Starkey, like his later namesake, seems to have always been pessimistically inclined.

for wool, swine more wholesome of flesh, and goats more gainful
to their keepers than here with us in England?"[1]   The extinction of harmful wild animals, especially wolves, gave both the
beasts and their guardians a more peaceful life, and fewer herdsmen were needed in consequence.   Murrain was still prevalent,
and where there was an outbreak it was considered usual and
kindly to put the head of a defunct beast upon a long pole and
set it in a hedge beside the highway, to warn drovers that the
district was infected.[2]

The *sheep* was the most valued of all animals at this time
and the greatest source of wealth on account of its wool; but
besides the use of its fleece, mutton and manure—

> His tallow makes the candels white
> To burne and serve us day and night,
> His skinne doth pleasure divers waies,
> To write, to weare at all assaies,
> His guts thereof do make whelestrings,
> They use his bones to other things,
> His hornes some shepherds will not loose,
> Because therewith they patch their shoes.[3]

Good wethers at the end of the century weighed from 40 to
60 lb., and their fleeces from 4 to 6 lb., an astonishing improvement since the last century.   The price of the best was 20s. and
the worst 6s.[4]   The milking capacity of the ewes must have
correspondingly increased, for Tusser equates the milk of five
ewes to that of one cow, as compared with Walter of Henley's
ten to one.   The practice of milking ewes seems to have been
rather out of favour in the first half of the century, for Fitzherbert (1523) notes that in the Peak district the lambs were
weaned at 12 weeks because ewe's milk was used there, whereas
it was more usual to leave them with their mothers 16 or 18
weeks.   Mascall also states that "the poor husband in many
places where they do use to milk their ewes" weaned the lambs
at 12 weeks, but that they were never so good.   Tusser, however, weaned his lambs on May 1st and advocated milking the
ewes from then until August 1st.[5]   Before lambing time they
were brought into good pasture near the farm-house, well

[1] Harrison, i. 128; ii. 1.

[2] Fitzherbert: *Husbandry*, 53.

[3] L. Mascall: *The First Book of Cattell*, 202.

[4] Jacob Rathgeb (1592) quoted in Harrison's *England*, lxxxiii.   But
judging by Shakespeare's *Winter's Tale* a little over 2 lb. was an ordinary
weight for a fleece.   (Act IV. Sc. 2, notes.)

[5] Fitzherbert: *Husbandry*, 44.   L. Mascall, op. cit., 235.   Tusser, 150.

sheltered and without brambles to tear the wool. The time-honoured expedient of making a sheep take to another's lamb by covering it with the skin of her own dead one was already used, another contrivance being to tie a dog within sight of her.

The sheep were often put out in folds in May, preferably on the rye-ground, and moved daily, but some thought that folding encouraged vermin, and preferred merely to drive in stakes where they wished the sheep to graze, as the animals would go to them to rub themselves. In this way the expense of hurdles was saved and the sheep could shelter themselves by the hedges in bad weather. Some only folded their sheep in July and August, and tied dogs at the four corners of the fold, or "set up shows of dead dogs' heads" to scare away hurtful beasts. In June the sheep were washed in a running stream, dried in the sun and shorn after two days, the yearling lambs being left until the end of the month or the beginning of July. At the sheep-shearing feast, a merry business cheered with song and pipe, nosegays were given to the shearers; meat, wafers, cakes and warden pies coloured with saffron were devoured, and sugar, currants, rice, prunes and many spices were used to compile other dishes.[1]

With the great popularity of sheep-farming enormous flocks came to be owned by single farmers, "some having 24,000, some 20,000, 10,000, 6,000, 5,000 and some more and some less," as the Act of 1534 states. By this Act, however, no one person might own more than 2,000 sheep, but such was the popularity of sheep-farming that the law was often evaded by assigning this number to each member of a family. The sheep were mostly white, inclined to be large-boned and leggy; the rams especially were high and long of body, and by some farmers were preferred without horns.[2] In the illustrations to Spenser's *Shepheardes Calendar* many of the sheep appear to be horned, as if that breed was still popular. They suffered a great deal from scab, and rot, a disease which some writers considered due to rank pasture, damp summers and "rowtie fogs," but in which Sir Thomas More saw a vengeance for the evils of enclosure. In winter, when the pasture failed, they were fed on hay, tares, barley straw and pease haulms, mistletoe, and ivy, elm and ash leaves (carefully gathered in autumn and dried), sometimes barley and peas ground together, or dried peas and acorns, and occasionally "three-leaved grass," green or dry, which looks like an early mention

[1] Shakespeare: *Winter's Tale*, Act IV, Sc. 2. Tusser, 271.
[2] L. Mascall, op. cit., 205, 239, 243.

of clover.[1]   But with the best that could be done for them in the early months of the year they were apt to be in a very poor condition.

The *shepherd*, says Mascall, should be wise in governing his flock with gentleness, and should throw nothing at them to frighten them.   Although on some sheep-walks the flocks seem to have fed independently without the attention of herdsmen, and although with certain writers there was a pleasant idea that shepherds enjoyed a leisured existence with plenty of time to cultivate their minds in a refined manner, in reality they

FIG. 57.—SHEPHERD (from Spenser's *Shepheardes Calendar*)

seem to have had a busy life.   The shepherd is exhorted to drive his sheep westwards in the morning and eastwards at night, so that they should always feed with their heads away from the sun; he is not to stray from his flock and is discouraged from resting, for "if he do not go he ought to stand, and to sit very seldom."   His indispensable needs were a hook, a knife, shears and a tar box or bottle, for the anointing of sores, and a dog "to bark when he would have him bark, and to run and to leave running when he would."   The crook shown in the illustrations to Spenser's *Shepheardes Calendar* of 1579 has a long staff with a well-curved hook at one end and a curious implement at the other which might be a slightly curved blade for cutting leaves and twigs for the sheep to eat.   The shepherd was recommended to have a little board beside the

[1] L. Mascall, op. cit., 215-16.  Tusser, 89.

fold, whereon he could lay a sheep cleanly when its sores needed dressing, and sometimes he had a little cabin on wheels. His tar was made into an ointment by mixing with oil, butter, or swine or poultry grease.[1] For marking sheep the tar was mixed with pitch, two-thirds of tar to one-third pitch.

The shepherds described in Barclay's *Eclogues* seem to be independent sheepowners, of varying degrees of prosperity, not hired servants. The rich shepherd Codrus wears a cloak with one or two hoods, and furred mittens of "curres" skin; he has

FIG. 58.—SHEPHERD AND BROKEN BAGPIPES (from Spenser's *Shepheardes Calendar*)

meat in his wallet and ale or wine in his flask, but the poorer ones, such as the aged Corinx, whose "patched cockers skant reached to his knee," carried bread and cheese and ale, and a wooden spoon stuck in his felt hat. Dairy produce and brown bread, with a little fish in Lent, was the usual fare of these men, eked out with vegetables. Coridon, who prided himself on having shot four thrushes and a sparrow with his bow in an hour to increase his larder, had a weakness for garlic and leeks, but realized that "great Lordes may not abide the stinke." They spent toilsome days tending the flocks that were their livelihood, repairing their folds and "cotes," guarding them from thieves at night, and procuring fodder for the winter. The sheep were only housed in times of great cold, and then the shepherds lay up to the chin in hay, "pleasantly and hote"

[1] L. Mascall, op. cit., 215, 231, 236. Fitzherbert: *Husbandry*, 15, 45, 46.

and entertained each other.   Each shepherd carried either the
straight "oaten pipe" or the bagpipes, and could play and dance,
for however poor "yet could he pipe and finger well a drone."

In many places, especially in the West and where pasture
was rough and scanty, and only a poor livelihood could be
picked up by the animals, *goats* were kept, but they were not
encouraged in the vicinity of houses on acccunt of the damage
to herbs and trees.   They were said to give three times as much
milk as a sheep, and their milk, cheese and kids were "very
profitable for the sick."   They were not folded, but if housed
it was recommended that the floor should be of stone or gravel,
as their natural heat was such that they did not need litter.
In winter they were fed on elm leaves, and seeds and branches,
and their houses cleaned out daily.   A hundred was considered
a large enough herd.[1]

Unlike the stunted cattle of mediaeval times as pictured in
the Luttrell Psalter, *oxen* by the end of this century were of a
good size, with long, spreading horns.   Rathgeb in describing
the "beautiful oxen and cows" of England notes their very large
horns, although they were not as large in body as those of Bur-
gundy, being low and heavy, and for the most part black.
Harrison, with his usual rosy-hued outlook, boasts that English
cattle "for greatness of bone and sweetness of flesh yield to none
other," and states that they often measured a yard between the
tips of their horns, and were as tall as a man of medium stature;
but these were probably exceptional beasts, for he admits that
the general breed was not well looked after.   He also says that
graziers anointed the budding tips of the horns with honey,
which "modifieth the natural hardnesse of that substance and
thereby maketh them to grow into a notable greatnesse."
Oxen were still prized for power of draught rather than production of beef.   Mascall considered that black and red oxen were
best for labour, the brown or "grizled" next, and the white
worst.   At ten years old they were fed up and sold to be eaten.[2]

"Let cows be beetlebrowed and stern of look," says one writer,
and animals of this forbidding aspect were considered as meet
for the plough as oxen or horses.   As a rule, however, their
chief function was the dairy, and sometimes a good milch cow
was bought specially for the winter, and another for Lent.
Some improvement in milking capacity must have taken place
since mediaeval times, for Tusser's dairy season extended
from the beginning of April to the end of November, instead

[1] Harrison, ii. 8.   L. Mascall, 250-3.   Fitzherbert: *Husbandry*, App. 139.
[2] Harrison, lxxxii. 1, 135, ii. 3.   Mascall, op. cit., 52.

of the old period of May to Michaelmas. At the end of August the old cows and ewes were no longer milked, but were fattened for selling or home killing. Those cows which needed housing in the cold weather were let out once a day for exercise and drink, and during this season the cattle were fed on all sorts of straw, beginning with rye, the least appetizing, and coming last to hay. About January, when supplies of fodder were dwindling, "browse" (or branches) were cut for them, and after having their tongues rubbed with salt they were dosed all round with verjuice if at all weak. The calves were looked after by the housewife and weaned at fifty days, and in May they were put out to grass in a carefully sheltered close.

The "keeper of cattle" was exhorted to rub and comb the oxen daily, and wash their feet; he was to keep their stalls clean and keep out poultry and hogs, as their feathers and droppings were thought to breed murrain. When past work the oxen were fattened with boiled barley or pulse, hay or elm leaves, and "boyled coleworts with bran." The village herdsman was commonly paid 2d. a quarter for the care of each beast when driven out to pasture.[1]

Henry VIII paid considerable attention to the breeding of *horses,* and observing the wretched little animals, descendants of the mediaeval stotts and affers, pastured upon the village commons, attempted to improve matters by statute. In 1535 all owners of enclosed ground of a mile in compass were to keep two mares able to bear foals of 13 "handfuls" at least, and they were only to be kept with horses of 14 "handfuls," which shows how small the ordinary working horse must have been. In 1541, in all counties except those of the North and extreme West, it was enacted that horses under 15 hands might not be turned out on the public wastes, and mares incapable of bearing reasonable-sized foals were to be killed and buried. It is improbable, how-ever, that these statutes were ever rigidly enforced. "Plain countrymen" still used the old inferior breed for draught and carrying purposes, but now, apparently for the first time, the farmer began to have the use of some of the old English "great horses," hitherto reserved for the warriors with their heavy equipment. In this century only the inferior animals of the breed and some of the mares found their way to the farm-yard; it was not until after the Civil Wars of the seventeenth century and the disuse of protective armour that the proud war-horse became the shire-horse of the farm wagon.

[1] Mascall, 61. Fitzherbert: *Husbandry,* 77. App., 139. Tusser, 60, 88, 90, 93, 156, 192.

But even so the infiltration of the superior breed began to tell. Harrison states that packhorses would carry 4 cwt. "without any hurt or hindrance," and the cart- or plough-horses were so strong that five or six of them would draw "3,000 weight (? 30 cwt.) of the greatest tale with ease for a long journey"; and Blundeville notes that the mares "will endure great labour in their wagons, in which I have seen two or three mares to go lightly away with such a burthen as is almost incredible." Carriers seem to have been specially particular in selecting fine animals, for the same writer says, "I have seen somtyme drawing in their carts better proportioned horses than I have knowne to be fynely kept in stables as jewels for the saddle." Foreign writers of the end of the century observed that in England horses were abundant, and though low and small were swift. The English "great horse" of this time was a fine animal, "though not of such huge greatness as in other places of the main."[1]

Mascall considered that a good horse should be "with a merry look and wild of countenance," and preferred brown-bay with a golden mouth. More than one white foot was not favoured. "If thou have a foal with four white feet keep him not a day," with three white feet "put him soon away," with two "send him to thy friend," but with only one "keep him to his lives' end." Tusser kept up the mediaeval practice of bleeding his horses at Christmas-time, the old date being St. Stephen's Day, and was careful to see that they had good food after it.

The *carter* or horsekeeper was exhorted to be patient and moderate in the use of his animals, and to use "fierce words" more than stripes. He was to rub and comb them every morning, and in summer to wash their feet after work, with cold water, or still better with wine or ale and butter, and to rub their legs with butter or oil to strengthen and supple the sinews. They were to be fitted with harness of the right size, and this and all the cart furniture was to be kept in good repair, and the harness hung up "for fear of dogs and cats." Bells were hung upon the "back wantie" to delight the horses, and the animals were to be placed in the stable according to their liking for each other. The carter had also to be able to shoe the horses at need, and was to carry with him on a journey hammer, nails and shoes in case of emergency. Lastly, he was to clean the stable out every morning, litter them at night and sleep with his charges

[1] Sir Walter Gilbey: *The Great Horse*, 34, 35. Harrison, i. lxxxiii. and v. (Rathgeb and Kentzner); ii. 3-4, 153.

lest they should break loose and fight, and, as in earlier times, "look advisedly and warily to his candle."[1]

Lean horses were fattened on dried wheat and barley given as a mash with ale and wine. If too fat they were given bran mixed with honey and warm water.

Harrison stated confidently that there were no *asses* and therefore no *mules* in the country, but Mascall (writing at the same time) had a good deal to say about them. "Moyls" he considered stronger than the contemporary horses, and said they were able to carry 5 or 6 cwt., and travel thirty miles a day therewith. He also makes the rather curious statement that they were somewhat "too high" for ploughing and that therefore a team of oxen was commonly used to break up the ground before them. It is difficult to believe that they were larger than the other animals employed for this purpose.

*Asses* were "commonly of a mouse-dun coloured hair," but that was not thought a good colour for mules.[2]

*Pigs* abounded, and according to Rathgeb were larger than in any other country. It was customary to ring their snouts and sometimes to yoke those which fed at large, to prevent them from uprooting the pasture and breaking the hedges, but this was not necessary while they were in the woods for the mast. The masting season in Fitzherbert's time lasted from Michaelmas to Martinmas, a period of six weeks, and 1d. or ½d. per head was still paid for the privilege. In most country places the village herd was still collected daily by the swineherd, "who commonly gathereth them together by his noise and cry and leadeth them forth to feed abroad in the fields," but they were carefully "styed uppe" by each owner from sunset to sunrise. Sties were also provided for the sows to farrow in, and eight piglings were considered enough for each sow to bring up; some farmers preferring five, or only three if very fine ones were desired.[3] Hogs or boars were styed at Michaelmas "and lodged on the bare planks of an uneasy cote" to harden the fat and make good brawn, which was the first dish at the farmer's dinner table from November to February. For this they were fed on barley, malt, oats and peas, with the washings of ale barrels or whey to drink, or by some with bran mixed with whey. The dairymaid was expected to keep their cabin clean. At other times, when they did not pick up their own

---

[1] Mascall, 107, 117-20, 169, 185, 187.

[2] Ibid., 110, 117. Harrison, ii. 3.

[3] Tusser, 95, 104. Mascall, 262, 264, 273. *Arch.*, xlvi. 377, 379, 384 (Ct. Rolls of Scotter). Harrison, lxxxiii. Fitzherbert: *Surveyinge*, 19.

living, they were fed on the waste matter from the brewhouses, or on acorns stored in cisterns with water, or dried and kept in vats.   Andrew Boorde considered the English pig an unclean beast which "doth lie upon fylthy and stynkynge soils," whereas in Germany they were kept clean and swam in rivers once or twice a day.   Tusser slaughtered his fat pigs at any time from Hallowtide (November 1st) to Shrovetide, as required. Small boys disported themselves at pig-killing time by blowing the bladders up "great and thin" and inserting dried peas and beans to make a pleasing rattle.   The swine suffered a good deal from "measles," though not so much, apparently, as the sheep from rot.   "Measeled hogs," as Tusser callously remarked, could always be sold to the Flemings.[1]

As might be expected in England, *dogs* were particularly good, especially the breeds used for sport, but according to Harrison those "of the homely kind are either shepherd's curs or mastiffs."   The shepherd's dog we have already dealt with. The mastiff, used to guard the farm from intruders from the earliest times, was "a huge dog, stubborn, ugly, eager, burthenous of body, terrible and fearful to behold."   Besides this alarming animal there were "prick-eared curs and others, only meet to give warning at night," and the "black and red little curs" of the poachers.   Tusser considered one dog enough for a household, and discouraged too many "ravening curs" which ate up good food.   The farmer was cautioned to keep the farm dogs from tearing the pigs' ears or killing them, and from preying upon the lambs during Lent, when their diet, as well as that of the household, was meagre.   "If a dog have fleas," says Mascall, "wash him with beaten cummin, 'ellebory' and water, or with juice of wild cucumbers; or anoint him with the leas of olive oil."[2]

It was as necessary as ever to have a guardian for the granary, and the *cat,* says Heresbach, "is a household servant to be cherished . . . for avoyding the mischiefs of ratts and mice."   Tusser agreed that one cat should be kept for this purpose, and considered that she earned her living, but he was very careful to keep her out of the dairy.[3]

Vast numbers of *rabbits* were still kept in enclosed warrens, from which, however, they were very liable to escape.   They seem to have been a different breed from the present, for we read that they were valued for their *black* fur even more than

[1] Tusser, 20, 32, 38, 46.  Barclay, *Fifth Eclogue.*  A. Boorde: *Dyetary,* 272.
[2] Mascall, 298.  Tusser, 22, 23, 49, 125, 139, 253.  Harrison, ii. 44; lxxxiv.
[3] C. Heresbach: *Four Books of Husbandry.*  Trans. B. Googe, 156a.

for the larder; but young rabbits or conies were much eaten. Tusser preserved them from slaughter from January to March.[1]

*Fowls* of a yellowish colour, as in Chaucer's day, were still preferred. Mascall liked his cocks to have golden neck feathers, and "changeable yellow" plumage, and his hens tawny or russet. Those with a tuft of feathers on the head he considered "reasonable good," and partly black fowls next best to the tawny, but grey or white he did not approve of. Grey hens, however, were evidently valued in humble cots, for Barclay represents his shepherd Coridon as saying:

> My wives gray hen one egge layde every day,
> My wife fed her well to cause her two to lay,
> But when she was fat then layde she none at all.[2]

Chickens which were hatched later than the middle of June were not considered profitable by Mascall. He put all newly hatched chicks in a sieve and perfumed them delicately with rosemary, as a preventive against the pip. But the most interesting feature of his work on poultry is the mention of an early attempt at incubation. He packed the eggs in hen droppings and bags of feathers and hatched them out "in an oven always of temperate heat," turning them often, and washing them in lukewarm water on the eighteenth day. As early as 1516 Sir Thomas More in his *Utopia* seems to look upon this practice as an ideal one, for there, he says, hens do not sit on eggs, "but by keepynge theym in a certayne equall heate they brynge lyfe into them and hatche theym"; but Mascall, in 1581, was the first writer to describe the process in a practical manner. To preserve eggs he kept them in dry salt or brine, instead of the "ancient way" in straw or bran. His fowlhouse was built at the eastern end of his house, and enclosed with stone at the bottom to keep out vermin; it had a window to the east, that the inmates might know when it was time to bestir themselves. Inside there were perches and nests, the latter sometimes baskets hung on poles, for the hens, and below them pens and nests for the ducks and geese separately. The birds had their wings clipped to prevent their straying; the house was kept clean, and the window carefully shut at night. He did not consider hens were good for laying

[1] Harrison, lxxxii. 304. Tusser, 87, 104.
[2] Barclay, *Second Eclogue.*

O

purposes after three years. They were fed on barley, vetches, peas, millet or any offal corn and house scraps.[1]

Then as now *geese* were considered as good as a watchdog, and were valued for their feathers even more than their flesh; feather beds were very popular with the increasing luxury of the times, the wings being used by the housewife as dusters. In a few country places the gooseherd still survived, and led forth his charges to the field, carrying a rattle of paper or parchment "the noise whereof cometh no sooner to their ears than they fall to gaggling, and hasten to go with him." At Scotter, Lincolnshire, a low, unenclosed piece of ground known as the "car" was reserved for them, and the owners were to clip or pull their wing feathers to prevent them from straying.[2]

*Ducks* were apostrophized as "grosse, greedye and filthie feeders" but were universally kept; and *swans* were also very plentiful. Strong nests were made for the latter out of reach of flood water, and they were fattened for the table in houses or in the yard, being a much esteemed dish at festivals.[3]

The great addition to the poultry-yard at this time was the *turkey*. This bird seems to have been first discovered by the Spaniards in Mexico in 1518, and was introduced by them to Europe.[4] Mascall states that it was brought to England about 1537-8, an appropriate time for the appearance of this pompous and choleric bird, which might be thought to have a certain affinity with the reigning monarch. Mention is also made of it in Cranmer's *Constitution* of 1541. Tusser refers to turkeys in 1557, as if they were commonly kept, but they were evidently a danger to the kitchen garden, as both he and Mascall agreed. The latter regarded the "turquie" as "a right cofer for oates and a sack for corne, a gulfe, a swallower of barns, a devourer of much meat," and although he admitted the delicacy of its flesh, considered it heavy and hard of digestion and preferred that of the peacock. Heresbach referred to their inability to stand cold and wet, and kept them in winter in a warm, dry place; he fed them in the same way as peacocks and

---

[1] Mascall: *Husbandlye Ordring and Governmente of Poultrie*, ch. 1-26. He also gravely states that if eggs are coloured and put under a hen the chickens will be of that colour! Tusser, 262.

[2] *Arch.*, xlvi. 378. Harrison, ii. 15. Tusser, 264. Mascall: *Poultrie*, c. 33.

[3] Mascall, ibid., c. 41 and 56. Tusser, 110. Harrison, lxxxiii.

[4] Heresbach states that the turkey was not seen in Cleves before 1530. Op. cit., 166a.

provided them with ladders to their perches eight or ten feet from the ground.[1]

*Guinea-fowls* are said to have been rare, and are only mentioned once[2]; but *peacocks* were still commonly kept. Mascall considered them a "strange bird to feede and to governe, for they hardly bee so familiar with any person as other birds will."

Every farm-yard had a house for *pigeons,* and they flourished in such multitudes that they had become hurtful to the crops and "great devourers of corn." Still they were considered a necessity for the table, and their droppings were much valued, especially for the hop-yard. Tusser repaired his dovehouse carefully in January, cleaned it out and refrained from killing the birds during that month, and fed them from December to February. Mascall (quoting Stephanus) recommended the farmer to perfume his dovecote delicately with "genoper" or rosemary, and sometimes with frankincense or lavender to make the birds "love the dovehouse more."[3]

Many other birds were caught and fattened in cages for the larder.

According to Harrison *bees* were plentiful, and in some "uplandish" places there were one or two hundred hives. They were considered very profitable, for:

> He that hath sheep, swine and hive
> Sleep he, wake he, he may thrive.

The hives, which were often of the capacity of four or five pecks, were commonly made of rye straw "wattled about with bramble quarters," or of wicker plastered with clay. They were set on the warmest side of the house, in the garden or orchard, in a dry spot safe from mice and moths, sheltered from the north and north-east, and with the mouth of the hive towards the sun. Tusser fed his bees in winter, if necessary, with honey, rosemary and water, and "burnt them up" after five years for the honey and wax.[4]

*Agriculture* at the beginning of the century was still in

---

[1] Tusser, 109. Mascall: *Poultrie,* ch. 44-5. *Encyclopaedia Britannica,* "The Turkey," which states that there were two breeds, the southern "mexicana" and the northern "americana."

[2] Fitzherbert, 145. He refers to them as "Ginny or turkie-cocks," but as the latter seem to have been fairly common by that date, it was probably a confusion of terms. This occurs in the Appendix by I. R., 1598. Mascall: *Poultrie,* ch. 50.

[3] Mascall, ibid., ch. 61. Tusser, 87. Harrison, ii. 15.

[4] Tusser, 19, 64. Fitzherbert, 75. Harrison, ii. 37.

a depressed condition, attributed by some to the decay of the monasteries, those great pioneers of mediaeval farming, and their subsequent dissolution. To counteract this, however, there was the growing prosperity of the independent yeoman farmers, and as early as 1523 the renewed interest in agriculture was shown by the appearance of Fitzherbert's *Boke of Husbondrye,* the first treatise written on the subject since that of Walter of Henley, two hundred years before. The spirit of enterprise so typical of the sixteenth century made itself felt in agriculture, not in any strikingly new departures, but in a revival of energy, especially towards the end of the century. By that time population was increasing a little more, and although enclosure for sheep-farming was still going on, land which had benefited by a long rest was being ploughed up again for arable, and, in the words of Harrison, men were more "painefull, skilfull and carefull" in their efforts to win profit from the hard-worked soil.

The farmer's calendar for the year remained roughly the same, being still without the root crops which later opened up new possibilities; but we find in both Fitzherbert and Tusser that it was now customary to work in an extra ploughing of the fallow land during the summer. Instead of the three mediaeval ploughings—"warectum" in April "rebinatium" in July, and "hyvernagium" in the autumn before seed sowing — these farmers did their first fallowing in April, a "twy-fallow" or first stirring in May or June, and a "thry-fallow" or second stirring in July or August, besides the autumn ploughing.[1]

The plough was still a cumbrous affair, mainly of wood, with "clouts" of iron to strengthen it, and share and coulter of steel. Different sorts were used according to the nature of the soil, some having wheels and others not. In Kent the "turnwrest" plough had already been invented, in which the mould-board could be turned to the other side, at the end of the furrow, so as to make the sods fall the same way going as coming.[2] Tusser mentions a plough-chain. Both oxen and horses were used for drawing the plough; Fitzherbert put forward much the same arguments in favour of the ox as Walter of Henley; Sir Thomas More in his *Utopia* advocated oxen alone for ploughing and draught; Tusser had teams of both, and used light sedge-collars for his plough-horses; Harrison considered cows as useful as either.

---

[1] Fitzherbert, 26, 32, 39. Tusser, 17, 35, 155, 171. Tusser was usually a month earlier in the operations than Fitzherbert.

[2] Fitzherbert, 2.

Wheat, rye, peas and beans were often sown with a plough following to turn the furrow over on to the seed, and a child was sometimes employed to go before the plough scattering the seed, but, observed Fitzherbert doubtfully, "me seemeth that chylde oughte to have moche dyscretion." This writer still followed the mediaeval rule of two bushels of wheat or rye to the acre, but otherwise varied from it a little, sowing only three bushels of oats and two of peas, and as much as four

FIG. 59.—PLOUGHING (from Fitzherbert's *Husbondrye*, 1525)

bushels of beans, and four or five of barley.[1] He often mixed saltpetre or black dregs of oil with the seed. Tusser sowed oats as early as January, and barley as late as May. Seed was selected by means of the casting shovel, the best and heaviest seeds flying the farthest if evenly cast, but sometimes children were employed to pick out the choicest seeds in the barn.

[1] Fitzherbert, 21-3, 40, 136. Tusser, 9, 189.

Towards the end of the century the practice of dibbling corn by hand or "pricking in" was evidently tried, for Sir Hugh Platt, writing in 1601, recounted the implements which had been formerly used for this and rejected. He related that when man got beyond making a hole for the seed with his fingers an instrument like a rake with a handle on the back was used, having teeth to make a dozen or more holes. Then a board was contrived, about 3 feet long and 1 foot wide, with holes at regular intervals through which pins, 3 or 4 inches long and as thick as the finger, could be thrust. The last device, which he said was already rejected, was to have the pins set into the board. This painstaking method of sowing was, however, more advocated in the following century.

After ordinary sowing, girls with slings and boys with bows and arrows were set to scare away the birds and "cry alarum," and the ground was harrowed with an ox harrow first and a finer horse harrow afterwards. In very stony soils the tynes of the harrow were made of ashwood as it was considered to wear better than iron. Particular care was taken to break up very finely the ground on which barley was sown. If the harrow failed it was beaten with mattocks, and rolled after a shower. Wheat also was sometimes rolled.

Weeding was not left until July, according to the mediaeval custom, but the winter corn was weeded in May and the spring crops in June. A forked stick and weeding hook, the blade an inch wide, set on a staff a yard long, was still used for this in dry weather, but in damp the hook was replaced by a pair of wooden tongs or "crotch," nicked at the side to pull up weeds by the roots.

As far as can be estimated the yield of hay was fairly good. "Land meadows" were said to produce not more than one good wainload per acre, but low meadows commonly two loads or more, and sometimes three, which, if the "load" was the same as the modern one of a little under a ton, does not compare unfavourably with the present day.[1] Aftermath was not considered so wholesome for the cattle as the first crop.

Harvest was conducted in the same way as formerly. After the barley had been mown, with a scythe and a "cradle" attached, the ground was raked over with a "great rake with iron teeth, made fast about a man's neck with a string," in order that none should be lost. Googe, in his translation of Heresbach, mentions an early form of reaping machine as used

[1] Harrison's estimate of the draught powers of sixteenth-century cart horses is high (see above, p. 212). Harrison, ii. 133.

in the Low Countries, " a lowe kinde of carre with a couple of wheeles and the frunt armed with sharp syckles, whiche forced by the beaste through the corne, did cut down al before it," but we have no record of its being adopted in England.

"Harvest Home" was a great ceremony. Hentzner, in 1598, describes one near Windsor at which the last load of corn was crowned with flowers, "having besides an image richly dressed, by which perhaps they would signify Ceres," and this was escorted to the barn with much shouting. Tusser gave his men "good cheer in the hall all harvest time long,"and each plough-man was given a goose at the harvest supper.

The wains in which the crops were brought home to the barn were built of oak, with strong, iron-bound wheels for ordinary ground, or broad, untyred wooden wheels for use where the soil was marshy and soft. There were also muck-wains and tumbrels and marling wains; and others, but more usually sleds, for turf. Carts were made of ash for lightness, and the axle-trees and nathes were greased with fat or soap mixed with snails. Rathgeb stated that wagons were not used for transporting "goods," but two-wheeled carts drawn by five or six strong horses, and so large that they carried as much as a wagon. Tusser used elm, ash and thorn for carts and ploughs, hazel for forks, willow for rakes, and holly and thorn for flails. Holly and ash were also used for ladders, and poplar for bowls, troughs and dishes.[1]

The grinding of corn continued to give rise to controversy between lord and tenant, and Fitzherbert notes the variety of tolls exacted by the miller: the better the mill the higher the toll. The windmill underwent various improvements, and the tower-mill, in which only the top storey or cap had to be turned to make the sails catch the wind, was evolved. But the "quernes that goo with hande" still survived, and in some places tenants were allowed to grind small quantities with them.[2]

Draining of the fields had made no advance during this period, and no new method of dressing the soil was evolved, although marling was somewhat revived. Harrison claimed to have a "kind of white marl that if it be cast over a piece of land but once in three-score years it shall not need of any further compesting." Tusser manured his fields before

---

[1] Harrison, lxx. Tusser, ii. 138, 242. Fitzherbert, 14, 15. H. Hall: *Society in the Elizabethan Age*, 53. Mascall, 119.

[2] Bennett and Elton: *History of Corn Milling*, i. 221; ii. 193. Fitzherbert: *Surveyinge*, 21.

ploughing for seed and fallowing, but after twy-fallowing and thry-fallowing; and also manured the hay meadows.

The yield of crops, if Harrison's figures may be trusted, had considerably improved since the early fourteenth century. In Elizabeth's time, in "mean and indifferent years," rye or wheat are said to have yielded 16 to 20 bushels, barley 36 bushels, oats 32 to 40 bushels, and other crops "after the same proportion," the average being less in the North but more in the South;[1] so that the increase of care in tilling the soil was evidently having its effect. The only new field crop was *buckwheat* or "brank," which Heresbach observed was brought from Russia and North Germany "not long since." Tusser advocated it as "comforting to the land" and useful for fattening all kinds of stock. It was sown in May, two bushels to an acre, on ground that had had wheat or barley, and was mown in July, or sometimes ploughed in to enrich the soil. More varieties of the usual crops are mentioned; wheat was white, red, mixed (called main), Turkey or Purkey, grey, flaxen, pollard, English, and peak; barley was sprot, longear, and bear or big; oats, red, black and rough; peas, white, green, grey and runcival. The most usual mixtures were miscelin or meslin (wheat and rye) and bullimong (oats, peas and vetches). The disease known as "smut" first appeared in wheat during the first half of this period, but did not attract much attention until the next century.

In some parishes it was the custom to sow a portion of public land with peas for the poor, each tenant contributing a portion of seed. At reaping time a bell was rung for their gathering, and no one was allowed to gather without a licence.[2]

The growing of *hops*, introduced from the Low Countries, became popular towards the middle of the century. Fitzherbert does not mention them (1523), but Tusser included their cultivation among his activities (1557), and Harrison in the last quarter of the century noted their increase "of late years," so that "few farmers have not gardens of their own." Tusser dug and weeded his hopyard in January and dressed it with droppings from the dovecote; he set the plants in February or March, 5 feet apart, and harvested them in August, cutting the strings and taking the poles right up before picking off the hops

---

[1] Harrison, ii. 134. The mediaeval yield, according to the anonymous *Husbandrie* and the accounts of Merton College, was: Wheat 10 bushels, rye 11 to 14 bushels, barley 16 to 32, oats 10 to 16, peas and beans 10 to 13½ (see above, p. 167).

[2] *Arch.*, xliv. 377-8, 384.

into the hop-manger. They were dried in the kiln, and stored preferably "in canvas or cloth," but sometimes in hogsheads.

*Vines* were very little grown because imported wine was better and cheaper than that made at home; and it is said that only 25,000 tun a year was produced in the country.[1]

The sowing and gathering of flax and hemp had now become one of the numerous duties of the housewife; the fimble or female hemp she kept for spinning, and handed over the carl or male hemp to the men for shoe-thread, ropes and halters. Flax growing was enjoined by statute, everyone owning 60 acres of ground being supposed to plant a quarter-acre with flax, but according to Harrison this useful plant was "contemptuously rejected."

*Woad and madder* were also neglected, although the latter became a little more popular towards the end of the century.

*Saffron* was grown in East Anglia (notably round Saffron Walden) and in Gloucestershire and the West. In September the flowers were picked, the three red stamens saved and the rest cast away; the stamens were then dried "on little kelles (kilns) covered with streined canvas, over a soft fire," pressed into cakes, "bagged up" and sold. In a good year an acre yielded 20 lb. of dried saffron. The roots were taken up every few years and cleaned, and the new heads separated and replanted in July or August, "between the two S. Marie's days" (July 22nd and August 15th).[2]

Quickset hedges were planted of whitethorn, crab, holly and hazel in wooded country, or of oak, ash and elm in open country, but never of blackthorn as that was hurtful to sheep's wool. After ten or twelve years these hedges were "plashed or pleached," the branches being cut half asunder, bent down and intertwined. Trouble was also taken to "increase bushy ground," especially with young oaks. Fitzherbert gathered "akehornes" and planted them in "earth pottes," setting them in the ground in February or March, but Tusser planted them at once and fenced them carefully round. Ash keys, nuts and others were similarly treated.

*Gardening* received a great stimulus in the reign of Henry VIII, and the farm-house garden profited with the rest, especially as regards vegetables. "The poor man," says Harrison, "thinks himself lucky if he may have an acre of ground assigned to him whereon to keep a cow and set cabbages, radishes, parsneps, carrets, melons, pompons or such-

---

[1] Harrison, i. 149; ii. 135.
[2] Ibid., ii. 50-5. A saffron grower was called a croker.

like stuffe," on which his family lived; and Tusser gives long and imposing lists of what should be grown. He left the work of gardening to the housewife and her maidens, and during March and April she was exhorted to garden from morning till night. The garden was to have a south or south-west aspect, and the plants grown were classified as "seeds and herbs for the kitchen," including marigolds and violets among many usual and unusual plants: "Herbs and roots for sallads and sauce," "Herbs and roots to boil or butter," including the usual root vegetables of the present day; "Strewing herbs," still necessary to counteract the lack of cleanliness, "Herbs, branches and flowers for windows and pots," a charming list of old-fashioned flowers; "Herbs to still in summer" for sweet water and cakes, and "Herbs for physic," altogether a quantity calculated to keep the household well employed. Although the expression used above suggests that the choicest flowers, and those which had no particular practical use, were confined to "windows and pots,"[1] the number of those grown for other purposes would have ensured a gay display of colour and redeemed the garden plots from the appearance of a mere vegetable garden. The sixteenth century was the age of great and stately gardens among the well-to-do, but the busy farmer and his harried wife would have neither time nor space for such conceits.

The most noticeable features among the vegetables are the popularity of melons, "pompions" and gourds, of onions, leeks and garlic, and the use of globe artichokes, the heads of which were boiled in beef broth. A cauliflower is pictured in Gerard's *Herbal*, but does not seem to have been common. The potato was introduced by Sir Walter Raleigh in 1586, but it was long before it came into general use. Strawberries were transplanted from the woods and in time of frost covered with straw supported on "crotches and bows."

Orchards now contained apricots, nectarines, bullaces, damsons, red currants, figs, and a kind of peach called the "melo-cotone." Other fruits mentioned are barberries, whortleberries, raspberries, gooseberries, cornel-cherries, filberts, and various sorts of plums: cornet plums, pear-plums, wheat plums, green plums (? greengages); also almonds, and apples and pears of innumerable kinds.[2]

The usual village and manorial officials and servants, of

---

[1] The expression might be taken to mean flowers grown for cutting and placing in pots and windows.

[2] Tusser, 85. Bacon: *Of Gardens*. Amherst: *History of Gardening*, 94, etc.

course, continued where needed, but in this period we hear more of the hired farm labourer, whose services in Elizabeth's time were obtained at the annual "Statute Fairs" held for ten days or a fortnight before Martinmas. A "chief servant in husbandry" at that time was expected to be able to plough, sow, mow, thresh, make a rick, thatch, hedge, and kill and dress hogs, sheep and calves; a "common servant" could mow, sow, thresh and load a cart, but could not expertly make a rick; while a "mean servant" could drive a plough, pitch to a cart and thresh, but neither of the last two expertly.[1] The housewife's maidservants, hired at the same time, had a busy life. They were exhorted to rise at 3 o'clock and either do their mending or "go brewing," but in any case must not lie later than 4 o'clock in summer and 5 o'clock in winter. They were kept to work with the threat of the holly wand, but "such servants are oftenest painfull and good, That sing in their labour as birds in the wood." They were sent to bed after supper.

The farmer's fare at this time was plentiful, but on much the same lines as before, according to the seasons. During the winter "Martilmas beef," dried and smoked in the chimney or "smoky house," pork, brawn, bacon and "souse" (pickled ears, feet, etc.) were varied with fish on Fridays, Saturdays and Wednesdays, and "white meat" (i.e. dairy produce). In Lent there were herrings and salt fish, but according to Harrison almost every household, "even of the meanest bowres," had one or more fish ponds in which they reared tench, carp, bream, roach, dace, eels, etc., for the table. Veal and bacon was the fare at Easter, and "grassbeef" and peas at St. John Baptist (Midsummer). At Michaelmas came fresh herring, and old ewes fattened "and such old things." Those who were able added fowls, pigeons and birds of all sorts, kids and rabbits to the bill of fare. Brawn, we are told, was always the first dish at dinner from November to February, and fruit pies, and raw fruit were also a good deal eaten.

The farm-house breakfast consisted of "pottage" (made of liquor from boiled meat, with vegetables and oatmeal) with a morsel of meat, dealt out by the housewife. Dinner was at noon, and the family and the farm servants partook of it all together. It consisted usually of broth and bread, with brawn or some other dish, but not more than "three dishes well dressed." The housewife set about getting supper when the hens went to roost, usually about seven or eight, and served it

[1] Curtler: *Short History of English Agriculture*, 109, quoting Rutland Act, 1564.

"with good cheer," roast meat being expected by the ploughmen on Sundays and Thursdays.

Of "white meat," eggs were eaten most often roasted or poached, and "tyred with a lytel salte and suger." The home-made cheeses were of several kinds: "green" or new, with the whey not thoroughly pressed out; "soft," neither too new nor too old; "hard"; "spermyse," a kind made with curds and the juice of herbs; and "rewene," which was considered best of all. Both cow's and ewe's milk was drunk, and "clowtyd" cream made.[1]

Bread was of many sorts, but only gentry and townsmen ate the three best sorts of wheaten bread—manchet, cheat and ravelled: country folk ate brown bread, either of wholemeal wheat, or more usually of "meslin" or "miscelin," mixed wheat and rye, or rye and barley; only "yll people" mingled wheat and barley. The very poor, with their fare of eggs, cheese and vegetables, ate bread of rye or barley only, and as in mediaeval times were sometimes driven to making it of beans, peas, oats, acorns, tares or lentils.

Fresh, home-brewed ale, made of malt, water and yeast, according to Andrew Boorde, "for an Englyshe man is a naturall drynke," but should not be drunk less than five days old. The new drink, beer, made of malt, hops and water, was coming to be "moche used in Englande to the detryment of many Englysshemen." Cider and perry were common in Kent, Sussex and the West, metheglin (of honey, water and herbs) was drunk in Wales, and in Essex and other places there was a "kind of swish swash" made of honey and water "which homely country wives, putting some pepper and a little other spice among, call mead." Whey and buttermilk were also drunk. Andrew Boorde noted the liking of the Welsh for roasted cheese and metheglin, and the bad cooking of the Cornish, doubtless due to the fact that their chief fuel was furze and turf. Their ale was white and thick, "lyke wash, as pygges had wrestled dryn."[2]

Special feasts were not wanting. At Christmas they had "good bread and good drink, a good fire in the hall, brawn, pudding and souse, and good mustard withal," and also beef, mutton, pork, shred pies (mince), veal, goose, capon, turkey, cheese, apples and nuts, with "joly carols."[3] This is the first appearance of the Christmas turkey.

[1] Tusser, xxxvi, 191, 249, 259. Harrison, i. 150, 166; ii. 10. A. Boorde: Dyetary, 262-6.
[2] A. Boorde: Intr. of Knowledge, 122. Dyetary, 256. Harrison, i. 155 on.
[3] Tusser, 73.

Plough Monday after the Reformation became merely an opportunity for collecting money for conviviality, and there was a good supper in the evening. On Shrove Tuesday there was the entertainment known as "threshing the fat hen," followed by a collation of boiled fowl and bacon with many pancakes and fritters. In addition to the sheep-shearing and Harvest Home feasts already mentioned, at the conclusion of the autumn wheat sowing the workers were given seed cake, pasties and "furmenty."[1] On the feast of the dedication of the parish church there was a "wake-day," before which vigil used to be kept in the church all night, but it was later adjourned to the kitchen, where watch was kept before the oven which was filled with "flawns"[2] to be eaten next day. There were also village festivals such as "bride-ales," at which the goodman of the house found the drinks and the guests brought their own victuals, and all were "verie friendlie at their tables . . . that it would do a man good to be in companie among them."[3]

The standard of comfort in farm-houses of any prosperity had very much risen during this period, and Harrison has much to say about the "amendment of lodging." Formerly a mattress or flock bed was considered luxurious, the usual bed consisting of a straw pallet covered by a sheet, with coverlets of "dagswain or hopharlots," and a good round log for pillow; while as for servants, "if they had any sheet above them it was well, for seldom had they any under their bodies to keep them from the pricking straws that ran oft through the canvas of the pallet and rased their hardened hides." By Elizabeth's time, however, the prosperous farmer owned several feather beds and pillows, many blankets and coverlids, and even tapestry and silk hangings. Fine napery adorned his table, and instead of utensils of "treen" (wood) his cupboard was set out with a "fair garnish" of pewter,[4] and often a salt-cellar, wine bowl and spoons of silver. Glass was also becoming fashionable, and the poorer folk used vessels of home-made glass made of fern and burnt stone, but these were very brittle and consequently apt to "breed much strife toward such as have the charge of them."

[1] Or frumenty, a drink made of wheat and milk.
[2] "A sort of flat custard or pie." (Webster.)
[3] Harrison, i. 250. Tusser, 270 on.
[4] Harrison gives the composition of pewter as 30 lb. of kettle brass to 1,000 lb. of tin, and 3 or 4 lb. of tin glass. A "garnish" consisted of 12 plates, 12 dishes and 12 saucers. Op. cit., i. 147, 240-1.

They did not despise games either, for:

> The sturdie plowman, lustie, strong and bolde
> Overcommeth the winter with driving the footeball,
> Forgetting labour and many a grevous fall.

This sport, known also as "camping," was allowed on the meadows in December, as it was supposed to settle the roots of the grass and destroy the moss.[1]  Whipping-tops appeared among the juvenile population in spring as they do now.

The manifold activities of the farmer's wife and her maidens seem to have reached their climax during this period, for the standard of living was higher, and as yet many things were manufactured at home which a little later were bought or sent out to be done.  "Seldom does the husband thrive without the leave of his wife" was a saying of the time, and the farmer with a lazy or weakly partner must have laboured under a grave disadvantage.  Many departments were under her control, her responsibilities on a large farm would be considerable, and her opinion sought and deferred to by all.  The busy housewife rose early and swept the house, an operation which, according to Andrew Boorde, "ought not to be done as long as any honest man is within the precynt of the house, for the dust doth putryfy the ayre, making it dence."  Since she had no tea-leaves to help her in this respect her husband doubtless took care to be out still earlier.  Next she "dressed up her dishboard," in other words decked out her dresser with what the household possessed of pewter and earthenware dishes, tidied up, milked the cows, strained the milk and fed the calves, and then proceeded to "take up her children and array them."  Then there was breakfast, dinner and supper to get for the household and the farm hands, who all dined together, one of the children no doubt being put to turn the spit when there was roast meat twice a week.  Afterwards there were many dishes and pots to wash, though the housewife was exhorted not to "scour for pride" as it wore out the utensils.  The bread, butter and cheese were made at home, and also the various forms of drink, notably ale or beer.  The malt for this was usually made during the winter, of the best barley, steeped three days and nights, spread upon the floor to shoot, and lastly taken to the kiln to be dried by gentle heat with straw or wood as fuel.  During this operation frequent turning was needed, and the maid in charge was recommended to sing at her task to keep herself awake.  Some housewives ground their own malt with a

[1] Tusser, 64.  Barclay: *Eclogue V.*

quern, but many sent it to the mill to be done, measuring it carefully before and after the process. Brewing was an elaborate business, although the new drink, beer, in which hops were added to the malt, kept longer, and necessitated less frequent brewings.[1] In some places it was customary to make cider, perry, metheglin and mead, as we have seen.

There were clothes to be washed, with soap made of such undesirable materials as hogs' dung or other sorts of refuse, "than the which," we are told, "there is none more unkindlie savor." No wonder lavender and other sweet-scented herbs were needed afterwards. Fine linen and hempen sheets were becoming very general in farm-houses, and the housewife had to keep careful count of them and the blankets and coverlids, and mark and mend them. But besides this she still had to manufacture them from the beginning; it was her business to sow the flax and hemp, prepare the stalks when grown, and spin them into thread for linen and "hempen homespun," on a wheel which during the second half of the century was at length provided with a treadle. Similarly with the wool brought to her from their own sheep, she had to sort, card, dye and spin it. The distaff, observes Fitzherbert pleasantly, is always "ready for a pastime, and stoppeth a gap." One wonders when such a gap could have occurred. Perhaps the actual weaving was not always done at home, but certainly there would be the making up of the linen, hempen, and woollen cloth into sheets, "broad-cloths," towels, shirts, smocks (now first mentioned) and other garments. Knitted stockings now appeared, which would give the housewife occupation to fill yet more gaps. She used alder bark to dye them black.[2]

Hemp and rushes also had to be prepared for candlewicks, and the candles themselves made of tallow, in a tube-shaped mould.

Every morning and evening there were hogs and cows to be fed, besides the milking; the poultry had to be attended to and the eggs collected, and the needs of the household dog and cat satisfied. The goose feathers were carefully kept and made into feather beds, and in season there was much preparation of brine and salting of meat.

In summer there was much to be done in the stillroom, a department now coming to the fore in well-regulated households. Besides the herbs used for distilling sweet-waters, syrups and essences for flavouring cakes there were numbers used for

---

[1] For a detailed recipe for the making of beer see Harrison, i. 158 on.
[2] Harrison, ii. 242.

home medicines, and the good housewife had much lore in physic. There was water of fumitory for the liver, cool herbs for the ague, acqua composita, vinegar, rosewater and treacle "to comfort the heart," and many such decoctions. She and her maids also gathered the mustard seed for household use and dried it in an upper chamber before grinding. All sorts of seeds were carefully gathered, and exchanged with the neighbours.

But besides all these household occupations and the over-sight of the maidens, the busy wife was exhorted during March and April to garden from morning to night, at busy seasons to help make the hay, shear the corn and winnow it (in contrast to the more thoughtful mediaeval custom which left the house-wife at home), and if necessity arose to help her husband fill the dung-cart, drive the plough and load hay and corn. It was often her part also to take the dairy produce, poultry, pigs and corn to market, buy necessary goods for the household and render careful account to her husband.

All things considered she must have felt the force of Fitz-herbert's remark, "May Fortune sometime that thou shalt have so many things to do that thou shalt not well knowe where is best to begyn," followed by the sage advice to "leave that last which will best wait." Thankful indeed she must have been to lock the doors, "save the fire," and go to bed, at nine in winter and ten in summer.[1] One hopes that in that vigorous and light-hearted age she was able to face her life with the true spirit of adventure, and so carry it through.

[1] Fitzherbert: *Husbandry*, 93-8. Tusser, 247-69, etc.

# CHAPTER X
## THE SEVENTEENTH CENTURY

### REFERENCES

| | | |
|---|---|---|
| G. Markham, C. G. H. | = | *Cheap and Good Husbandry.* |
| G. Markham, E. H. | = | *The English Housewife.* |
| G. Markham, F. H. | = | *Farewell to Husbandry.* |
| Best | = | Henry Best: *Rural Economy in Yorks in* 1641. (Surtees Soc.) |

P

# CHAPTER X

## THE SEVENTEENTH CENTURY

THE seventeenth century was a period rich in new ideas for the furtherance of agriculture, but poorly provided with the means of carrying them out.

Towns were growing, and their markets were ready to absorb any surplus agricultural products, for immediate use, for storage in municipal granaries against scarce years, and even for export in times of plenty. The increasing demands of London caused still further developments; by degrees the corn supply of the country, instead of merely going to the nearest local centre for distribution, was drawn towards the capital and a single centralized market. Thus a metropolitan market was established which fixed the prices for local trade over a wide area, and farming, which at first had been a matter of providing sustenance for its own manorial group, and had advanced from that to supplying the markets of the nearest towns, now became a commerical enterprise promising considerable profit. By the time of the Restoration, London was able not only to obtain enough corn to supply her growing population, but even to organize an export trade and open up foreign markets.[1]

The adventurous spirit of the Elizabethan Age was late in penetrating to such a comparatively prosaic field of action as the farm-yard, but under the incentive of those new conditions it did succeed in reaching even the minds of agriculturalists. Land was taken up by men who meant to make farming pay as a business proposition, and the result was a flood of experiments and brilliant suggestions foreshadowing the real and practical advance of the following century; thus we hear of the Dutch idea of feeding animals on root crops during the winter—garden vegetables transferred to the fields—and rumours of artificial grasses to augment the hay crop, the use of new materials for manuring the fields, and fresh methods of agriculture which did not bear real fruit until at least a hundred years later. For many were the disadvantages of the age. The Civil War with its attendant evils checked the growing prosperity; the danger of pestilence, famine, or at least scarcity, was always present;

[1] Gras: *Evolution of the English Corn Market*, 123, 125, 256.

roads were mere mud tracks, and England, although utilizing tidal rivers and such others as were navigable, did not begin to improve them or to develop a canal system until late in the century. But the real stumbling-block to progress was the ever-present shortage of labour, by which all but the very small farmer, however desirous of carrying out up-to-date improvements, was hampered at every turn. England was still thinly populated, and in country districts especially little increase took place, because the surplus drifted to the unhealthy towns. Bad housing, smallpox, typhus, and other fevers, insanitary conditions, and the lack of milk, vegetables, and fresh food during the winter, were partly the cause of this; disadvantages which all fell heaviest on the family of the wage-earning farm hand. Most of the farm work was done by single men and women living in, and receiving food and clothing as part of their wage, and if, when married, they lived in cottages of their own, they fared badly after the rough but plentiful provision of the farmhouse, for the wages of the day labourer were not enough to support a family properly. The assessment of wages by Justices of the Peace, instituted under Elizabeth and most characteristic of this century, was probably designed to check demands for exorbitant wages, but rather defeated its own ends by fixing them at a rate disadvantageous to marriage and the natural increase of the population. From poor nourishment and lack of midwives infant mortality and the death rate in childbirth were appalling, and comparatively few children grew up into healthy lads and lasses.[1] Thus the supply of labour remained short, and it was not until the middle of the following century, when conditions improved and the population increased, that pioneer farmers were able to make real headway and profit by the suggestions of this period.

In the meantime, it was the small farmer who prospered, who was able to work his land with the assistance of his family, and was little dependent on hired labour. Fortunately these sturdy yeomen were a very numerous class, and in their little farms scattered all over the country lived a healthy and industrious life.

After the rage for rebuilding so characteristic of the sixteenth century most farmers probably found themselves well housed, and those who had occasion to erect new houses went on building in the same style for a considerable time. In the west and other wooded parts timber construction continued well into the eighteenth century, although in some districts

[1] Alice Clark: *Working Life of Women in the Seventeenth Century*, 73-87.

oak was growing scarce.  Worlidge, writing after the Restoration, advocated brick as the best building material, and recommended that the farm-house should be sheltered from the wind by trees, and should look out upon a flower garden, instead of the less pleasantly scented yard; a considerable advance in taste.  In the larger houses considerable development took place, the most marked change being that the hall, that essential feature of the house since early days, was losing its status owing to the multiplication of other living-rooms.  When for symmetrical reasons the doorway was placed in the centre of its wall, instead of being at one end, shut off by the "screens," comfort departed, and it rapidly became a mere vestibule leading to the other apartments.  But with this and subsequent alterations in the planning and appearance of large houses we are not much concerned; in farm-houses the great kitchen and living-room with its wide, open chimney and elaborate appurtenances of racks, spits and pot cranes long perpetuated the memory of the mediaeval hall.  The roof for the most part was of a single span, and the upper rooms in a line opening the one from the other, the traditional arrangement being that the master and mistress of the household should sleep in the strategic position at the head of the stairs, in the centre of the house, the daughters in the rooms on the one side, with the maidservants in the inner chambers beyond, and the sons and the menservants stretching away on the other side, so that all were well supervised.

Officials known as "chimney-peepers," whose business was to see that chimneys were properly swept and the thatched roofs not endangered, occur in court rolls of the seventeenth century, but not afterwards.  Towards the end of this period iron baskets began to be used in some fireplaces, marking an intermediate step towards the modern grate.[1]

The barns were built of much the same materials as the farm-houses.  Henry Best's great barn at Elmswell, built in 1607, was of timber framework, 126 feet long by 33 feet broad, the sides filled in originally with wattle and clay, but later with brick, and the roof thatched.  It was 27 feet high to the roof ridge, had interior props at intervals of 18 feet, and was furnished with large folding doors for the wagons to enter. The earth floor was made by raking and watering it thoroughly, and after a fortnight beating it smooth with broad, flat pieces of wood.  Best also had extensive "helmes" or cart-sheds into

---

[1] G. Slater: *English Peasantry and Enclosure*, 22 (Stratton and Grimstone, Dorset).  Gotch: *Growth of the English House*, 76, 143, 151, 282.

which his wains and carts were put after harvest for the winter. The "long helme" in the stackyard held four wains, two carts with the bodies taken off the wheels and put in sideways, and long ladders and wheelbarrows besides. There was also a "helme" in the foreyard for three wains and three "coupes" (four-wheeled carts). The wheels were raised on stones to keep them dry. There were still "hovels" in which to stack beans, and it was recommended that the ricks in the yard should be raised on frames to prevent the corn from being rotted by the damp or devoured by vermin. These frames were to have four pedestals (or six for a long rick) of wood or stone, 3 feet high and tapering upwards; squares of wood were placed on the top of each, connected by a strong framework, with smaller poles laid across to support the rick. If not thatched, scarecrows, dead birds or "clapmills" were placed on the tops of the ricks, but usually they were thatched carefully, and sometimes ashes and lime sprinkled over them to fly up into the faces of marauding birds. Thatchers were busy on the farm before haymaking and after harvest attending to ricks, houses and out-buildings, and in some places mud walls with thatched tops were still found as in mediaeval times. Two women helped the thatcher, one to draw out the straw into "bottles," and the other to serve him and temper the mortar. They were given three meals a day, at 8, 12 and 7, of butter, milk or porridge and cheese, with eggs, pies or bacon. The thatching straw instead of being piled up and watered every night was sometimes laid out in the yard for a while, so that "swine wrought in it," as this was supposed subsequently to keep away birds and vermin from the thatch. For merciful reasons thatchers stopped work about Martinmas (November 11th) "for it is an occupation that will not get a man heate in a frosty morning, sittinge on the toppe of an house wheare the winde commeth to him on every side."[1]

Another important building in the yard was the malt-house and kiln. It was considered most convenient to have this round, furnished with garners for the grain and cisterns under-neath to steep it in; other bins were provided for the dried malt, and a place for fuel near the kiln mouth. Markham states that the old form of kiln was square at the top with narrow strips of wood 4 inches apart laid across from the main beams, a structure very liable to catch fire from careless use; but the "French kiln" then in general use was made of brick or stone

[1] Best: *Farming Book*, 47 n., 107, 137-9, 146. Markham, F.H., 70-5, 94. Worlidge (I.W.): *Systema-Agriculturae*, 52.

which did away with this danger. In the West Country small households often had an arrangement by which the kiln was built at the back of the kitchen chimney, communicating with it by a square hole which could be opened or closed at will by an iron plate, thus utilizing the kitchen fire instead of a separate one in the kiln-house.[1]

As regards the farmer's *stock*, negligence in the breeding of oxen and sheep is still complained of in 1651, but *horses* fared rather better. Already, as we saw in the last chapter, the "misfits" and superfluous members of the "Great horse" breed were finding their way to the farm-yard, with beneficial results, but towards the end of the seventeenth century, after the Civil War and the final disuse of body armour consequent on the increased use of gunpowder, these heavy horses were no longer needed to carry the warrior and his equipment, and were turned to humbler uses. The proud, but of necessity somewhat lumbering, war-horse was set to draw coaches through the mire of the execrable roads, and thence descending to the cart and the plough became the ancestor of the mighty shire horses of to-day. The Old English black or "Great horse" had been bred with some care, stallions from Naples, Germany, Hungary, Flanders and Friesland being imported to improve the breed, which was not quite so large as that of the Continent. It was descended from the ancient large breed of Upper Europe, originally dun or white, but which, early blended with the North African horse, typically bay with white markings, produced a strain usually black or grey, but sometimes chestnut or brown.[2] Thus during the century under consideration, in horse breeding as in agricultural practice, the foundation was laid for the pioneers of the following century to develop.

But early in the seventeenth century sturdy useful animals were evidently fairly plentiful. Packhorses, says Markham, should be "exceeding strong of body and limbs, but not tall," with broad backs, thick withers to avoid saddle galls, and a long stride to get well over the ground at a walking pace. Horses for carts and ploughs were best of ordinary height that they might be easily matched, "of good strong proportion, bigge brested, large bodied, and strong limbed, by nature rather inclined to crave the whip than to draw more then is needful." He fed them on hay, chaff and peas, the last sometimes mixed with oat hulls or chopped straw, giving them "warm grains"

[1] Markham, E.H., 160-2.

[2] W. Ridgeway: *Origin and Influence of the Thoroughbred Horse*, 364, 373.

and salt once a week. Most farmers, notably Worlidge, kept
to the mediaeval practice of bleeding the horses in December,
but Markham doubled this and while young and strong bled
them twice a year, at the beginning of spring and autumn,
after which they had a week's rest and were given a
"comfortable drench," usually two spoonfuls of "diapente" in
a quart of strong ale.  In the heat of summer their heads were

FIG. 60.—TOPSELL'S HORSE

perfumed with frankincense and they were encouraged to swim
in the water.[1]

*Oxen*, instead of being referred to in the mass, now began
to be distinguished by the breeds raised in different localities.
Thus at the beginning of the century the long-horned breed
of Yorkshire, Derby, Lancashire and Stafford were considered
the best.  These animals were all black, with very large white
horns tipped with black, and were "of stately shape, bigge,
round, and well buckled together in every member, short

[1] Markham, C.G.H., 5-10.

joynted and most comely to the eye." Their cows gave particularly good calves.

The Lincolnshire cattle had horns "little and crooked," were pied with white, and had "tall bodies, long and large, lean and thin thighed and strong hoofed," and with admixture from the Continent were the ancestors of the modern shorthorn. They were considered the best for labour and draught. Somerset and Gloucester produced "blood-red" cattle, also referred to as shorthorned but later, evidently after the evolution of the true shorthorn, classed as "middle-horned," which were especially good for milking.

Draught oxen were still spoken of in 1669 as "worthy beasts in great request with the husbandman, the oxe being useful at

FIG. 61.—TOPSELL'S COW

his cart and plough."[1] They were fed chiefly on straw, sometimes mixed with hay.

The prevailing idea was that all cattle, even the dairy cows, should be as large as possible because their ultimate destination was the butcher, and the quantity of meat they produced was regarded rather than the quality. Markham considered that a good milker should have a crumpled horn and a thin neck, but the standard of milking capacity was not very high, for he states that 2 gallons at a milking (presumably 4 gallons a day) was rare, $1\frac{1}{2}$ gallons was considered good, and a gallon certain not to be found fault with. Milking times were between 5 and

[1] I.W.: *Syst. Agric.*, 150.

6 a.m. and 6 and 7 p.m.; some also milked in the middle of the day, but this was not recommended. The milkmaid was constrained to set about the business of milking with the tenderest consideration for the feelings of her charge, "to do nothing rashly or suddenly about the cow which may afright or amaze her, but as she came gently, so with all gentleness she shall depart." Her dairy was to be kept so clean that "a Prince's bed-chamber must not exceed it," but possibly in those days that would not be an alarmingly high standard to attain; the vessels were to be scalded and aired once a day, and the milk strained through a fine linen cloth. Butter was salted and preserved in earthenware pots, well pressed in, with salt at the bottom and the top; or in barrels, in which case the butter was pierced to the bottom with holes made by a small stick, and brine poured in. This was best done in May. As for the buttermilk which remained, the best use for it was "charitably to bestow it on the poor neighbours."

Some farmers took the newly born calves from their mothers at once and reared them "upon the finger" with skim milk, but others did not wean them until eight or ten weeks, or even let them run with the cows all the year. A writer of 1697 recommends that a cow should be "heartened" after calving with a pint of malmsey, the yolks of three eggs, and a pint of "sweet wort," which seems refreshment on a generous scale. The calves were put out to short sweet grass after Lady Day, first being accustomed to the light and liberty of the yard, so that they should not become dazed and run away.

We do not hear so much of "murrain" as in former times, but Worlidge described it as a swelling in the throat which choked the cattle.[1]

Like the oxen, *sheep* were coming to be known by the locality of their breed, but the quality of their wool was more considered than their mutton. In the former respect the small-boned, black-faced sheep of Hereford and Shropshire were the favourites, giving a "curious fine staple of wool." Worlidge states (in 1669) that those of the Herefordshire breed "about Leicester" bore the finest fleeces. Next came the Cotswold breed, and after them the large-boned sheep of the Midlands (Warwick, Leicester, Bucks, and parts of Worcester, Northants and Notts). These were of the best shape and their fleeces of the deepest staple but the wool was coarse. The

[1] Markham, C.G.H., 41-5. E.H., 140-8. A.S.: *The Husbandman, Farmer and Grasier's Compleat Instructor*, 3. Best, 117. I.W. (Worlidge): *Systema Agriculturae*, 100, 150.

largest sheep came from Lincolnshire, with their "legges and bellies long and naked," and their wool still coarser. Those of Yorkshire and the North were of "reasonable big bone" but with rough and hairy wool. The Welsh sheep were thought the worst, being little, and the wool of poor staple, for they were "praised onely in the dish, for they are the sweetest mutton."[1]

Markham and Best preferred both rams and ewes without horns, and long bushy tails were considered a good point, thin "candle-tayle" sheep being looked on with disfavour. Large and long-bodied rams (or tups) were favoured, with broad foreheads, and a "cheerful large eye." A good ewe had a "neck like a horse," large, thick shoulders and a broad back, "she

FIG. 62.—TOPSELL'S RAM

herselfe seeminge everyway rownde and full." Again these two writers agreed that she should have short legs, but "A.S.," writing in 1697, differed from them in advocating long, lean legs and smooth horns. She was considered to be in her prime from the fourth to the sixth shearing. Lambs were often timed to fall about Candlemas (February 2nd) as then they could be fattened for sale at Helenmas (May 3rd) but those on special pasture came later. A weak lamb was put in a basket in the chimney-corner, a good way from the fire, fed on cow's milk and covered over for the night. On open-field farms the ewes and lambs were put out on the low-lying pastures from Lady

[1] Markham, C.G.H., 65. I.W.: *Systema Agriculturae*, 150.

Day (March 25th) until folding time, which lasted from May till September, when they were folded principally on the land that was to grow rye or meslin.  The lambs were left with their mothers until Michaelmas unless it was the custom of the farmer to milk his ewes, when they were weaned at the beginning of May.  Sheep's milk and the cheese and whey made from it were still considered more nourishing than those of cows, but the practice of milking ewes was evidently on the decline.  Lambs were sometimes shorn during their first year, but it was considered better not to do so.

Henry Best usually washed his sheep during the first half of June and clipped them about the middle of the month, but he sometimes did it in May if the weather was favourable.  A fair, hot day was chosen for the washing, and the washers, standing in deep water and dealing with 120 sheep each in a day, were cheered by a "hot posset" of ale and milk with bread-crumbs and spice in it.  The loose wool was removed and carefully put in a "poake" by the shepherd.  Some days later the shearing started, a clipper doing 60 to 70 sheep a day, or an extra good hand 80 to 90.  Two fleece-winders were provided to five clippers, and there were two girls, one to collect the stray locks of wool, handfuls of which were given to the poor, and the other to keep the fire going under the tarpot with which the shorn sheep were marked.  The fleeces were stored in a room with a clean boarded floor and a good lock, piled in tens so that the tithe could easily be taken.  On this farm 3 lb. was considered a very good average weight for a fleece, 2½ lb. passable, and 2 lb. bad.  The ancient practice of pulling the fleece off by hand was now forbidden by statute, but had probably been long disused except in the most remote and backward spots.  The workers had ale, bread and cheese at noon, "and perhapps a chees cake," and when all was finished there was a dinner of five or six courses, of which roast lamb or mutton formed one.  It is recorded that Sir Moyle Finch at Watton (Yorks) had a piper for the occasion, who played to them upon the bagpipes all day for 6d.  Each shepherd on that estate received a bell wether's fleece in 1641 as in Saxon times.

After harvest the lambs (called hogs from their first midsummer to the next shearing time) were greased with a salve made of 2 gallons of tar to 8 lb. of tallow, thickened with a quartern of wheatmeal, to preserve them from scab and vermin. Weak lambs were housed at night during the winter, but most sheep seem to have remained out-of-doors, and were foddered with hay in hard or snowy weather when there was nothing to be picked up.[1]   Sheeprot, "which hath undone many an honest, simple man," was still prevalent, and remedies were sought for in vain.   The theory of Gabriel Plattes was that too much moisture dissolved the sheep's livers and made them "repleat with noxious and waterish humours," and that if they could not be removed to astringent salt marshes, where the rot did not occur, they should be fed through the winter on dry, hard hay.   The chief cause of the disorder he considered was immoderate rain in May or June, which produces a "frim and frothy grasse" very detrimental to the sheep.[2]

*Goats,* white, black, or "pyed," were still fairly generally kept in districts more suited to them than to sheep, chiefly for the delicate flesh of the kids, for their milk which was

FIG. 63.—TOPSELL'S GOAT

"esteemed the greatest nourisher of all liquid things whereon we feed, and the most comfortable to the stomach," and for their hair, because the ropes and other useful things made from it were said not to rot in water.   Sometimes it appears goats were kept in stables where there were a number of horses, mysteriously "to preserve them from many epidemical diseases."[3]

All writers of this period agree enthusiastically on the excellence of the *pig* as a food producer; and the variety of toothsome dishes he contributed were still the chief support of the farmer's family during the winter.   He was easy to feed, and in the dish was found "so lovely and so wholesome that all other faults may be borne with," and his "troublesome, noysome and unruly" manners were forgiven.   White pigs

---

[1] Markham, C.G.H., 66-69. Best, 1-30, 69-78, 94-7. A.S.: *Husbandman's Compleat Instructor,* 57.

[2] G. Plattes: *Discovery of Infinite Treasure,* 64-74.

[3] I.W.: *Syst Agric.,* 151. Markham, C.G.H., 82.

were considered the best breed, and those "all sanded" were well thought of: the pied variety were distrusted as "most apt to take meazels," although black-spotted ones were not rejected. Black pigs were unpopular and not common. Some improvement of breed must have taken place since mediaeval times, for at this time a good pig was described as being long and large of body, deep sided and deep bellied, with thick thighs and short legs; for the long-legged swine "but couseneth the eye, and is not so profitable to the butcher." Illustrations of the time show that they were still prick-eared but with a less high-ridged and bristled spine than the mediaeval species. The best pigs came from Leicester and Northants, owing to the abundance of pulse foods in those districts. Sows were expected to farrow three times a year, and twelve to sixteen piglings were common at a litter.[1] The usefulness of the pig in disposing of waste products and converting them into prospective bacon is dwelt upon, for, says Markham, "he is the husbandman's best scavenger and the huswife's most wholesome sinke, for his food and living is by that which would else rot in the yard and make it beastly." For the husbandman he cleared up pulse, chaff, barn dust, garbage and weeds, and for his wife "draffe" (the sweepings of the malthouse), swillings, whey and tub-washings. As the wood-lands decreased, less is heard of driving the swine out to mast in the autumn, but where available this was still done for six or eight weeks, and the fattening was afterwards completed in the sties on dry peas or beans. In other places they were fed at that season with hips and haws, sloes and crab-apples from the hedges, the poor people gathering a store of these for winter feeding. In Herefordshire particularly they were fed on elm-leaves, gathered up in bags. Red ochre mixed with their food once a fortnight was said to keep off the measles. They were allowed less liberty than of old, except where there were suitable feeding grounds, as they were thought to lose flesh with too much running about, and it was therefore more economical to keep them "penned in some court," both for quicker fattening and for the collection of manure. Blith, in the middle of the century, advocated a hogyard with troughs outside the fence, and holes through which the hogs should put their heads to feed, and so avoid scattering the food about and wasting it. Complaints were made in some parts that poor

[1] Markham, C.G.H., 87-97. A.S.: *Husbandman's Compleat Instructor*, 85-7.

men fed their pigs on young fish from the fish stews, thereby endangering the supply of fresh-water fish.[1]

*Asses* and *mules* were little kept, although it was pointed out that they were hardy animals, easy to feed, "enduring much labour and travel, and able to plough on fairly light ground." One objection put forward to the ass was that he destroyed trees by eating the bark.[2]

*Dogs* were found, as ever, very necessary servants on the farm, both to guard the premises and help the shepherd, although with the total disappearance of wolves size and strength was not so important in the sheep-dog. They were now fancied "rather long than thick" to favour speed, and were guided by the whistling, hissing or waving of the shepherd. They were preferred of a brindled or yellowish colour, and their tails were cut at about a month old.

The bandog or house-dog was still usually a brindled mastiff with a great head and a thundering voice, "his face representing Terror like that of a Lyon." Smaller and more alert dogs were also sometimes kept, to rouse the slower mastiffs by their yapping.[3]

The farm-yard *cat* was "of divers colours but for the most part gryseld, like to congealed yse," but judging by a contemporary illustration she was not unlike the common English tabby of to-day. It was already considered "if she put her feete beyond the crowne of her head that it is a presage of rain."[4]

Large numbers of *rabbits* or conies were kept, and were evidently profitable, particularly for their fur. Those having silver hairs mixed with black were now more popular than the plain black, but the latter were still preferred to white, pied, yellow, dun or grey. Many of the best skins were exported. Rabbits were also much used for food, being specially in season from Martinmas to Candlemas, when they would help out the monotonous winter fare of salt meat. The ordinary sort kept in warrens had hay and branches scattered there to help them in winter, but sometimes suffered from rot like the sheep. "Rich conies," i.e. specially fatted ones, were kept in small hutches of which the description has a very modern sound; they were 2 feet square and 1 foot high, divided into two rooms, of which the smaller one was dark, and the larger with "open

---

[1] S. Hartlib: *Legacie*, 98. W. Blith: *English Improver Improved*, 146. A. Speed: *Adam out of Eden*, 101. I.W.: *Syst. Agric.*, 151.
[2] I.W.: *Syst. Agric.*, 149. S. Hartlib: *Legacie*, 95. A.S., op. cit., 119-27, 135. E. Topsell: *Historie of Fourfooted Beastes*, 173.
[3] I.W.: *Syst. Agric.*, 151. A. S.: *Husbandman's Compleat Instructor*, 135.
[4] Topsell, op. cit., 103, 105.

windowes of wyar," through which the cony could feed from
a trough placed in front. These large, fat conies were considered worth five ordinary ones, their skins being worth 2s.
or 2s. 6d. instead of 2d. or 3d., and their flesh much superior.
They were fed on oats and hay, presented to them in cleft
sticks to avoid waste, and greens two or three times a fortnight,
and their boxes were cleaned out daily. The cony is described

FIG. 64.—TOPSELL'S CAT

as "a very timorous and frightful creature, and naturally subject
to melancholy."[1]

Cottagers and poor folk found *poultry* particularly profitable
as they were able to shift for themselves on any common piece
of ground during the greater part of the year, and they were
valued on every farm both for the market and for their droppings, which were considered a valuable manure. The favourite
colour for a cock had changed from yellow to red, "and if he
be a little knavish he is so much the better," but at the end of
the century he was preferred red "mixed with yellowish and a

[1] A.S.: *The Husbandman's Compleat Instructor*, 142. Markham, C.G.H.,
100-2. Harrison, ii. 271. (Fynes Morrison's *Itinerary*.)

sprinkling of black." The best hens were also red, with a high, thick tuft of feathers on the head. "The dung-hill Cock," says Markham, "is a fowle of all other birds the most manliest; stately and maiesticall, very tame and familiar with the man"; and the hen "should be valiant, vigilant and laborious," both for herself and her chickens. It was considered that poultry prospered better wandering loose, and not confined in yards or paved courts, for the cock "delighteth in open and liberall planes where he may lead forth his hennes into greene pastures and under hedges." But for the market and the table, hens and capons were fattened in pens or dark houses, on blood and bran, boiled root vegetables and bread ends soaked in beer or skim milk, or sometimes crammed with balls of wheat or barley meal and milk. To make hens lay well they were given buckwheat or hempseed. In 1633 eggs in London were three a penny, a considerable rise from the mediaeval score for a penny. For rearing chickens hens were set preferably in February, in the increase of the moon, on nineteen eggs, marked on one side to make sure that the hen turned them. The nest was perfumed with brimstone or rosemary, and later the chicks also with the latter, as in Elizabethan times. After two days the chicks were fed on oatmeal or white bread-crumbs, soaked in milk, and kept in the house a fortnight. Markham, writing early in the century, has nothing to say about incubating, but Stevenson, in 1661, states that "some will set eggs in stows or ovens in winter," but that they were little or no use in England and the chicks never good or profitable. Worlidge, however, a few years later, was a believer in the idea, and declared "you may with facility hatch three or four dozen of eggs in a lamp-furnace made of a few boards, onely by the heat of a candle or lamp." This seems a step nearer the modern incubator than the earlier oven-hatching, a practice which he ascribes to the Egyptians. He arranged for these chicks to hatch out at the same time as broods hatched in the natural manner, so that one hen, put with the chickens in a warm room, could mother a large number.

Markham considered the best way to preserve eggs was to bury them in a heap of old malt, as he said salt caused them to "waste and diminish," and straw or bran made them musty. His henhouse, like Mascall's had an eastern window, strongly latticed, perches aloft for the fowls and turkeys, with pins in the wall for them to climb up by, and pens on the ground for the other fowl, surmounted by hampers of straw for the hens to lay in. He recommended that the henhouse should be near the kitchen, brewhouse or kiln, "where it may have ayre of the fire and be perfumed with smoke, which to pullen is delightful and wholesome,"—a taste their ancestors doubtless acquired when they roosted "o'er the beam" in the mediaeval hall. One cock was kept to ten or more hens.[1]

White or grey *geese* were considered the best, pied not so good, and black worst. They were eaten at two seasons in the year, after harvest, presumably about Michaelmas, when the "elder" geese were fed on the stubble and afterwards fattened on oats, beans and barley-meal, and at Whitsun, before which the young or "green" geese were fed up for a month in pens or dark houses, on ground malt or oats and milk, or carrots chopped small. Worlidge put forward the idea that darkness conduced to the fattening of any creature, and described how Jews, having wrapped the goose up in a linen apron and stopped her ears with peas, to prevent any distraction from sight or hearing, fed her three times a day with pellets of ground malt or steeped barley. In some places geese were shorn of their feathers, but those to be used for quills were not taken until moulting or killing time.[2] *Ducks* were as universally kept as ever, being a good dish fattened like geese and very economical to feed, on "corne lost or other things of lesse profit."

The flesh of the *turkey* was so much appreciated, "either from the spit or in paste" that his defects were borne with, for he was destructive to the garden, devoured more corn than he was worth unless he had a large range on which to pick up food, and was very delicate, "being an extream chill bird." Turkey-cocks, "large, stout, proud and maiesticall," were yet considered melancholy fowls, on account of their doleful cry and their apparent anger at the sight of red colours, "being possest with a strong conceit that they are mocked."[3]

*Swans* and *peacocks* were by this time kept "for their beauty and magnificent Deportment," more than for the larder. The adult birds seem to have rather gone out of favour on the table, but fatted Cygnets were still considered a noble dish at great entertainments, and young peacocks were eaten. The chief use of the peacock, "a Bird of Understanding and Glory," was to keep the farm premises clear of snakes, adders, newts and toads.

[1] Markham, C.G.H., 110-21. I.W.: *Syst. Agric.*, 152. A. Speed: *Adam out of Eden*, 115. A.S., op. cit., 149. M. Stevenson, *Twelve Moneths*, 13.
[2] Markham, C.G.H., 122-4. I.W.: *Syst. Agric.*, 153. Harrison, ii. 273. (Fynes Morrison's *Itinerary*.)
[3] I.W.: *Syst. Agric.*, 153. Worlidge, like earlier writers, refers to them as "Turkeys or Guiney-hens or cocks" and seems to confuse the species. Markham, C.G.H., 125-6.

Q

Where Henry Best lived in Yorkshire a male swan was called a "cobbe" and a female a "penne," and if the two belonged to different owners the brood was divided between them, one cygnet going to the owner of the land on which the nest was built. Men called "swanners" collected the young birds about midsummer, marked them and cut the joint of the right wing, and they were brought home at Michaelmas to fatten like geese. Fat swans were sold at 10s. a pair, and unfattened at 5s.[1]

We are assured early in the century that no kingdom in the world had so many dove-houses, so that evidently the popularity of *pigeons* had not decreased. They only needed feeding in bad weather, when they were given dried peas, clean water, and sometimes a lump of salt. Careful farmers whitewashed the inner walls of the pigeon-house and cleaned it out once a week, the droppings being much valued as manure, and gave the birds dry sand and gravel in which to clean themselves. Worlidge somewhat unscrupulously recommended that the pigeon-holes should be washed with asafœtida boiled in water to scent the pigeons, so that wherever they visited, others "will be so well pleased therewith that they will bear them company home, to the great encrease of your stock."[2]

Partridges and pheasants were often reared under hens, and fed when young on ants' eggs. Best put his young partridges out under nut trees in the orchard, and at a month old cut a joint of one wing to prevent their flying away. They were fattened in a house with little boxes in the corners for them to hide in, and wheatsheaves and pans of water in the middle. Quails and many other wild birds were caught and fattened in cages for the table.[3]

Everyone was exhorted to keep *bees*, for they "aske nothing but an house rent-free to dwell in, and when they die they bequeath their riches to their landlords." Markham declared the bee "a creature gentle, loving and familiar about the man which hath the ordering of them," unless he had a strong, ill-smelling savour about him, when they were "curt and malicious." Hives were still commonly made of straw wreaths bound with bramble, or often of wattle, covered 3 inches thick on the outside with a plaster made of cow-dung mixed with sand or lime. If further covering against heat and cold was needed the straw hives were given a covering of this plaster, which dried entire so that it could be lifted off in one piece; and the wattle and daub

[1] Best, 122. Markham, C.G.H., 128-9. I.W.: *Syst. Agric.*, 154.
[2] I.W.: *Syst. Agric.*, 154. Markham, C.G.H., 130-1.
[3] Markham, C.G.H., 132-4. Best, 109. S. Hartlib: *Legacie*, 114.

hives had a threshed wheatsheaf with the ears tied together fitted over the top and held down by a hoop. They were sometimes "comely made of firboards" or other wood, octagonal, or round and hooped, and these could be furnished with shuttered glass windows through which to watch the proceedings within. Experiments were even made with hives entirely of glass. New hives were rubbed inside with balm and fennel or other sweet herbs, dipped in cream. They were placed on benches in a sheltered spot, in or near the orchard, preferably under some sort of roof or pent-house, and some farmers who had no convenient place out-of-doors set them in the lofts or upper rooms of their houses and let the bees go in and out through the window. Levett, the best authority of that century on bees, liked to face the mouths of his hives south, or failing that, west, but some liked them turned south-east, to catch the sun earlier in the morning. When the bees swarmed it was customary to "play them a fit of mirth upon a pan, kettle, bason or such-like instrument to gather them together," but Levett looked upon this as "a very ridiculous toy and a most absurd invention," the bees preferring peace and quiet. They were fed in winter, if necessary, with honey and rosewater, "sweet wort" from the brew-house, toast sopped in ale, or dry bean-flour. The washings and "offal" of the hives were used to make mead.[1]

Turning to *agriculture* we find abundance of experiments, both with new implements and new crops, which, even if they made little headway at the time, at least paved the way for future improvements in the same direction.

By the middle of the century a considerable variety of ploughs was available: there was the cumbrous old double-wheeled plough, of which the wheels, notably in Hertfordshire, were sometimes 18 or 20 inches high, the furrow wheel being in some cases larger than the other. Secondly there was a single-wheel plough, which would "admit of more lightness and nimbleness," and with this implement, made neat and small, it was declared possible for a man with one horse to plough an acre a day, instead of utilizing two men and a large team of beasts. There was also a "plain plough" without wheel or foot, and a sort called a "Dutch bastard," which is not described but seems also to have been wheelless. But besides these more ordinary varieties a double plough was known to Blith, which made two parallel furrows, and another was described by Hartlib about the

---

[1] S. Hartlib: *Legacie*, 151-5. Best, 61-7. I.W.: *Syst. Agric.*, 158-71. W. Lawson: *Country Housewife's Garden*, 75-81. J. Levett: *Ordering of Bees.*

same time in which the two blades were set one behind the other at different depths, in order to cut a single furrow 12 or 14 inches deep. In Norfolk a plough was used with a harrow attached behind, but this necessitated some contrivance for sowing at the same time, or else the seed was sown before and ploughed in. Worlidge adds to this list the double-wheeled plough called a "Turnwrest," noticed in the last century by Fitzherbert. There was also a turfing plough with a horizontal blade, to pare off the surface of poor or rough land preparatory to burning it to enrich the soil.[1] Markham still considered oxen best for drawing the plough, especially in rough and hilly country, because of their strength and endurance, as well as their usefulness to the butcher later on, but he admitted that a horse team was faster, doing $1\frac{1}{4}$ or $1\frac{1}{2}$ acres a day where an ox team did but 1 acre. He used four to eight oxen in a team, according to the nature of the soil; and in steep and rocky counties such as Cornwall, "where carts and wagons cannot go"—or at any rate did not in those days—he recommended that the oxen should be kept entirely for ploughing, so that the horses might be free for the carriage of hay, manure, corn, fuel, etc., on their backs, by pot, pannier, hook or pack saddle. The primitive and barbarous practice of attaching a light plough to the tail of the horse for purposes of traction was now made illegal by statute, although it had probably been long disused except in Ireland. Worlidge stated that he had heard of ploughs drawn by "Mastive dogs," and the promise of some driven by the wind, but he did not pay much heed to them.[2]

The four ploughings of the fallow advocated by Fitzherbert and Tusser were still customary, and even a fifth was sometimes added; cross-ploughing, i.e. with the furrows running at right angles to the former direction, being advocated for some of them. The latter, of course, was an improvement only possible on enclosed lands, and not on the narrow strips of the open fields. Ridging up the fields was still recommended, not only that the water might run off into the deep furrow, but because it was thought that the height and roughness of the ridge served to "break the fleeting winds" and temper them to the winter crops.[3] The ploughman was expected to rise at 4 a.m. and see to his horses or oxen, feed and rub them down and clean out their stalls. After an interval for his own breakfast at 6, he

[1] I.W.: *Syst. Agric.*, 206-7. W. Blith: *English Improver Improved*, 198-220.
[2] I.W.: *Syst. Agric.*, 207. Markham, F.H., 117-21.
[3] I.W.: *Syst. Agric.*, 32.

started ploughing at 7 and continued until 2 or 3 o'clock. The cattle were then unyoked, rubbed down and fed, and the ploughman had an interval of half an hour for dinner, after which he returned to his beasts, cleaned out their stalls and foddered them again. That done he went to the barn and

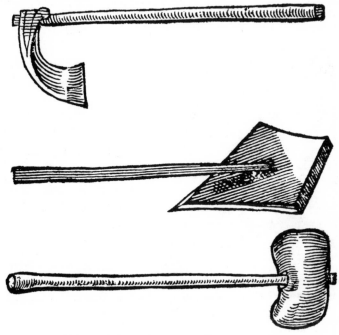

FIG. 65.—IMPLEMENTS FOR CLOD-BREAKING

prepared fodder for the next day, and went in to his own supper about 6. Until 8 o'clock he was occupied indoors by the fireside at a variety of tasks, such as mending shoes, knocking hemp and flax, picking and stamping apples, grinding malt on a quern or peeling rushes for candle wicks. At 8 he took his lantern and fed and littered down his cattle for the night and went to bed.[1]

Heavy clay fallow, after ploughing, was sometimes broken up with iron hacks or mattocks, or beaten to a smoother surface with beetles or mauls, and ground for barley was often harrowed and rolled to a fine surface. Rollers were made of solid rounded tree trunks.

[1] Markham, F.H., 115-16.

Seed corn was selected by dropping it into water and taking the heaviest grains, which sank to the bottom; or by rubbing the grain from the middle of the ear by hand. A new idea that seemed popular was to steep the seed in "liquor from the dung-heap," brine, or the lees of wine, beer and other drinks, as a simple and easy way of manuring it. It was thought that birds were less likely to eat it when sown, and that if swollen and slightly sprouting it grew quicker.[1]

FIG. 66.—MAXEY'S METHOD OF SETTING CORN. 1601

Best changed his barley and wheat seed every four or five crops. Sir Hugh Platt's contrivance for setting corn by hand by means of boards set with pins to drill holes in regular rows, we described in the last chapter, and this practice was adopted in some few places of England during the first ten years of the century, notably in Middlesex. It was much advocated by Maxey, who considered that an acre set in this way and heavily dunged gave three or four times as much yield as one sown in the ordinary way; half a bushel of seed being used instead of 2½ bushels.[2] Gabriel Plattes attempted to improve on this, in 1639, by inventing a machine by which the holes were still drilled by the workman's weight, but by an elaborate arrange-

[1] Markham, F.H., 12, 69. G. Plattes: *Discovery of Infinite Treasure*, 37. Sir Hugh Platt: *New and Admirable Art of Setting Corn*, ch. 6 and 7.

[2] Ed. Maxey: *A New Instruction of Plowing and Setting Corn.* Harrison, iv. 290. (*Coryat's Crudities.*)

ment of sockets and perforated boards the seed ran into the holes
thus made from above. A second "engine" was to follow it to
"lay the earth neatly in little furrows," but this does not seem
to have materialized. The first machine met with little favour,
for, says a writer of 1651, "I know there are many difficulties in
it, which he himself could never wade through."[1]  Both this
author and Worlidge, in 1669, considered that it was better to
hoe in corn by hand than set it in this laborious manner.
Worlidge gives an illustration of a new corn-drill, a machine on
wheels with a coulter in front, to cut a little furrow in the carefully prepared land, and a hopper behind from which the seed
fell into the furrow, and was covered over, as by a harrow, by a
piece of wood fixed at the back of the machine,[2] but apparently
this machine never actually came into use. But in spite of these
very advanced ideas the ordinary husbandman continued to sow
his corn broadcast in the old manner; 2 to 3 bushels an acre
being the usual amount sown.[3]  To protect the seed when sown
it was suggested that it might be sprinkled with the lees of bitter
oil, but as a rule a boy with his bow and arrows followed the
harrow as of old, "making a great noise and acclamation."
Thorough-going farmers kept on these "field-keepers" every day
from an hour before sunrise to half an hour after sunset until
the corn was well sprouted, "and though the endurance may
promise much pain and trouble, yet questionless the labour to
any free spirit is both easie and pleasant." We will hope that
the said boys were sufficiently "free spirited" to agree. Some
were allowed to have a musket or "Harquebush," which doubtless delighted their hearts, although no bullets were provided.
Some farmers instead of employing this cheerful company
stretched lines of pack thread with feathers "knit in" across the
fields, but one can hardly imagine this could have been done
over a wide area.[4]  In some places sheep were driven over the
newly sown fields to manure them.

Maxey seems to have been the first to suggest that wheat and
rye could be sown in the spring, on rich land, as well as in the
autumn according to immemorial custom; and this idea seems
to have been taken up as regards rye, for both Markham and
Worlidge include a sowing of March rye in their calendars.
Spring wheat was not adopted until a much later period.

The crops were harvested with the same old implements,

[1] S. Hartlib: *Legacie*, 8.
[2] I.W.: *Syst. Agric.*, 47-9.
[3] Best, 115. E. Maxey, op. cit.
[4] Markham, F.H., 74-5.

the saw-edged shearing hook for wheat and rye, the scythe, with cradle attached, for oats and barley, and the hook for peas; the inventive brains of the period not having turned in that direction.

According to Sir Hugh Platt, in 1601, a yield of 4 quarters of wheat per acre contented the ordinary farmer, and indeed might do so now, but he goes on to state that many obtained 6 quarters, a few 7; and if 10 or 12, then "hee acknowledgeth himselfe to have receyved an extraordinary favour and blessing from the Heavens,"—as well he might. Norden recorded that the careful husbandry of the West produced 4 to 10 quarters of corn per acre, which seems remarkably good, and the experimenters in new methods of cultivation would of course have us believe that they could produce a good deal more. Maxey, wishing to depreciate the usual method, said the ordinary yield was 2 to 2½ quarters, the same figures that Harrison gave, late in the previous century, and the truth probably lay between his too low figures and the very optimistic ones of Sir Hugh Platt.[1]

When the last of the grain was gathered in and all the peas pulled, there was the usual Harvest Supper, which on Henry Best's farm consisted of puddings, bacon or boiled beef, meat or apple pies, "and then creame brought in platters and everyone a spoone," followed by hot cakes and ale. Some, we are told, cut up their cake and put it in the cream, to make "creame-potte or creame-kitte."

Best sent his corn to market about Martinmas, on the backs of packhorses, in strings of four or six fastened head to tail, and the unfortunate drivers in charge had to rise about 3 a.m. to reach the market town by 9 o'clock. Corn for the household use was sent to the miller to be ground as needed, a "cadger" calling for it and bringing it back, and if one bushel of corn did not produce 6 pecks of meal the miller was changed. The miller got a ¼ peck of corn from each bushel as his due, and the fine flour was sieved at home.

Corn in the granary was considered liable to heat, and some farmers mixed chaff, beans, iron, or stones with it to prevent this; another expedient being to run it through a small hole in the floor into another chamber below, like the sand through an hour glass.[2]

Peas, beans and vetches, and sometimes oats, rye and meslin or mixed corn, were kept in closed casks. Beans were dried in

[1] Sir Hugh Platt: *New and Admirable Art*, etc., ch. 8.    T. Norden: *Surveyor's Dialogue*, 232. E. Maxey, op. cit.

[2] I.W.: *Syst. Agric.*, 50. 93-100. Best, 93, 100, 103, 105.

the kiln if threshed before March, and oats were similarly treated before being ground for household use; the oatmeal was then kept in closed casks or garners, well trodden in, "as hard as you can."

The crops evidently suffered a good deal from smut and mildew, and to prevent the former it was recommended that the land should be sprinkled with chalk or lime before sowing; "sooty" corn was washed, dried in the kiln and well aired, and could then be used, but not for seed. It was supposed that dew had a bad effect upon the corn if not washed off by rain, and hence, on mornings of heavy dew, two men were instructed to walk up and down the furrows with a line stretched between them knocking the dew off the ears. Best mentions a disease called "slaine," similar to smut but more tenacious.[1]

Treatment of the soil with various sorts of dressings made great strides during this century. More distinction was drawn between different sorts of soil and their appropriate treatment, notably by John Evelyn, who pointed out in his *Terra* that each sort of earth must be improved by applications of a contrary nature; thus if wet and heavy it should receive a deposit of sand or ashes, and if cold and dry a dressing of hot, moist compost was best. The general idea seemed to be that clay lands should receive sand and lime to improve them, and light sandy ground a dressing of marl. The use of this time-honoured substance, "a very gallant thing" for the fields and "as strong and cheerful as ever it was before," was much revived during this period, and was considered to benefit the soil where it was laid for ten or fifteen years. Chalk, lime, fuller's earth, sea sand, river soil and stagnant mud were also applied where suitable, and even the ordinary farm-yard manure was apportioned with some care, swine's dung being considered the coolest to the hot lands, and pigeon and poultry droppings, having most heat, to the cold soils. Cow-dung was still used by the very poor as fuel, but the products of the stable, byre and sheephouse were considered suitable for any sort of land, and were spread upon the arable fields twelve or fourteen loads to the acre and immediately ploughed in. Speed recommended a brick-enclosed manure heap, with a paved floor, that the liquid might drain off into a tank; and Evelyn advocated the careful making of "compost" in pits prepared for the purpose. The pit he described was long and deep, paved or lined with chalk and clay and covered from rain, and into it drained all the channels from the outhouses. Stable litter was

[1] G. Plattes, op. cit., 60. Markham, F.H., 81-5. S. Hartlib: *Legacie*, 15, 18. Best, 99.

laid in the bottom and covered with mould and successive layers of rotten fruit, garden stuff, pigeon and poultry droppings, sheep's dung, earth, cow-dung, ashes, soot, fern, sawdust, ditch scourings and anything else that came convenient, heaped up to the top of the pit. It was then left to mature for two years, after which the layers were separated and mixed with fresh earth until it became "comparatively sweet and agreeable to the scent." He also had a pit for liquid manure, under a pergola or covered shed, into which were cast rank weeds, acid plants, cabbage stalks and all manner of savoury morsels, mixed with pond water, the liquid from the pickling tubs and other matters not even "comparatively sweet." We hope it was in a retired spot. Speed used a handy pond for the same purpose, and having stirred in marl and added manure and a variety of sweepings, drew the liquid from it into barrels and took them on a cart to his fields, where it was distributed by an engine "such as is used in London when houses are on fire"—presumably a kind of pump and sprinkler. Others recommended a per-forated hogshead for sprinkling brine over the ground.[1]

Besides these more obvious dressings a large variety of new substances were laid upon the fields to enrich them: rags, wool-clippings, hair, horn shavings, seaweed, some kind of shelly soil called "snayle cod," bones, oyster shells, malt dust, oat hulls, fern, the rotted bottoms of hayricks, the grounds of the malt vat, the dregs of beer, ale and wine, meat and fish broths, and any fatty or oily liquor that was left over, were all pressed into service, as well as leaf mould, bark, rotten vegetables, waste soap ashes, saltpetre, the ashes from iron ore, "sal armonaike," offal of beasts, fish bones, fish that was past its best, and rotten pilchards after the oil had been extracted. Many and strange must have been the country scents in those days. The unfor-tunate "spreaders of muck" were mostly women, boys and girls.[2]

A method of preparing rough and barren ground for culti-vation was to pare off the surface, pile it into heaps and burn it, with whatever heath, bracken or bushy stuff grew on it, then scatter the ashes and plough them in. This in many places seems to have had a beneficial effect.

Enclosed farms naturally had a great advantage over the open-field ones in the fact that the arable and pasture did not need to be permanently separated, but could be changed over in their uses, that each might have a rest, and it was naturally

[1] J. Evelyn: *Terra*. A. Speed: *Adam out of Eden*, 122-30.
[2] Best, 140. W. Blith: *English Improver Improved*, 147. Markham, F.H, 48. Sir Hugh Platt, op. cit., ch. 6. I.W.: *Syst. Agric.*, 64-9.

on these farms that those improvements were adopted which
the slow-moving farmers could be persuaded to try. / Interest
was now aroused in the draining and irrigation of the land.
Blith in 1652 advocated drainage by deep trenches filled at the
bottom with faggots or large stones and turfed over until
shallow enough not to endanger the cattle. Worlidge mentions
a special trenching plough for this purpose, with a long, sharp
blade to make a deep cut preparatory to digging the trenches;
and Blith gives three varieties. Trenching spades had either
one or both sides turned up into a fin, while those for turfing
were shaped like the spade on a pack of cards. Ordinary spades
according to Blith's illustration, still had only the lower part of
the blade made of iron, in the mediaeval fashion, and were
either square or rounded.

Irrigation of the meadows and pastures was recommended
where possible by "drowning" or flooding them during the
winter months, and in 1669 it was stated that "this is of late
become one of the most universal and advantageous improvements in England within these few years." It was accomplished
most easily by directing a stream over them, by means of dams;
or otherwise by pumping the water up into a system of trenches,
which, branching out from the main trench, grew more
numerous and shallow until the water was thoroughly distributed over the ground, a counter system of drainage trenches
at the bottom of the field carrying the water off again when
required. Gabriel Plattes in 1639 wished to utilize the Persian
water-wheel, which raised water in earthen pots emptying themselves into a trough on a higher level; and Blith also com-
mended this idea. Worlidge gave an engaging illustration of a
"Persian Wheel," which, however, raised the water by blades,
not pots; and he, like Blith, also depicted a little windmill used
for the same purpose very much of the same appearance as
those used at the present day. There is, however, little reason
to suppose that these devices came into general use, and he
complained that much water was occupied by water-mills which
might have been used for profitable irrigation.[1]

Ordinary haymaking took place in late June or early July,
and "in sundry places" there was a second mowing later in
the year, but this was not common. Henry Best's hay barn
or "leath" held twenty-six loads well trodden down and packed
so close up to the ridge that "a catte can hardly goe betwixt the

[1] G. Plattes: *Discovery of Infinite Treasure*, 32. Markham, F.H., 65-6.
W. Blith, op. cit., 20-68. I.W.: *Syst. Agric.*, 16-21.

hey and the ridge of the howse." The rest was stacked on the grounds where the sheep were to be fed during the winter.[1]

Some experiments were tried with carts. Greater steadiness was wished for on the exceedingly uneven roads, and in order to lower the body of the cart for this purpose, and yet retain the advantage of the large wheels, which caused less joggling than small ones, the bed of the cart or wagon was sometimes sunk below the level of the axles, with the tail turned up over the hind axle. A three-wheeled cart drawn by one horse is also mentioned.[2]

Farm servants were hired just before Martinmas at the fairs, and the rates of wages for the district were given out publicly at the same time. Three classes of servants appear in these assessments. First there were the regular farm hands who lived in and received a certain yearly wage, with sometimes an allowance for "livery." These were usually the bailiff (on a large farm), the foreman or "chief hind," the ploughman and carter, each of whom commonly received £4 to £5 a year; the shepherd, receiving £3 to £5 according to age and capacity; and the "common hind" and other assistants £2 to £3.[3] A foreman was expected to be able to take charge, sow, mow, stack peas, "go well with your horses," and be used to marketing and the like.

Secondly there were the day workers, who sometimes received food and sometimes not, their wages varying accordingly, and less being given in winter than in summer, which can hardly have been helpful to a family budget. A common labourer could sometimes get as much as 10d. a day without his food, or half as much with it. Lastly there were the "task workers" paid by the acre or the piece, according to the nature of their work. Extra harvest work, hedging, ditching, repairs to ploughs and carts, and other jobs were done by these men; also all the operations necessary for the wood-pile, so important in every farm-yard, the heavier wood being dealt with by a "pair of sawyers" at fixed rates.

A noticeable point is that many farm workers received as good wages as skilled artisans, such as masons, carpenters and

---

[1] Best, 36. Harrison, iv. 310. (*Coryat's Crudities.*)

[2] I.W.: *Syst. Agric.*, 208. Speed, op. cit., 85.

[3] Wages varied a good deal in various counties, being particularly high in Suffolk and much below the average in Derby; but they rose everywhere as the century advanced. *Quarter Sessions Records*, Somerset, Devon, Wilts, etc. Cox: *Derbyshire Annals*, ii. Thorold Rogers: *History of Agriculture and Prices*, vol. vi.

plumbers, which shows how highly the scarce agricultural labourer was valued.

Women were hired by the day for mowing, reaping and hay-making, at rates a good deal lower than men, and in some parts clipped and tended sheep, spread manure and weeded corn.[1]

A ploughman with one team ploughed 1 acre or 1½ acres a day on stiff ground, or more on light soil. A good mower could do 3 or 4 acres of oats a day, in Yorkshire, but Markham expected rather less. He cut 1 or 2 acres of hay, according to the nature of the ground. A reaper did about an acre a day, or 8 to 10 stooks of twelve sheaves. One rod of ditching, 4 feet broad and 3 feet deep, or 2 rods of hedging, 5 feet high, was considered a day's work. Four bushels of wheat or rye were threshed in a day, 5 bushels of peas or beans, and 6 of barley or oats.[2]

In the new *crops* advocated during this period we have the first suggestions of that new order of things introduced by the cultivation of root crops and artificial grasses. The great success with which these crops were grown in the Low Countries was energetically pointed out by English writers, but there, except for a few enterprising exceptions, the matter remained until taken up by the pioneers of improved farming in the following century.

Speed in 1659 had much to say in eulogy of *turnips*. He noted that they were excellent for feeding and fattening all sorts of cattle, and that cows fed on them gave milk "with full vessels" three times a day throughout the year. He recommended them sliced for horses, and boiled for poultry to increase the egg supply, and to fatten pigs. He also fed swine on the liquor the roots were boiled in, and went so far as to state that boiled turnips worked into meal made good bread, which at any rate was useful to feed rabbits; and the same root "with some small addition" would make good cider. The tops, he suggested, should be mixed with rape or linseed cakes and grains to make "pottage" for cows. This first mention of the very advanced idea of cattle cake was borrowed from the Dutch, who bought up English rape seed, after the oil had been pressed out, to make cake for their beasts. Speed grew his turnips a little less than 6 inches apart, and sowed less than 10 quarts to an acre, getting two good crops a year on poor ground. Wor-

---

[1] Alice Clark, op. cit., 62. In one instance, in Wilts, the same wages are quoted for women and men for haymaking and "gripping of Lent corn" (Hist. MSS. Commission: *Various Collections*, i. p. 174).

[2] Best, 43, 50, 132. Markham, F. H., 112-14.

lidge in 1669 stated that turnips were still grown "usually in gardens," but did well in the fields for cattle and swine; and at the end of the century Meager had grasped the idea that they improved the ground for other crops in addition to their usefulness as food. He mentions two sorts, long and round, the latter being the more usual. The same writer recommended *carrots* for fattening swine.[1]

The enterprising Speed had many suggestions for new cattle foods. Potatoes and "pumpions," he declared, could be used in the same way as turnips, and Jerusalem artichokes for feeding poultry and swine; while an acre of cabbages was more beneficial to milch cattle than two acres of hay, and could be kept all the winter pulled up by the roots and set in sand in a cellar. Unfortunately all these useful plants remained merely garden vegetables for many a long year to come, although one would have thought the harassed farmer, confronted with the difficulty of keeping his stock in good condition through the winter, would gladly have seized upon any expedient offered.

Artificial grasses, even clover, seem to have met with little better success. From the middle of the century onwards the Dutch "trefoil or clover-grass" was much recommended, and also "St. Foyne" and "La Lucerne," which were described as French grasses. Clover was usually sown in March or April with the barley or oats, and mown after the corn was gathered in, but if sown by itself was mown in May, or early June, and again in August, unless left for seed or feeding on the ground. To extract the seed it was threshed twice and then "chaved" with a fine rake: it was sown 9 to 15 lb. to the acre, and continued without resowing for three to five years. It was considered good for all cattle, also turkeys and geese; and the seeds were given to cattle like oats, and to the poultry. The virtues of sainfoin and lucerne and their use on poor soil were also urged, but they were still not very common a century later. Worlidge mentions, in addition, "La Romayne" or French-tares, spurry seed, hop-clover, "Long grass from Wiltshire," and saxifrage, stating that where this last grew "Housewives say there is never ill butter or cheese." Speed even wished to grow sow-thistles for milch cows, and says they were formerly given to sows who had not enough milk to nourish their pigs.[2]

None of these new crops could be introduced in open-field

---

[1] A. Speed: *Adam out of Eden*, 19-29. I.W.: *Syst. Agric.*, 42. L. Meager: *Mystery of Husbandry*, 99.

[2] I.W.: *Syst. Agric.*, 25-30. A. Speed, op. cit., 34-41. W. Blith, op. cit., 177-85.

farming without the consent of all those in the village who held strips: probably no easy matter, for there would always be a certain number desiring to continue steadfastly in the ways of their forefathers. Nevertheless the revolutionary idea that the ancient rotation of crops could be interfered with had been suggested.

New sorts of wheat mentioned were Rivet (white and red), whole straw, chilter, ograve, sarasins, bearded, longreade and doddereade; and there was a kind of barley called "Rathripe" which ripened two or three weeks earlier than other sorts. Buckwheat was grown, particularly in Surrey, for feeding swine and poultry, and lentils were considered "excellent sweet fodder" for young cattle. Hemp and flax were rather neglected, rape and cole-seed were grown for extracting oil, and in some places woad, madder and weld or dyer's weed (a yellow dye) for home dyeing, and also saffron and liquorice. Hop-gardens were popular, but vineyards were almost extinct.[1]

*Orchards* flourished particularly in Kent, where many apples and cherries were grown, the trees being planted in lines 20 to 30 feet apart, and corn grown between them until they were matured. Apricots, peaches, nectarines, "melacotones" and cherries were trained against walls, "with tacks and other means," but William Lawson, the chief writer of this period on orchards, considered this practice bad for the trees, on the grounds that the wall interfered with the roots and caused them to die young. Lawson, writing in 1618, had not lost the spacious outlook of the Elizabethan age, and planted his orchard for future generations; forty or fifty years he considered but a small span in the life of fruit-trees. He was inclined to think that like other trees three hundred years was needed to bring them to their prime, and they might well continue in it for a similar period. Consequently he planted them at least 20 yards apart to give them plenty of room for expansion, and for the first twenty or thirty years permitted saffron, liquorice, flowers and herbs to be planted between them, but grass was not suffered to grow round their roots. He considered the perfect form of a fruit-tree to be low and widely branching all round. The fruit was to be gathered, with many precautions against bruising, at the full of the moon, and laid in a dry loft on dry straw, at first in heaps, but after ten days or a fortnight, wiped carefully, laid "thin abroad" and turned softly once a month.

Besides the fruit-trees in this spacious orchard there were roses, woodbine, purple cowslips, primroses and violets, with

[1] I.W.: *Syst. Agric.*, 36-41. Best, 99. Blith, op. cit., 221-47.

borders of raspberries, barberries, currants, and strawberries—
red, white and green.  A "store of bees in a warm and dry
beehouse, comely made of fir-boards," filled the air with their
drowsy hum, and "a brood of nightingales" and other birds,
especially robins and wrens, were thought good.  Humbler folk
planted their trees as many feet apart as Lawson did yards, and
some had a "little Nurcery" in which to rear plum and cherry
stones, apple and pear pips, and others, as well as acorns, ash
keys, etc., to replenish the dwindling numbers of the forest
trees.  Farmers were exhorted to plant trees along the hedgerows, both to renew the stock of timber and to give shelter to
the cattle.  Plattes had a quaint little device for guarding his
fruit-trees from harmful spring frosts: an earthern pot "like a
little still" was filled with half a peck of small coal (at the cost
of ½d.), lighted and covered with a tile with wet hay laid on
it, and one of these was hung by a cord in each fruit-tree, to
"give an aire all night."  Lawson guarded his cherries carefully
from birds "either with nets, noise, or other industry."  To
increase his stock of fruit-trees quickly he had a curious practice.  About the end of June a hand's breadth of bark was cut
from the bough of a tree, and stiff clay, 2 inches thick, applied
to the bare part.  A mixture of good rank mould and manure
was then plastered round the branch, above the clayed part, and
bound round "as big as a football," the whole being left until
the following February.  The branch was then sawn off just
below the clay and planted in good soil, as it was, the idea being
that the clay would have stopped the sap from rising and the
branch would put forth roots in the ball of mould and manure.[1]
Sir Hugh Platt was in favour of burying "dogges and cattes"
under the roots of fruit-trees to fertilize them.

In the *garden* the flowers and the vegetables were kept more
separate than formerly, lest "garden flowers shall suffer some
disgrace if among them you intermingle onions, parsnips, etc."
The herb garden, with herbs and flowers for cooking and
distilling, was also in a separate place, so that the necessary
picking of the herbs should not spoil the design of the flower
garden, which was often an elaborate matter.  "Daffadown-
dillies" and daisies were now considered more for ornament
than for use; there were July-flowers or gilly-flowers in nine or
ten colours, some as large as roses; also wall-July-flowers or
wallflowers and many tulips.  In the vegetable garden there

---

[1] W. Lawson: *Husbandman's Fruitful Orchard*, 95, etc.; *Country Housewife's Garden*, 93; *A New Orchard and Garden*. G. Plattes. op. cit., 12, 14.
S. Hartlib: *Legacie*, 19-21. Harrison, ii. 188.

were French and kidney beans, white, French, Hastings, and rounceval peas, Jerusalem artichokes newly introduced in 1617, and asparagus, which was cultivated much in the same way as at present, and left three to five years before cutting. Cauliflowers now became an ordinary vegetable, and potatoes, a luxury costing 2s. a pound in the reign of James I, by the end of the century had become fairly common in gardens, and were usually eaten either buttered or in milk, though they were said to make good bread, cake, paste and pies. Garlic was still universal, and the skirret, a root sweeter than a parsnip, was popular, as well as the common root vegetables, and a variety of cabbages.

Lawson considered that weeding could be done by any maids who were willing to take the opportunity of a shower of rain, but "withal," he added, "I advise the mistress either to be present herself, or to teach her maids to know herbs from weeds." Speed recommended that a horse's head should be set upon a pole in the vegetable garden to banish the flies that destroyed cabbages and other herbs.[1]

The busy housewife and her maids had much the same tasks as before, but rich farmers' wives were coming to content themselves with organizing the household under competent servants. These were hired at Martinmas with the other farm hands and had their wages similarly fixed. Thus a woman "skilful in ordering a house," or a good "cookemaid," or a servant "able to keep a dayrie and doth take charge of it soe as the same be of ten kine at least, and doth take charge of brewing and baking, by which is meant that they shall make good what they lose or spoil," were all paid about £2 to £2 10s. a year besides their food and clothing. Ordinary dairymaids, washmaids, and maltsters, such as were hired by the plain farmer's wife, commanded rather less, and younger damsels were paid in proportion.[2] They were usually required to be good milkers, and able to wash, brew, cook and bake. Doubtless their capacity as spinners was taken for granted, for we learn from Markham that the wool was still sorted and spun, both warp and weft, at home, the warp being spun close and hard twisted, for strength, and the weft looser. Some of the wool could be sent to the dyers if the housewife wished it, and the weaving was usually

[1] Speed, op. cit., 48, 165. I.W.: *Syst. Agric.*, 131, 145, 224, 244. W. Lawson: *Country Housewife's Garden*, 66-75. Markham, F.H., 106-7. Amherst: *History of Gardening*, 137.

[2] A. Clark, op. cit., 50. *Quarter Sessions Records*, Somerset, Yorks, Lincs, etc. Cox: *Derbyshire Annals*, ii.

R

sent out, or a journeyman weaver hired, but the really good housewife was expected to know how to weave if necessary. Hemp and flax for the linen was grown and prepared at home still, but was often given out to be spun and almost always to be woven, so that the home work was a little lightened. The webs came home to be bleached, and in March all the household linen which had grown yellow during the winter was spread out in the sun for this purpose.[1]    In this month also the housewife would be much occupied in rearing chicks. In May she began to be specially busy in the dairy, and this was considered the best month for potting butter. Churning was done before each market day, and butter was sold in Yorkshire in 1-lb., or 2-lb. cakes, at 5d. per pound in Lent and 2½d. in May. Rennet was made at home, from the stomach of a young calf. The kinds of cheese given by Markham differ from those mentioned in the last century; the best sort he considers the "new or morning milk cheese," made of new milk and yesterday's cream with hot water added to curdle it. "Nettle cheese," the finest summer variety, was laid on fresh nettles after draining from the brine and covered with them, the nettles being renewed and the cheeses turned every two days. The coarsest sort was "Flitten milk cheese," and lastly there was the winter or "eddish," which was made in the same way but owing to the season would not harden. On farms where it was the custom to take calves away from the cows at once there would be many to be fed from the dairy,[2] and pigs also to be looked after.

June was the month for distilling sweet waters and innumerable medicinal potions, and for drying and laying up herbs to be stored, and besides this the housewife was expected to understand the selection, preserving and "curing" of wines, and the mingling of such luscious drinks as "muscadine." Late in the season more homely wines were made from plums, mulberries, raspberries and currants, the last a "pleasant, sharp liquor."[3]

In September there were "cauly flowers" and other vegetables of that tribe to be sown in the garden, and hops to be gathered and dried ready for making the "drink by which the household is nourished and sustained." The grain for the October malting was carefully chosen by the housewife, the bedding of the kiln prepared with best rye straw covered with a hair-cloth, and fuel

---

[1] M. Stevenson: *Twelve Months*, 12.

[2] Markham, E.H., 145-52. Best, 105. Flitten was probably the same as Flotten, i.e. skimmed milk. Eddish was the late grass which grew after the mowing, and the cattle fed on this would produce the "eddish" cheese.

[3] Markham, E.H., 112-13. I.W.: *Syst. Agric.*, 98, 112, 233.

of wheat straw provided if possible, but failing that, other sorts of straw, rushes, furze, bracken, brushwood, or any other kind of wood. This done, there was the household supply of beer, ale, cider and perry to be brewed, although less of the two last seem to have been made than formerly. "March beer" had ground peas, wheat and oats added to the malt. Mead was still made from clarified honey, the washings of the hives, strained and fermented, flavoured with herbs and spices.[1]

Baking, of course, was a continual business, and the bake-house in large households was provided with bolting apparatus of "large pipes," and sieves of all sorts. Meslin (wheat and rye) seems still to have been the most usual bread for the farmer, the rye having dwindled to a proportion of not more than one-third, though Fynes Morrison in 1617 states that the husband-men ate brown bread made of barley and rye. The very poor mixed peas with rye or meslin, and Markham mentions a coarse kind of bread composed of 2 bushels of barley, 2 pecks of peas, 1 peck of wheat or rye, and 1 peck of malt, ground all together and sieved. Wheat flour was used for pies, but "folk's" piecrust was made of meslin, and "folk's puddings" of barley, not rye, because "it maketh them soe softe that they runne aboute the platters." At harvest they were made of wheat flour. Oatmeal was used for many dishes, especially in the north, the fine meal for thickening or making "pottage," for some kinds of bread, and mixed with blood and liver for a dish Markham fondly believed to be "Haggas or Haggus." With the coarse oatmeal were made "black puddings," of blood, suet and herbs, "white puddings" of cream, eggs, suet, currants and spices, and "Good Friday puddings," of eggs, milk, suet and pennyroyal; also the stuffing for geese, and a form of porridge mixed with butter, called by seamen "loblolly."[2]

Rice, although as we should expect "grown not much in our kingdom," was plentiful, and thought cheap at 8d. a pound. It was considered "most sweet, fresh and pleasant," and many wholesome dishes were made from it, "some thick, some thin, some baked, some boyled," the most popular being boiled rice with butter and sugar.[3] Several new vegetables, as we have seen, could be procured from the kitchen garden, and "sallets" were served with vinegar, oil and sugar. The ordinary farmer's wife, however, did not deal in very elaborate cookery, but she

[1] Markham, E.H., 153, 181. I.W.: *Syst. Agric.*, 175, 238. Harrison, ii. 188.

[2] Markham, E.H., 178-80, 187. Best, 104.

[3] Markham, F.H., 102-5.

was bidden to be "cleanly both in body and garments," and "not butter-fingered, sweet-toothed or fainthearted."    Her staple dishes seem to have been "collops," i.e. poached eggs served with slices of bacon, fish, beef or pork; "fricases" or fried dishes, "pottage" of mutton chopped up and boiled with oatmeal and herbs, and sometimes a kind of pudding, presumably of firm consistency, which was wrapped round a spit and roasted.[1]

The discreet housewife was recommended to have her garments "comely and strong, made as well to preserve the health as to adorn the person, altogether without toyish garnishes or the gloss of light colours, and as far from the vanity of new and fantastick fashions as neer to the comely imitation of modest matrons."    She wore a gown of home-made cloth, a kirtle of "some light stuff," with a linen apron, and a coif and a high felt hat on her head.    Her husband's clothes were also of home-spun.[2]    Markham's list of the "Husbandman's Recreations" would lead us to suppose that he was given to many kinds of sport; hunting of the stag, fox, hare, badger, otter, boar, and goat; hawking, coursing, shooting with the long-bow and cross-bow, bowling, angling, and cock fighting, but the hard-working yeoman farmer, ruminating in his slow mind on the reports of strange new experiments and improvements in agriculture, would probably have little time for such diversions as these.

[1] Markham, E.H., 50-73.
[2] Markham, E.H., 3.    Harrison, iv., 281.

# CHAPTER XI
## THE EIGHTEENTH CENTURY

### REFERENCES

M.Y. = William Marshall. *Rural Economy of Yorkshire,* 1788.

M.N. = William Marshall. *Rural Economy of Norfolk,* 1787.

M.G. = William Marshall. *Rural Economy of Gloucester,* 1789.

M.M. = William Marshall. *Rural Economy of Midlands,* 1790.

M.W. = William Marshall. *Rural Economy of West of England,* 1796.

M.S. = William Marshall. *Rural Economy of Southern Counties,* 1798.

Y.E. = Arthur Young. *Tour through Eastern Counties of England,* 1771.

Y.N. = Arthur Young. *Northern Tour,* 1770.

Y.S. = Arthur Young. *Six Weeks' Tour through Southern Counties,* 1768.

# CHAPTER XI

## THE EIGHTEENTH CENTURY

DURING this century there spread over England the greatest advance in the practice of farming that had taken place since agriculture was first established in this country. At last, under the stimulus of increasing population, expansion of markets, troublous times and the fresh needs resulting therefrom, the suggestions and tentative experiments of the seventeenth century were taken up and developed, and under the leadership of such great farmers as Lord Townshend, Bakewell of Dishley, Coke of Holkham, Arbuthnot and Duckett, in the latter part of the period brought about a revolution in the affairs of the farm-yard.

The great difficulty the farmer had had to face since the earliest times was how to feed his stock during the winter. The number of animals he could keep alive until the fresh grass came in the following spring was dependent on his crop of hay, straw and oats; the rest had to be killed and the meat salted down for winter use, or sold at the nearest market. But now this problem was solved at last by a series of root-crops and "artificial grasses" which filled up the gap, and for the first time enabled him to keep his full stock of sheep and cattle in good condition all the year round. When these new crops came into general use fresh and wonderful possibilities opened up before the enterprising farmer. More stock could be kept perennially upon the farm; more stock meant a larger supply of valuable manure; more manure enabled more ground to be fertilized, larger crops to be raised and pasture stimulated; hence the cattle thus reared could have their breed and condition cared for and improved, a matter for which there was crying need. New crops, better methods of cultivation, development of the science of stock breeding, all received attention; but these things of necessity brought great changes. The ancient open-field system of husbandry with its scattered strips of arable, its communal meadows and pasture land, its flocks and herds driven out together by the village herdsmen, was doomed. Its fixed and universal rotation of crops did not admit of the introduction of turnips, clover and sainfoin, both on account of the difficulty of getting everyone to agree to such innovations, and also because the customary autumn feeding of the cattle on the stubble fields would be interfered with. Attention to stock breeding was also impossible with the promiscuous herding on the village commons, and the consequent spreading of any disorders, as was shown by the miserable condition of those communal flocks and herds remaining at the end of the century, "a motley mixture of all the different breeds of sheep and cattle at present known in the island; many of which are diseased, deformed, small, and in every respect unworthy of being bred from."[1]

Already, as we have seen, much enclosure had been effected, at first chiefly for purposes of sheep-farming, but at the beginning of the eighteenth century about half the arable land of the country was still open.[2] As the century advanced more and more parishes were enclosed, this time not so much for grazing as for cultivation: during the last quarter the increase was especially rapid, and the process was practically completed by the Enclosure Acts of the beginning of the nineteenth century. The Industrial Revolution of the middle of the eighteenth century brought about a great increase of population in the growth of the manufacturing towns; fresh and large markets for his produce opened before the farmer, and the increased demand for food spurred him to fresh efforts.

Little change took place in the planning of the farmer's house, which by this time had become very much like that of to-day. Those rebuilt in the pleasant and dignified style of Queen Anne's reign would have large and well-proportioned sash windows, an innovation which often seems to have been inserted rather incongruously into more ancient houses to replace the old casements and give more air and light. Fire-grates, too, were a new contrivance for the sitting-rooms, but the old open hearth, with brick oven at the side heated by burning wood in it, long continued in the kitchen. Sometimes a garret was fitted up as a smoking-chamber, the smoke being let into it from the downstairs fire by means of a register in the chimney, thus avoiding too much heat. Scores of tongues as well as hams and "hung beef" were thus cured.[3] The typical Norfolk farmhouse had a kitchen and "backhouse" or scullery, a parlour, and sometimes a "keeping room" for the family to sit in apart from the servants.[4] The materials of which it was built varied according to the resources of the district. Stone was used in

---

[1] Ernle: *English Farming*, 159, quoting *Farmer's Magazine*.
[2] Ibid., 154.
[3] R. Bradley: *Country Housewife and Lady's Director*, Part II, 30.
[4] M.N., i. 81.

parts where it was easily obtainable, such as Yorkshire and the North, and Northants; in the West, the Cotswolds, most of Somerset, Cornwall and West Devon.  Half-timbered houses were still customary in Kent and parts of Gloucester and Somerset, the panels filled in with plaster or brick, or, as a later development, covered with weather-tiles in Kent or weather-boarding in Gloucester.  Brick alone was gaining ground, and was in general use in the Midlands, Norfolk, and parts of the South, and was being adopted in Kent.  In Devon and Cornwall some earth building or "cob" survived,[1] roofed with thatch, and is still found at the present day.  Tiles were the most common form of roofing, varied by occasional thatch, all over the country except in the West where they gave place to slate.  In Gloucestershire a knobbed tile had come into use instead of the common sort hung by a pin.

The use of oak for house-timbering was going out, although it lingered, together with elm in some parts.  More commonly only door and window frames and sometimes wall-plates and upper floors were made of it, the modern unbeautiful deal which was growing more and more fashionable, being used for the rest. For ground floors bricks or tiles were most often used, and in Yorkshire cottage floors were often made of a mortar of lime and sand.  Ordinary diamond-shaped window-panes cost 1d. each, while the new-fangled squares for sash windows cost from 6d. to 1s.  But building was altogether wonderfully cheap at the beginning of the century, since we are told "a fine front built with Free Stone, very curious, will come to about 3d. per foot."[2]

Towards the end of the eighteenth century some attention was paid to the planning of the farm-yard with a view to the greatest convenience and saving of labour, instead of placing the various buildings in a haphazard manner.  Thus the main points on which stress was laid by Young and Marshall were: (1) that the barn should stand between the rick-yard and the cattle-sheds, so that the stalls could be supplied with hay and straw through openings in the wall with little trouble; (2) that the dairy should be situated conveniently near the cowsheds, calfhouse and hogsties, for the purpose of feeding the two latter, also that the hogtroughs should be fed, if possible, by spouts leading from the brewhouse, dairy, and scullery; (3) that the valuable liquid manure of the farm-yard should not be wasted or allowed to run off, but should be drained into pits or

---

[1] For a description of this process of building see M.W., i. 60.
[2] Giles Jacob: *Country Gentleman's Vade Mecum*, 32-5.

cisterns and absorbed with straw; (4) that water should be run from a pond or stream into cisterns in the yard for the animals to drink from. But these for the most part remained counsels of perfection. The model farm-yard which most excited the admiration of Young when on his famous tours was that of Lord Darlington at Staindrop, a spacious range of buildings which paid due attention to all these rules.[1] Humbler establishments arranged their buildings with an eye to suitable shelter for the beasts; thus in Yorkshire they were usually built round three sides of a square with the open side to the south, although, as Marshall pointed out, an L-shaped arrangement with the angle set to the north would do almost equally well. Small farms sometimes condensed their buildings into one oblong block, with the dwelling-house at one end and the stable and cattle houses at the other, with barn and hayhouse over; an arrangement which gave little shelter in an exposed situation.[2]

Barns varied in size according to the method of threshing and dressing the corn prevalent in the county. In the North where the corn was threshed bound, and winnowed with a machine fan, a barn with a threshing floor 15 feet by 10 feet was considered large enough for two men. In Norfolk where corn was threshed loose and cleaned by casting right across the floor from one porch to the other, 24 feet by 18 feet was considered a good-sized threshing floor, and it was in this county that the largest barns were found. The use of large barn doors was declining, a small door of which the bottom half could be shut to keep out poultry and pigs being found more useful and economical; and in the interior of some barns partial partition walls took the place of supporting beams.

In the West barns were mostly small, except for those surviving from monastic times. Marshall in 1789 noted one of these still in use, having a chamber over the porch with a fireplace and chimney, opening into a gallery running along inside the barn from which the barnward could overlook the work of the men below.

Timber and weather-boarding was a very common mode of construction in many parts, with the roof tiled, but in some counties, notably Kent, still thatched. In Thanet the great thatched roofs, taking an immense quantity of straw, sloped down to cover a range of sheds on each side. Tiles were naturally considered safer from the risk of fire.

[1] Y.N., ii. 477 on.
[2] M.Y., i. 118-19.

The foundations were sometimes of brick, or at other times the whole structure was raised on short pillars to keep out rats and mice. In the West the barn was often built on sloping ground, with a doorway on the higher level, through which the packhorses, which carried home the crops, shot their loads. Barn floors were of clay, brick, stone, or planks, the most elaborate, in the Midlands, being made of bricks set edgewise in mortar, the interstices carefully filled in, and a plank floor laid on the top.[1]

*Granaries* were sometimes built over the barn, with a trap-door and tackle to raise and lower the sacks, a height of 10 feet being allowed below; but this was considered unhealthy for the thresher, as the dust from threshing was confined instead of escaping into the high roof. Built over the porches granaries had not this disadvantage, but they were more commonly met with over cart-sheds, or simply raised some 4 feet above the ground.[2]

In the West the ancient span of 16½ feet was still adhered to for the *cattle-shed* or oxhouse. This allowed 2 feet for the thickness of the walls, 9 feet for the depth of the stalls, a 3-foot trough or manger, and a passage along the heads of the stalls from which to fill the mangers. The roof was supported by oaken posts on stone blocks, each pair of oxen having a stall 7 feet wide with a post between them. In the Midlands the cowsheds referred to by Marshall as "modern" were arranged in the same way, with the addition of water-troughs, the posts between each pair of cows being shaped like a K without the lower limb. A farmer at Retford in 1771 had an elaborate oxhouse for fattening cattle, the stalls for each pair being 8 feet or 6 feet wide, for large or small beasts. The square manger at the head was filled with hay through a hole from the stackyard, and was flanked by stone troughs, one for oil-cake and the other for water, which was pumped through a pipe from a cistern outside. The doors had sliding shutters for ventilation, and in a small room at the end the oil-cake was broken on a wooden anvil and warmed on a stove.[3]

But except for the fatting cattle, the calves, and the oxen in winter, the beasts were usually kept in open yards, either with sheds open in front (called in the West "Linhays") to feed and shelter in, or merely surrounded with bundles of firewood

[1] M.M., ii. 23.

[2] M.Y., i. 124. M.G., ii. 19. *Tusser Redivivus.* June. J. Mortimer: *Whole Art of Husbandry*, 113.

[3] M.W., ii. 241. M.M., i. 30. Y.E., i. 405.

to give a little protection. The old-fashioned "hovel" under the beanstack, enclosed with upright slabs or faggots, was disappearing. Cart-sheds, which Marshall considered should be near the main approach to the farm, in Norfolk were often furnished with a bank of earth at the back, 18 inches high, on which to rest the tails of the tilted-up carts.[1]

Ricks, as before, were frequently raised on stages, and Nourse considered that preserved from vermin in this way and well thatched the corn kept better than when threshed and put in a granary. In the Midlands a brick parapet with the coping projecting outwards was sometimes built to contain ricks, with a brick floor inside. It was apparently in the Maidstone district of Kent that farmers first thought of covering stacks with sail-cloth laid over a rope stretched between two posts. The posts were stepped in cart-wheels weighted with stones, and the raising of the cloth gave ventilation and prevented pools of water from forming on the outside of the cover.[2]

In the farm-yard *stock* an extraordinary change for the better took place, due to the fact that for the first time the modern principles of breeding were at length carefully evolved by the efforts of Bakewell of Dishley and other farmers of his ilk. This improvement is shown by a comparison of weights of cattle sold at Smithfield at the beginning and end of the century. In 1710, beeves 370 lb., calves 50 lb., sheep 28 lb., lambs 18 lb. In 1795, beeves 800 lb., calves 143 lb., sheep 80 lb., lambs 50 lb.[3]

Hitherto, as we have seen, size, irrespective of shape, was the criterion of cattle; those most esteemed were large, long-bodied, big-boned, coarse and flat-sided, but the principles introduced by Bakewell produced a very different sort of beast, middle-sized, small-boned, round-carcased, kindly looking, and inclined to be fat.[4] The first of these new requirements was utility of form, a carcase in which the valuable roasting joints were developed at the expense of the coarser boiling joints and waste parts, the bones also being reduced as occupying unnecessary weight and space. Secondly, the animal to be profitable must come to early maturity, and must have a propensity to fatten in a short time in readiness for the butcher, thus economizing in the amount of food it consumed. Also the texture of its flesh must be fine grained and its form symmetri-

[1] M.M., i. 32. M.N., i. 85. M.W., i. 59. *Tusser Redivivus.* June.
[2] M.S., i. 167. M.M., i. 33. Tim Nourse: *Campania Foelix*, 42.
[3] Cunningham: *Growth of English Industry and Commerce*, iii. 706.
[4] G. Culley: *Observations on Livestock*, 57.

cal, the ideal shape for both cattle and sheep being that of a hogshead, "truly circular with small and as short legs as possible, upon the plain principle that the value lies in the barrel and not in the legs."[1]    All these qualities being recognized as hereditary, Bakewell established the practice of in-and-in breeding from the best members of the same stock, instead of the former custom of importing good males of a different breed, and he conducted careful experiments in the feeding of his beasts to discover the best diet for fattening them, and also how to maintain the largest quantity of stock on a given amount of ground. But the animals produced on these lines by Bakewell were predominantly grazier's beasts, little attention being paid to their milking capacity and still less to their powers of draught. The Dishley cattle were derived from an improved breed of the old long-horned black cattle of Lancashire, Yorkshire and the neighbouring counties, the best and largest of which were found at the beginning of the century in the North Riding, near Ripon. Originally all black, with occasional white faces, they became in time pied with white, the best variety being the Craven breed, developed by Webster of Canley, from whose famous herd Bakewell derived his stock, notably his bull "Twopenny." The Craven breed, however, were unsuitable for draught owing to their horns, often a yard long, which were liable to gore each other or become entangled in the hedges or banks, and break either the horn or the neck of the animal.    It was this breed which supplied the mediaeval hunting horn, and as the use of that implement declined there was little reason to cherish that shape of horn.    Their colour varied a good deal from the original black, being described as "brindle, finchbacked and pyed," sometimes brown with a white streak, the lighter ones being preferred.[2]    Long-horned cattle were also found in North Wilts and parts of Hampshire, Somerset and Gloucester, and a small variety on the heaths of Surrey.[3]

The limitations of the long-horned breed caused the short-horned cattle of the East Coast to come into prominence in their turn.  This seems to have been a breed new to England, probably introduced from the Continent during the seventeenth century; they are described by Mortimer in 1707 as the "long-legged, short-horned Dutch breed of Lincolnshire and Kent," and in colour were red mixed with white.  He considered them the best dairy cows, giving two gallons at each milking.  By the end of

[1] Y.E., i. 112.
[2] For a detailed description of this breed see M.M., i. 277.
[3] M.G., ii. 75, 96, 152.  Y.E., iii. 416.  M.S., ii. 85.

the century improved breeding caused them to yield six gallons a day commonly and sometimes as much as nine.[1] The short-horned cattle extended up the East Coast from South Lincoln-shire to the Scottish border, and became best known as the Holderness breed. These were at first large-boned and clumsy beasts with heavy hind quarters, but twenty years of careful breeding improved their flesh and reduced their hind quarters, although with a consequent increase of danger in calving. A typical working ox measured 4 feet 11 inches high at the withers,

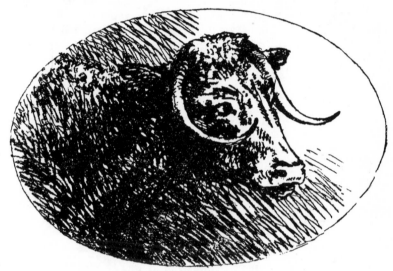

FIG. 67.—LONGHORN COW

and a cow 4 feet 5 inches. The Teeswater variety of this breed became the most popular, having clubbed, down-hanging horns; and a little later the Durham shorthorns, bred from the Holder-ness and Teeswater varieties combined, became famous both as milkers and graziers' beasts under Charles and Robert Colling of Ketton, their bull "Hubback" being the ancestor of many nineteenth-century shorthorns.[2]

In the West there were middle-horned cattle, of the colour described as "blood-red," found at their finest in Gloucester and Hereford, where they had an irregular white line along the spine,

[1] Mortimer: *Whole Art of Husbandry*, 166. Culley: *Observations on Livestock*, 40. Young, oddly enough, says that they were not good milkers. Y.N., ii. 133.

[2] M.Y., ii. 175-86. Culley, op. cit., 40. Ernle: *English Farming*, 188.

and the face and neck almost black.  The Hereford, however, acquired sometime during this century a white face also, and a white band underneath the body, probably from admixture with the Welsh.  These were large animals, with a long full carcase, but fine bones, good milkers and especially good draught animals. The cows also fatted well, and in the opinion of Marshall they were the best cattle in England.  In Devonshire they were a little smaller and all red, the best being found in the north of the county, where they were specially good for draught, being the fastest walkers, but better for fattening than milking. Farther west they became coarser, with larger and more upright horns. Middle-horned cattle very similar to these but rather heavier in build were used in the weald of Sussex, the best specimens being found in the Petworth district.  A curious variety called "sheet cows," "observable in Gentlemen's seats in various parts of the island" is recorded by Marshall, being like the Devon cows with, as it were, a white sheet round the barrel.[1]  The native breed of Norfolk was also middle-horned, blood-red with white or mottled faces, but smaller in size than the Hereford variety, small-boned, short-legged and round-barrelled.[2] But in Norfolk and Suffolk the cattle most prevalent were the Scotch cattle, mostly black or dark brindled polled Galloway "runts," with some Kyloes, Highland and Skye, which were brought south in the autumn by Scottish drovers, sold at St. Faith's Fair in October, and fattened on the turnips of East Anglia for the Smithfield market. This trade in drover's cattle had been growing ever since the Union of the crowns of England and Scotland.  A breed of hornless dun cattle was found in Suffolk, which were particularly good milkers and butter producers.[3]

Alderney and Guernsey cows were kept round Exeter, and in the Isle of Wight, but Culley considered them too delicate and tender ever to be much attended to by British farmers, and says they were only kept by the nobility and gentry, for their "exceeding rich milk to support the luxury of the tea-table."  A good many Welsh cattle were kept in Kent, in the Maidstone district.[4]

Great changes in cattle-feeding were made possible by the introduction of turnips and artificial grasses, and it was in the districts where these were adopted that the greatest improvement in stock naturally took place. The increased demand for butcher's meat consequent on the growth of the large industrial centres

[1] M.W., ii. 100.
[2] M.G., i. 213; ii. 192.  M.W., i. 235-8.  M.S., ii. 134, 194.  M.N., 323.
[3] Culley, op. cit., 65.
[4] Ibid., 72.  M.W., ii. 22.  M.S., ii. 281.  M.S., i. 321.

made careful methods of feeding and fattening well worth while. Oilcake (made of the bran of linseed or rape) was another useful food adopted during this century, prized partly for its fattening properties, but chiefly because the animals fed on it gave particularly rich manure. Marshall writing in 1789 stated that it had been in use for half a century, but in Gloucestershire had only been known twenty or thirty years.[1] Young reported its use among a few enlightened farmers in the Midlands, but it was far from being generally used all over the kingdom, although recommended by Speed in the middle of the last century. Stall-fattening became popular at this time, an elaborate example of a house for this purpose at Retford having already been described. This farmer, Mr. Moody, believed in lack of ventilation for accelerating the process of fattening, and for the first fortnight did not even open the door holes, at the end of which time the animals had sweated off their hair and a fresh coat began to come. The cowmen who attended on them in that atmosphere were hardly to be envied. There and in Gloucestershire cake was given two or three times a day, in the latter county alternated with hay, but in Gloucester the bullocks often had little yards attached to their stalls, in which they could get some air. But the more usual method, as in Norfolk, was to fatten bullocks in the fields on turnips, sometimes finished off with rye grass and clover. Even with these foods prejudices had to be overcome, an early idea being that cattle would overeat themselves on clover and burst, or would choke themselves on pieces of turnip. We are asked to believe that in Norfolk at the beginning of the century it was usual for a man to attend on the cows with a piece of rope knotted at one end, with which if necessary he shoved the refractory piece of turnip down the animal's gullet.[2]

Bull calves particularly were often left with their mothers all the year, but except in the Sussex Weald it was more usual to take away the heifer calves either at once or within a week or two and rear them by hand on skim milk and porridge, and later on grass and hay.

The yield of milk was naturally improving with the increased attention to stock-breeding, although at first dairy cows did not receive the most attention. Young gives as the average yield 5 gallons a day in districts covered by his Northern Tour and 4 gallons in his Eastern.[3] A more detailed examination of his

[1] M.G., i. 233-4.
[2] D. Hillman: *Tusser Redivivus*, Dec. 11.
[3] These names, however, are quite misleading as to locality, as his Eastern Tour wandered from Doncaster to Somerset and his Northern journey began in the Home Counties.

figures shows that 4 gallons was a very usual yield, though in some localities less was obtained, while the improved shorthorn breeds of Durham and Northumberland often gave from 5 to 9 gallons.

Work in the dairy for mistress and maids still began usually at 4 a.m., milking hours in the West being between 3 and 5 a.m. and 3 and 4 p.m. In some parts of Gloucestershire the milk was carried from the cow pastures to the dairy in a barrel-cart drawn by a horse. In this, the chief dairy district, the dairy customarily opened on a small yard or garden with a well in it, and benches for washing and drying the utensils. It was a large cool place with a stone floor, shelves of elm or ash, a window to the north, an outer door north-east or north-west and a door communicating with the kitchen to the south. The vessels were of wood in that district, but in some parts of Wiltshire and Yorkshire leaden vessels with an escape pipe for drawing off the milk under the cream had been adopted.

The labour of butter-making was lightened by the introduction of the horizontal barrel-churn, which was considerably easier to work than the old-fashioned upright kind. Ellis in 1750 refers to the barrel-churn as "new invented" and known in but few counties, but by the end of the century it was in common use in the north, in Norfolk and Suffolk and in Gloucestershire.[1] In the West of England, especially in Devon and Cornwall, the ancient method of raising the cream by gentle heat was still universal. The milk was put in large shallow vessels of brass or earthenware and stood on the wood embers of the hearth, or sometimes on charcoal stoves, for hours until thickened into clots. Nowadays tin pans are used, set in an outer one, of tin or iron, in which the water is heated round them. Marshall considers it the survival of a pristine method of preserving the cream when cows were few, in order to collect enough for butter-making, but local tradition ascribes its introduction to the early Phoenician traders, and indeed similar cream is made in Babylon to-day. To make butter the cream was thrown into a large bowl and beaten by hand until the butter came, churns not being used at all. The disadvantages of this method were the liability of the cream to collect the "more volatile parts of the fuel" if a careless hand stirred the fire, and a certain smoky flavour and lack of waxy texture in the butter. At the beginning of the century it still lingered in Somerset, but by the end had retreated to West Devon and Cornwall,

[1] W. Ellis: *Country Housewife's Companion*, 309.

where the true delicious scalded cream is still to be found in many parts, although unfortunately rapidly disappearing before the modern "separator".[1] Separated cream is also scalded but has not the same flavour.

In the dairy districts of Yorkshire the butter was inspected and graded by "searchers" into three classes and "grease." Classes 1 and 2 went to the London market and the grease to the wool industry. In parts of the Midlands where the flavour of the cream suffered from the rancidness of turnips or the bitterness of barley straw, this was remedied by being poured on to hot water and skimmed off again when cool.[2]

In many districts, especially in the cheese-making counties, whey butter was made, the whey being set for cream after the removal of the curd, and either allowed to stand or scalded by gentle heat to make it rise. Butter was then made of it in the usual way, and if of good quality sold for only a penny or two less than ordinary butter.

Rennet for cheese-making was made by soaking the calves' "vells" or stomachs in salt whey and brine, the "ill flavour" being removed by the addition of sweetbrier, dogrose and bramble leaves, and a lemon stuck with ¼ oz. of cloves.[3]

Cheeses were still classified by the counties in which they were made, but now the method used in one county spread to the district round and carried the name of its origin with it. Thus the mild and waxlike cheese known in its best quality as Gloucester, or sometimes in other qualities as Warwickshire, was made not only in those counties but also in Somerset, Wilts, Berks, Oxford, Worcester, Leicester, Derby, Stafford and even Yorkshire. Similarly Cheshire cheese, of a dry, loose texture, was made there and throughout North Wales. These two sorts were market cheeses of the eighteenth century, the excellence of Gloucester cheese being due to careful management in the dairy. It was usually made of the skim milk of the evening added to the full milk of the next morning, a thermometer being used to ensure the addition of the rennet at a temperature of 85°. The curd was cut up with a triple cheese knife, scalded, squeezed with the hands into the vat and many times turned and pressed. It was sometimes coloured with a preparation of annotto.[4] Towards the end of the century a curd-mill was coming into use in Yorkshire for the final break-

[1] M.W., i. 244-7. W. Ellis, op. cit., 321.
[2] M.Y., ii. 202. M.M., i., 316.
[3] M.N., ii. 108. M.G., i., 294
[4] M.G., i. 286-312; ii. 183.

S

ing of the curd. This consisted of two rollers in a box, turned by a crank and fed by a hopper, the upper roller being set with iron spikes an inch long and the lower with bevel-headed nails.[1] Other counties still had their cheeses, that of Suffolk having the worst reputation, being described as "so hard that pigs grunt at it, dogs bark at it, but none dare bite it."[2] The famous Stilton cheese also made its appearance early in this century. Marshall, writing in 1789, declares its first maker to have been Mrs. Paulet of Wymondham, who was still living then, but as Defoe mentions Stilton cheese in his *Tours*, sixty-seven years earlier, with comments on its well-developed maggots, the good lady must have begun making it young. This, however, was a luxury cheese, made of cream, and not for the ordinary farmer's consumption.[3]

In the dairy districts it was becoming customary to keep a proportionate number of pigs to the size of the dairy to consume the "swillings" economically, i.e. the whey and buttermilk, which were otherwise wasted or bestowed on poor neighbours.

The breeds of *sheep* increased and improved during this period as much as other cattle, and in the second half of the century with them too the quality of their meat became the first consideration, their wool, although still valued, taking a second place. But it is not until the last thirty years of the century that the names of new and improved breeds begin to appear. Ellis adds to the formerly well-known kinds (Herefords, Gloucester and Cotswolds, Midlands, Lincolnshire and Welsh) the Romney Marsh breed, large hornless animals with close curled wool and long legs; the Oxfordshire, not unlike the Lincolns, and a Dorsetshire variety, horned and white-faced with short white legs and fine close wool. The great advantage of the Dorset breed was its propensity for early lambing. The lambs were usually dropped about a month before Christmas and carefully reared in "little dark cabins," or in pens opening into a large suckling room, where their own mothers, alternately with foster mothers, were brought to them at intervals of four hours. Sometimes they were fattened off on oats, peas or meal, and were sold for the London market in April and May. The ewes were kept in good condition on the best pastures available, and by progressive farmers fed on oilcake, turnips and cabbages, besides hay. This "house-lamb" breeding extended from Dorset eastwards to the Isle of Wight and West Sussex. A

[1] M.Y., ii. 208.
[2] Ernle: *English Farming*, 193.
[3] M.M., i. 320.

slightly heavier variety of the same sheep was used in Wiltshire, and Dorset ewes were bought up by farmers of the Home Counties, those domiciled in Hertfordshire growing larger and coarser than in other counties.[1] Ellis mentions also a kind of sheep called "Turkey," large heavy white beasts with black spots and broad tails sometimes weighing as much as 14½ lb. But of these we hear no more.[2] Young, Marshall and Culley add many other species. The Herefords became best known by the Ryland breed, described as giving sweet mutton but finer wool, and on account of their fine short fleeces it was the custom to keep them in sheepcotes at night, low buildings with racks against the walls for fodder, which must have acquired a wonderfully unsavoury atmosphere since they were only cleaned out once or twice a year. In the Midlands arose the famous breed of "New Leicesters," begun by Allom of Clifton, and developed by Bakewell. These animals, the most fashionable of their time as graziers' beasts, were light-boned with bodies so broad that when well fattened they were wider than they were deep, and almost as broad as long. Their heads were long and hornless and their fleeces long.

In Yorkshire the enormous "mud" or "mug" sheep of the Tees valley, white faced and hornless, and standing so high on their legs that they resembled small cattle, were so much improved that the Teeswater breed became the outstanding one of the district. This breed was particularly valuable for its propensity for having twin lambs. They were sometimes crossed with the short-legged Swaledales, which would tend to make them less striking in appearance.[3] On the Moorlands of Yorkshire and the North the ancient type of mountain sheep remained; a small active, hardy race with wide curling horns, black or mottled faces, a "fierce, wild-looking eye," and long coarse wool. These are the sheep now found all over the Highlands of Scotland, to which they were introduced from the north of England in the late eighteenth century.[4]

Culley mentions in the North the Herdwicks, a small, hardy, hornless breed with short matted fleeces, found in the rocky parts of Cumberland; and also the Cheviots, a valuable and fine wooled race which is still to the fore.[5]

The small-horned sheep of Norfolk seem to have been akin

[1] W. Ellis: *Complete System of Experienced Improvements on Sheep*, 251-2, 263. Culley, op. cit., 136. Y.E., iii. 331-3. M.S., 200.
[2] W. Ellis: *Husbandry* (abridged), ii. 101-16.
[3] M.Y., ii. 213. Y.N., ii. 240. Culley, op. cit., 121.
[4] Culley, 143. M.Y., 215, 271.
[5] Culley, 147-50.

to the moorland breed, though some were a good deal larger and although still horned had short fleeces. According to Culley they were not much use except for folding. Exmoor also had a small, horned, long-wooled breed with white faces.

Another still famous breed which came into notice for the first time were the Southdowns of Sussex, middle-sized sheep with heavy hind quarters, thick short heads and necks without horns, grey faces and a very fine close wool. Ellman of Glynde built up this breed as Bakewell did the new Leicesters, crossing them with a Merino strain, and they were taken up a little later by Coke of Holkham, who gave attention to the expansion of their chests. Merinos were also imported from Spain by George III in 1792, and proved a valuable cross for short-wooled sheep.

The milking of ewes was very much on the decline, and by the end of the century only survived in a few parts of the North. Culley states that Cheviot sheep were milked to make cheese for eight or ten weeks after weaning the lambs, and Young notes the same practice at Wooler in Northumberland, but it was recognized as detrimental. The butter made from ewe's milk was used, mixed with tar, for salving the sheep, with the idea of keeping them warm in bad weather, and free from scab. In parts of the Midlands an ointment of butter and flowers of sulphur was used to prevent fly.[1] On the Yorkshire Wolds sheepfolds (for hornless sheep) were sometimes made of network of small cord, being light and easy to move, and a very modern-looking rack on wheels for feeding sheep is recorded by Young in Lincolnshire.[2]

The average fleece among long-wooled sheep towards the close of this century was about 8 to 9 lb., and among the short-wools about 2 lb. The mountain breeds averaged about 3 lb. Sheep-shearing festivities seemed to be dying out, as little is heard of them in this century, except in the very sophisticated form instituted by Mr. Coke at Holkham, at which this simple and necessary process of farming was elaborated into an agricultural show with the distribution of silver trophies, and the homely feast of the farm hands in Shakespeare's days degenerated into a banquet given to hundreds of visitors at the great house. A peculiarity among farmers of the West of England was that they did not wash their sheep before shearing. The

[1] Y.N., iii. 65. M.M., i. 403.
[2] M.Y., ii. 251. Y.N., i. 97.
[3] These are taken from Culley's lists, slightly moderated by Marshall's figures.

result was a heavier fleece, but as the wool fetched a lower price there can have been little advantage in that respect.[1]

Among the *pigs* the chief event was the arrival between 1770 and 1780 of the Chinese variety. These animals, of which there seem to have been large and small breeds, were renowned for their rapidity in fattening and their early maturity, qualities

FIG. 68.—OLD ENGLISH HOG

prized in every farm-yard species in that century. Their broad fleshy carcases, short thick heads with erect ears, and very short legs formed a most valuable admixture for the large and somewhat lank native breeds with their long legs and snouts, and many of the English breeds thus gained in compactness of form, a cross with the Berkshire giving the most satisfactory results, in spite of some diminution in size and fertility. Some complained of their mischievous ways, since only walls or extra good palings could keep them from wandering where they should not go. The old English large white breed remained chiefly in Yorkshire and Lancashire and in the West, "very plain, thin, awkward hogs," with long legs, large ears and wattles hanging from their throats; but their popularity was giving way before the Berkshires, which were common in almost every part. The latter are described as reddish or sandy with black spots, having

[1] M.W., i. 261; ii. 125.

large lop ears, short legs, and a tendency to fat, although small-boned. Even in Yorkshire and the Vale of Gloucester, ancient strongholds of the gaunt white pigs, the Berkshires were penetrating, and bringing with them a certain admixture of the useful Chinese blood. This mixed breed was described as "cadish and quiet" and had the useful quality of eating "noxious

FIG. 69.—CHINESE PIG

weeds" such as docks. A variety of the Berkshires found especially in the Downs of Kent and Surrey was the "tunback," so called from their rounded backs, probably due to the oriental admixture. Young noted the prevalence of Chinese pigs from London to Berkshire, and in the Midlands the up-to-date farmers kept a half-bred Chinese variety called "tonkey."[1] Marshall records these and a species of black hogs with "whole feet" like an ass. In the Vale of Gloucester there was a breed of Shropshire hogs, usually white with long bodies and huge slouching ears "which almost trail upon the ground to make way for their noses." A similar breed was found in Northants, almost blinded by their ears, which at least had the advantage of making them remarkably gentle.[2]

[1] A corruption of Tonquin, their place of origin.
[2] M.M., i. 327. Y.S., i. 125, 150. M.Y., ii. 210. Culley, 172-5. Youatt: *The Pig*, 53-63.

The sound and economical practice of keeping a suitable number of hogs to be fattened on the waste products of the dairy—buttermilk and whey—was gaining ground, especially in dairy counties. In East Anglia the farm-yards had hog cisterns, handy to the kitchen, dairy, and hoghouse, well lined and roofed over, in which the dairy and brewery swillings were collected, with some boiled turnips and bran or meal added. The sows and young pigs were fed from it in summer, and the rest of the swine in winter, by which time it must have been well flavoured and scented. In Suffolk and part of Essex hogs were fed on clover, but not apparently in other parts. Young conducted a careful series of experiments in feeding these animals, and considered clover and lucerne in the field good summer food; for winter he found pollard (wheatmeal) mixed with skim milk gave the best results all round, though boiled carrots were the most fattening, and potatoes were also good. In Sussex the swine were fed on the marshes from May to the middle of September, in charge of a swineherd, but in a wet summer many died of "rot." The sows even farrowed in the fields and were kept there on herbage.[1]

As a variety from the ordinary pigsty Young noted some in Derbyshire with a convenient stream running through, and Defoe states that in Lincoln, owing to the prevalence of monastic ruins, the sties were built in ecclesiastical style with stone walls and arched windows and doors. Rubbing posts were often provided against which the pigs could relieve their hides of superfluous matter and acquire a certain polish.

The era of the great shire *horse* as a farm animal now opens. The old black English horses of what was known as the "Fen breed" were much improved early in the century by the importation of six mares from Zeeland sent over by Lord Chesterfield, and Derbyshire taking the lead in the careful breeding of this stock, Leicester and other Midland counties soon followed. No longer called upon to charge swiftly in battle the shire horses developed a staid and dignified deliberation of movement which provoked the criticism of Marshall, who referred to them disrespectfully as the "black snail breed" and considered them "better calculated for eating than working, and whose tendency is to render their drivers as sluggish as themselves." They were, however, pretty universally used throughout the Midlands and South, in the Vale of Gloucester, where their colour sometimes inclined to brown or tan, and even to some extent in West Devon. The largest specimens of the breed, the London dray horses, reached 18 hands, and it is stated that "there is no

[1] Y.E., iii. 407. M.S., ii. 203, 242.

country can bring a parallel to the strength and size of our horses destined for the draught, as there are instances of single horses that are able to draw the weight of 3 tons."[1]

Bakewell effected improvements in this animal, as with his other stock at Dishley, and having strength and activity in his mind rather than height or weight, produced a somewhat smaller horse with a short thick body and short clean legs, and with a hardier constitution than the very large species, which Culley complained tired more quickly and were more subject to disorders than middle-sized beasts.[2]  The Clydesdale, the famous Scottish breed of cart horse akin to the Shire, had already become prominent in that country, but was not much seen in England until the following century.

In East Anglia the Suffolk Punch was the popular breed, "a very plain made horse" but strong, active, and excellent for draught.  They were usually bay-sorrel in colour, with a white blaze on the face, not more than 15½ hands high, with a round barrel and short clean legs, rather lower in front than behind, "half horse, half hog" in appearance, but stronger and hardier than the black horses.  They improved a good deal as the century went on, fresh blood being introduced in 1773 from a Lincolnshire trotting horse, which gave them a little more action.[3]

In some districts, however, the heavy cart horse was not favoured, and the Yorkshire breed of Cleveland Bays, originally coach horses but later used for farm draught, came very much to the fore, as strong, active animals.  It was said that three of them would draw 1½ tons of coal 60 miles in twenty-four hours with two or three baits. In the Cotswolds also there was a revolt against "elephants" for the plough, and a kind of strong coach horse probably akin to the Cleveland was used.[4]  In this part the old custom of keeping a goat in the stable as a preventive against staggers still lingered, and Marshall suggests that staggers being a nervous disorder the smell of the goat might have a beneficial effect on the horse's nerves. In the hilly country of Devon and Cornwall the small local breeds, Dartmoor and Exmoor ponies, were much used, with somewhat larger animals for packhorses, which were still almost the only means of carriage in South Devon, the Vale of Exeter and throughout Cornwall.

[1] Sir Walter Gilbey: *The Great Horse*, 55, quoting *Sporting Mag.*, 1796.
[2] M.M., i. 257.
[3] W. Ridgeway: *Thoroughbred Horse*, 374.  Culley, 27.  M.N., i. 42.
[4] Culley, 25.  M.Y., ii. 157.  M.G., ii. 32.

Team horses were usually put into stables in November or into yards with covered sheds. The ancient practice of bleeding horses at Christmas and midsummer is mentioned by Hillman in 1710, but after that seems to have fallen into disuse. The pride which was taken in a team of good cart horses gave rise to much ornament in harness and trappings, and also to the practice of "bell-teams," a method of giving warning of their approach, very useful in the deep and narrow lanes of Surrey, Sussex and Kent. Culley describes a Berkshire team of six stallions in tandem, the leader having six bells, the second five, and the others each one less.[1] In Sussex the four-horse teams until late in the nineteenth century had sets of sixteen bells, tuned in two octaves, the leader having the five smallest bells, the shaft horse the three largest, and the middle horses four bells each. The flocks of Southdown sheep also had sets of twenty-two or twenty-six little brass bells, which must have made a merry chime over the wide stretches of the Downs.[2]

*Asses* were recommended by Mortimer at the beginning of the century as "useful to the poor" for light ploughing, carrying burdens, turning the mill, fetching water, and giving good milk, their hides also making durable shoes; and in South Devon at the end of the century they were a good deal used as pack animals.[3] Culley records a Spanish ass at Beverley 14¾ hands high, but says these animals were much abused and overworked. Mules were not much kept.

*Goats*, except for their occasional presence in stables, seem to have been little kept, except in Cornwall, where there were considerable numbers, herds of as many as a hundred being kept for milking, and for the sale of their flesh and that of the kids in the market.[4]

The shepherd's *dog* was usually expected to be active and swift, although in the opinion of some a lame shepherd and a lazy dog were best because they did not overdrive the sheep. In the Downlands of Sussex the quick and intelligent dogs kept the sheep from the open corn-lands "with great caution and severe labour," and their feet became so worn and full of corns that they were rarely able to stand full work for more than three years. In West Devon and East Cornwall they were particularly well trained, being tall and shaggy with very short tails, mostly "grizzled" but some dun-coloured. Ellis states that

[1] M.S., i. 393. Culley, 31.
[2] I am indebted for this information to Miss Horn.
[3] M.W., i. 284. J. Mortimer: *Whole Art of Husbandry*, 163.
[4] M.W., i. 316.

most sheep-dogs were shaggy, to stand the cold and exposure of their manner of life, but some were large, half-rough lurchers, and some smooth-coated. All should have sufficient courage to face rams. They began their training at six months old, and were fed on barley- or wheatmeal mixed with skim milk, warm water or "pot liquor."[1] So much damage was done by uncontrolled and mad dogs that there was a demand arising for their taxation as they were a curse to the sheep-farmer in particular.

*Cats*, as Darwin has observed, encourage the growth of clover, because they eat the field-mice which eat the humble-bees which fertilize the plants. But as the eighteenth century advanced a much more serious matter arose for their attention and that of the farmer in the unfortunate arrival from the East of the common rat. These very undesirable animals first appeared in Russia about 1727, and entered this country by ship some three years later, quickly becoming an increasing evil to the farmer in spite of the fact that they waged a war of gradual extermination on the old English black rats formerly in possession. The sole virtue of the new variety was their lack of sociability. The home-bred black rat had favoured the houses and haunts of man and had endowed him liberally with the deadly plague-flea, but the brown rat preferred barns, outhouses and sewers, and consequently the plague departed. The cats in charge of the barn and stackyard had an increasingly busy life, although aided by traps. One farmer near Maidstone adopted the expedient of feeding his rats and mice generously with a paste of wheatflower, sugar, and oil of carraway, and when they were "accustomed to assembling for it" either mixed in arsenic or took them wholesale in a trap.[2]

Hares and pheasants were a great pest to the farmer, especially in Norfolk, where the hares nibbled the turnips, the pheasants ate the corn and both attacked the clover. There were still dovehouses in some places, but in the West pigeons were not numerous and there were no pheasants.[3]

*Rabbit* warrens were still profitable, both for the fur of the animals and their flesh, in the Vale of Pickering (Yorks) a large warren being let for as much as £300 a year. The pelt of the common grey was used for felt and that of the black with silver hairs dressed for fur. For eating they were sometimes fattened in an enclosure of turnips within the warren. Their close season

[1] W. Ellis: *Improvements on Sheep*, 10. M.S., ii. 380. M.W., i. 258. M.Y., i. 346.
[2] M.S., i. 111.
[3] M.W., i. 164. M.N., i. 172-3.

began at Candlemas.[1] In West Sussex we are told that *deer* were sold in the market like ordinary cattle.[2]

Marshall noted that *bees* were very plentiful in Yorkshire, especially in the Moorlands, but their honey was brown and strong flavoured. Norfolk also produced a good deal of honey, but of an inferior quality, and throughout Devon and Cornwall there were hardly any bees.

Among the *fowls* the breed we hear of most is the Dorking (or Darking), large white fowls, or sometimes yellowish with white legs, having a double claw behind. They were fattened highly in the surrounding county, the capons to as much as four to six pounds, and sold at the poultry market of Dorking, which was the best in England. Other breeds mentioned are the Poland and Hamburg; the Hertfordshire "dunghill fowls" were in good repute, and the game hen was said to lay most eggs but they were small in size. In the West the hens did not lay well, and this was attributed to their roosting in the open air, frequently in trees, which did not seem to answer so well as when they were housed in the warmth of smoky cottages, or even in the livery stables of London. Fresh recommendations for preserving eggs were to lay them in running water, or to varnish them.[3]

*Geese* abounded, particularly on the Lincolnshire Fens, and in Somerset, in the neighbourhood of Taunton and Sedgemoor. Myriads were kept there, mainly for their feathers, which were pulled several times in the course of the summer, so that by September the poor birds were "looking piteously melancholy," though they were said to suffer no inconvenience by the process. They were kept on the moor all winter and in time of frost fed on beans. In Lincolnshire they were plucked five times a year, at intervals of six weeks, beginning at Lady Day. Droves of geese were also sent to the London market from Norfolk and Lincolnshire, feeding upon the stubble fields on the way. Ellis recommended geese cross-bred between the English and Portuguese. They were said to fatten well on carrots cut small.[4]

In addition to the common white *duck* there were the crook-bill, the Muscovy (the largest sort), and, in 1750 recently become popular, the Normandy. "A good parcel of ducks," says Ellis, "will do great service in a turnip field where they are seized by

[1] Culley, 218. M.Y., ii. 227, 256.

[2] M.S., ii. 207.

[3] M.W., i. 265. M.S., ii. 416. W. Ellis: *Country Housewife's Companion*, 153-8, 168.

[4] M.W., ii. 84, 372. Defoe's *Tours*.

the black caterpillar."[1] Aylesbury was already famed for duck-raising.

Two sorts of *turkeys* are mentioned, the common Norfolk or Suffolk, and the blue Virginian. Great numbers of them were bred in East Anglia, even by little farmers and cottagers, and droves of large, fine birds, 300 to 1,000 together, were conducted to London, proceeding stoutly upon their feet. A daunting experience for the young to meet an army of turkey-cocks upon the march. A little later carts of four tiers were invented to carry them and the geese and chickens to market in a more convenient and rapid manner. Nettles were to be avoided near turkey-houses, lest they should sting their heads to death.[2]

Little is heard of *peacocks,* but a writer of the beginning of the century recommends that this bird should be kept from the garden by "a little sharp cur that will bark, tease him about as long as he can stand, at least till he takes his flight, and he will come no more there."[3]

In the farmer's *implements* a good deal of development took place. Among *ploughs* there was still much room for improvement, for the majority were still amazingly cumbrous, although towards the end of the century lighter and improved forms began to come into use. Of the varieties mentioned in the last chapter the heavy old wheeled plough particularly associated with Hertfordshire, a solid structure of wood with a beam 10 feet long and 6 inches square at the coulter-hole and a proportionately heavy share, was still used in the Home Counties, and slightly differing varieties in East Anglia and the South. In Norfolk it was guided by one handle and a rein and was distinguished by having the mould-board of wrought or cast iron instead of wood, which was an innovation. Cast-iron shares are said to have been invented at Norwich during the latter half of this century.[4]

Equally cumbrous was the old Kentish turnwrest plough, having "as much timber as would build a Highland cart"; a log was dragged behind to act as a shoe, and the solid beam rose high above it. The share had no fin. A somewhat lighter form was used on the South Downs, the shifting of the mould-board to make the furrows fall the same way going backwards and forwards being particularly suitable to hilly country.[5] In

[1] Ellis, op. cit., 162.
[2] Ellis, op. cit., 159-60. M.N., i. 375. Defoe's *Tours.*
[3] *Tusser Redivivus*, February 8.
[4] M.N., i. 53. Y.S., i. 107-8.
[5] M.S., i. 59; ii. 368.

West Sussex and other parts of the South a plough with a single small wheel in front, and a rising beam, was used, and the old, long, heavy form of the single wheel plough survived in the Cotswolds and surrounding districts, and in some parts of the Midlands.[1]

The swing, or wheelless plough, was common in different forms in the North, the Midlands, and parts of the South. An ancient and clumsy variety with a foot was used in the Sussex Weald and part of Surrey, and a curiously antiquated form throughout Devon and the West. Marshall describes the latter as the plough customarily portrayed by heralds and sign-painters, and hence considered it a survival of Norman or

FIG. 70.—MORTIMER'S PLOUGH (1707)

Armorican origin. It was ill-formed and long in the body, the mould-board being high over the keel, and without a curved breast to turn the soil over, the result being merely a rut and not a properly turned furrow. It was known as a "sule."[2] A more modern and scientific form was the short, light Dutch or York-shire plough (doubtless a development of the "Dutch bastard" of the last century), which was also a swing plough, but con-structed upon principles brought from Holland, and was common in Yorkshire and neighbouring parts of the Midlands. It had a "winding" or curved mould-board which, as in the modern plough, turned the sod well over, whereas the old straight or even slightly convex mould-board merely pushed the sod aside; a most important improvement in methods of tillage. The mould-board was still often of wood, but this material was coming to be reinforced or superseded by cast-iron plates. The form of this plough made in Rotherham and known by that name, was particularly popular in the Midlands.[3]

In the Home Counties the swing plough underwent more improvements than any other variety. A light form with a

[1] M.G., ii. 35.  M.S., ii. 172, 234.
[2] M.W., i. 126; ii. 144.
[3] M.Y., i. 257.  M.M., i. 104.

curved mould board was used in Bucks, and the enterprising Mr. Arbuthnot, of Ravensbury, Surrey, attached a notched regulator to the end of the plough-beam, by which the height of the draught-line was scientifically adjusted to the height of the horse drawing the plough.    A further step towards the modern implement was the invention of an *iron* swing plough, by Mr. John Brand, of Lawford, Essex.[1]

Circular coulters, like a little wheel, were sometimes used, mention of them being made in Lincoln and Gloucester, but the ordinary knife form was evidently more usual.[2]    Double and even triple-bladed ploughs were used occasionally, the latter notably by Mr. Duckett, of Petersham, Surrey, who also

FIG. 71.—ARBUTHNOT'S PLOUGH, SHOWING CHANGE IN SHAPE OF MOULD-BOARD.    (ABOUT 1771)

had a double-shared trenching-plough, drawn by four horses, the second blade making a deeper cut behind the first.    A still heavier variety was used for cutting drains by Mr. Arbuthnot, the other famous Surrey farmer, which cut a drain 16 inches deep with its two blades, one following up the other, and was drawn by eight horses.[3]

A plough with double mould-board, for making a trench or earthing up rows of plants, is mentioned as early as 1727, and was later made so that the mould-boards could be expanded at will to regulate the width of the trench.[4]    A useful paring plough, with a horizontal blade 14 inches wide to raze weeds, was the Berkshire shim, and besides the plain "stricking-plough" for cutting seed furrows there was a "drag-plough" for cutting a series of them, by coulters set on crossbars in a triangular frame, with a small wheel at each corner.[5]

[1] Y.E., ii. 211.
[2] Y.S., i. 149.  J. Mortimer: *Whole Art of Husbandry*, 39.
[3] Y.E., ii. 244, 518.
[4] R. Bradley: *Complete Body of Husbandry*, 35.  Y.E., ii. 502.  M.M., i. 106.
[5] M.S., i. 63; ii. 323.  Y.E., ii. 500.

Jethro Tull, the apostle of new agricultural methods which we shall presently describe, was the first to invent, at the beginning of the century, a really practical drilling-machine in which the seed was delivered through notched barrels, into regular furrows.[1] Originally designed for sainfoin and other light seeds, machines of this description came to be used a good deal for turnip seed, additional hoppers being sometimes added to distribute manure with the seed, some harrowing arrangement being added behind, and the whole contrivance drawn by a man or boy. But drill-ploughs, drawn by horses, were also used for the sowing of ordinary corn crops, and were made with from one to four blades, each provided with a hopper for the seed.[2]

Also connected with Jethro Tull and his methods were the new-fangled horse-hoes, to till the growing crops between their carefully drilled rows. These were made with from three to six stout blades like pointed hoes, but lighter sorts such as the "nidget" of Kent and the "tormentor" of Devon, had several rows of small triangular blades.[3] There were also weeding harrows, with one or more rows of teeth like a reversed rake, drawn by a horse and guided by handles. Heavy ox-harrows were still sometimes used for ordinary purposes and were occasionally provided with shafts and wheels, which was a step towards the modern "cultivator." Horse-rakes were used for cleaning fallows, and in Devon a heavy wooden-toothed ox-rake called a "drudge" collected the bits of pared sward for sod-burning.[4]

Harrows and ploughs were in many parts moved to and from the place where they were required on sledges, especially where the ground was soft. These were either of the ordinary shape with runners, or simply formed from the forked branch of a tree, with a triangular frame raised above it, and the staple for draught let into the stump. This sort could also be used as a snow-plough or road-maker. In many of the hilly districts of Wales and the West wheeled carts were still almost unknown, sledges or packhorses being used for farm transport. A peculiarity of Devon was the "Gurry Butt" or dung sledge, made like a truck, about 18 inches deep, on runners. If the ground was too rough and steep even for sledges, hay, straw, and

[1] Worlidge had put forward this idea in the last century but his machine was never actually made.
[2] Y.E., ii. 246, 510; iii. 83. Y.N., i. 317.
[3] Y.N., i. 95, 317. Y.E., i. 297; iii. 83. M.S., i. 63.
[4] Y.E., ii. 263. M.M., i. 108. M.W., i. 128.

similar matters were conveyed on horseback packed between crooks or bows; and manure, sand, and suchlike in "pots" or strong panniers with the bottom letting down at need like a falling door.

In Yorkshire a sledge drawn as it were sideways, and well weighted, was used for smoothing the surface of the meadows or spreading dung on them.[1]

Rollers of wood, tapering slightly at each end, or as a new invention set with spikes, were used for "reducing at once the surly glebe into a fine condition" suitable for tillage, especially for barley.[2]

Wagons varied much in size according to district. The men of Sussex negotiated their execrable roads with immense vehicles making a track 6 feet wide between their ruts, the wheels often being 6 inches broad. No wonder a team with chiming bells was needed to announce the approach of such a carriage. In the Midlands also they were large and clumsy, 14 to 15 feet long in the body and weighing over a ton, but only 5 feet 2 or 3 inches between the wheels. Kentish wagons were smaller, and those of Yorkshire "paltry," carrying only 40 to 50 bushels, as compared with the 90 to 100 bushels of other parts. But the best sort was found in Gloucester, and those parts of the West where wheeled carriages had penetrated. Their strong point was a wide, shallow body with side-rails arching outwards over the back wheels, thus lowering the bed of the wagon without sacrificing the steadying size of the wheels; a principle advocated by Worlidge in the last century. Marshall mentions a Turnpike Act which restricted the width of vehicles between the wheels to 4 feet 6 inches, which he considered too narrow.[3]

The old two-wheeled ox-cart or wain was common in Hereford, Somerset and farther west, the Cornish wain being a simple cart without sides but with two strong bows bending over the wheels to keep the load off them.[4] The Norfolk carts are specially mentioned as having the shafts continuous with the body and not hinged, which suggests that this was unusual. Some were constructed so that they could be converted into wagons by the addition of a detachable pair of wheels in front, thus making them steadier for loading.[5]

On the borders of Somerset and Devon horse-drawn barrows were used for carting manure, with three solid iron-bound

[1] M.W., i. 124.  M.Y., i. 262; ii. 121.  M.S., ii. 137.
[2] W. Ellis: *Husbandry* (abridged), i. 307.  Y.E., i. 293, etc
[3] M.S., i. 58; ii. 136.  M.M., i. 103.  Y.N., i. 251.  M.G., i. 57.
[4] M.W., i. 123, 310.  M.G., ii. 196.
[5] M.N., i. 51.  Y.E., ii. 141.

wheels of thick plank, about 2 feet in diameter, very like the carts described by Speed in 1659.[1] According to Marshall wheeled farm vehicles of any sort were unknown in most of Devon and Cornwall until about 1770, and even when he wrote in 1791 few had been introduced, packhorses still being the prevailing means of carriage, with sledges for harvest and a few other purposes.[2]

Carts and wagons were usually constructed of oak or elm.

The ancient *ox team* was retreating more and more before the improved breeds of farm horses, and, in some places, because of the hardness of better roads. By the last quarter of the century horses reigned supreme throughout the Midlands, the Home Counties and East Anglia, besides more outlying parts such as Westmorland and South Cumberland, the Vale of Gloucester, Thanet, and the West Downland of Sussex and Hampshire. Oxen still predominated in Wales, Devon and Cornwall, Hereford and most of Somerset; a good many were used in Durham, Northumberland, and the North and East Ridings of Yorkshire, in parts of Gloucester and Dorset, in Sussex and the Weald of Kent; and a few were still found in Surrey, Derbyshire, Wilts and the Isle of Wight. A very common practice was to mix the teams, a pair of horses, for instance, being harnessed in front of a pair of oxen, doubtless with the idea of hastening their pace, but this was not the custom in Sussex. In that county teams of eight to fourteen oxen were used, on the theory that to exert their full strength would retard their growth. Four or five horses were common for a team, harnessed in various ways, from four abreast to a string in tandem.

Oxen were as a rule yoked in the old way, occasionally with a sliding yoke to regulate the distance between them, but a few instances are mentioned of their being harnessed, three to five in tandem with collars and chains, and this was advocated as increasing their speed.[3] In Kent and Sussex they were usually provided with wicker or net muzzles and in Devon were still shod, a business to which it was well to train them young, as they were thrown on their backs and their bound feet held by a forked pole, of which the other end was firmly planted in the ground. In Devon they were still cheered on when ploughing by a perpetual chant "like a cathedral service," but in Sussex for some curious reason they are said to have been exhorted in the

[1] M.W., i. 384.
[2] M.W., i. 117.
[3] M.G., ii. 30. Y.E., i. 172. M.S., ii. 139.

T

"Yorkshire language."[1]  The tuneful Devon ploughman was not too energetic, even where his ground was hampered by large stones, against which his plough was apt to damage itself and he to be thrown forward with his face against it by the sudden jerk. Instead of removing these obstacles he got over the difficulty by fastening the draught chain to the plough-beam not with the customary iron bolt, but with a wooden pin, which broke with the shock of these collisions, and thus the plough-neck and the ploughman's teeth were saved from disaster.[2]  Writers of the time complain that an unnecessary number of both horses and oxen was nearly always used, eight being a fairly frequent team, and as many as twelve oxen not unknown, thus causing waste both in beasts and drivers.

The ridging up of ploughed lands was still a good deal advocated, and was approved by Tull, on the grounds that the water was drained off into the furrows, and that a certain amount of shelter was given by the ridges.  It was also argued that ridging increased the surface of the ground, and that thus the roots had more room and the ears more air; but this was not really of much advantage except to pasture-land, where the grass did especially well in the furrows, and the ridges were drier for the beasts to lie on.  The ridges varied a good deal in height and width in different places. In Sussex and Devon they were mostly shallow and narrow, but in parts of the Midlands were as much as 3 feet high.  Gloucestershire beat even this, for they were often 4 feet high, to a width of 15 feet, so that when the corn was fully grown a small man walking in the furrow disappeared from view.  In the past they had been still higher, and we are asked to believe that a certain man who hired six or seven teams and sent them to a field to plough, when he arrived there himself was unable to perceive any of them, as they were all at work in the furrows, so turned away supposing he had come to the wrong place.  However, by the time Marshall wrote, the prevailing ridges in that county were only from 2 to 2½ feet high, and 8 yards wide.  In East Norfolk the ridges were carefully directed north and south to get the maximum of sun.[3]  But a number of farmers objected to the practice, both because in wet seasons the furrows were apt to become stagnant canals of water, merely wasting the ground, and also because they made cross-ploughing impossible.  On level fields this cross-ploughing or pin-fallowing, by which the ground at each fresh turning was ploughed at

[1] M.S., ii. 136.  M.W., i. 119.
[2] M.W., i. 140.
[3] M.G., i. 76-7.  Y.N., i. 57.  Jethro Tull: *Horse-hoeing Husbandry*, 243.

right angles to the former direction, was becoming popular, as giving a better tillage, and occasionally the furrows were run from angle to angle of the field. Four ploughings before sowing wheat were quite common, and sometimes as many as five or six were given.

In the early part of the century Jethro Tull endeavoured to impress upon the farmer two principles which were to have a lasting effect on agriculture. These were, first, the necessity of repeated tillage of crops by hoeing *after* ploughing as well as before, both to keep them clear of weeds, and also by breaking the surface at intervals to bring fresh nourishment to the roots; and second, the drilling of seed in regular rows, a practice which not only made systematic attention with the hoe possible, but was also much more economical of seed than the ordinary broadcasting. Manuring he disapproved of, because it encouraged weeds, and also for the somewhat fastidious reason that plants "bred up and fattened amongst these toads and corruption" could hardly be good to eat; but he maintained that with diligent tillage and cleaning the ground would continue to give good crops without needing to lie fallow.[1] Hence the introduction of turnips as a "fallow crop," which if well tended bore out the truth of his assertions. His prejudice against manure was not accepted, but drilling and hoeing were found to be particularly adapted to the growing of turnips. With wheat the system was not so successful and was little adopted, except by such super-farmers as Coke of Holkham and his peers, although Young, writing in 1771, states that there was "a certain amount" of it, and that it was common in Kent, where they drilled their rows much closer than Tull advocated.[2]

In East Norfolk exclusively the farmers adopted a habit, towards the end of the century, of dibbling their wheat (as they had always done their peas) with egg-shaped dibblers, one in each hand, three droppers with the seed following the dibbler, and dropping two or three seeds into each hole, after which they were covered with a bush harrow.[3] Young noticed in Yorkshire an ingenious contrivance for sowing in drills; a skin bag hung round the man's neck from which the seed was dropped through a funnel, its discharge being regulated by the turning of a handle.[4]

But broadcast sowing and hand weeding continued the

---

[1] Jethro Tull: *Horse-hoeing Husbandry*, 50, etc.
[2] Y.E., iv. 196, 215. Marshall, however, does not notice it in that county.
[3] M.N., ii. 36-40.
[4] Y.N., i. 209-10.

common practice, and for the most part it was only in the turnip field that the mediaeval weeding hook and forked stick was supplanted by the hand-hoe or the horse-drawn implement. In some places the seed was still steeped in brine and afterwards dusted with lime to guard against smut.

Very varying quantities of all seeds were sown, and many new rotations of crops were tried in all but the most backward districts and where the old open fields remained; the general principle among enlightened farmers being to insert a crop of roots, pulse, or grass between every two crops of corn. The best of the new rotations was the famous "Norfolk course," of wheat, turnips, barley, clover (or clover and rye grass), and back to wheat again, without any interval of fallow, the turnip crop if well hoed answering that purpose. The clover could be varied by other artificial grasses, and the wheat by oats or peas, and variations of this admirable course were adopted in many parts of the kingdom. Thanet farmers worked on the principle of cleaning their lands before spring corn and artificial grasses rather than before wheat. East Kent had a "round tilth" of wheat, barley, beans, wheat, which answered quite well when the beans were drilled and well hoed. But the old rotation, regardless of roots and grasses, lingered in many places, and from Dorset westwards it was common to grow wheat, barley and oats after each other with some years of pasture or sometimes clover and rye grass, before returning to wheat.[1]

Barley and oats were universally mown with the scythe, and in some parts wheat also, although it was more usual to reap that crop with the "shearing hook." In the North the corn was mown with an inward motion, and gathered into sheaves by the "cradle" on the scythe; in the Midlands and South it was mown outwards, with or without a cradle, dried in swathes and afterwards raked into cocks. In the West all were bound into sheaves owing to the use of packhorses instead of wagons. The laden packhorses were driven in strings to the barn to be unloaded, and galloped gaily back to the field by their boy, to the dismay of peaceful travellers in the road. In that part the harvest was generally ten days or a fortnight later than the rest of England owing to the greater rainfall.[2] In damp seasons the corn was

[1] Y.S., i. 285.  M.N., i. 132.  M.S., ii. 13.  M.W., i. 137.

[2] It should be remembered when considering dates of the month in this century that in 1752 the Calendar was changed in England from the old style to the new. September 3rd of that year was accounted September 14th, a sudden curtailment of autumn which caused considerable disturbance at the time. After 1752 dates were the same as at present: before that they occurred twelve days later in the season than now.

often put up in little stacks in the field called "Arrish mows," and there the "Harvest holla," or signal for rejoicing, was given after the last wheat was cut, instead of after the carrying was completed, as in the rest of England. In Norfolk regular paid harvesters were employed, who played cards in the barn when it was wet, but moved with great dispatch when at work. In the Midlands proceedings were not hurried, and all labourers who had been employed throughout the summer were given a load of coals at the end. Gleaning was still allowed, but there were many complaints that it ought to be confined to the really poor and needy.[1] In Gloucestershire to save time and housework there was no regular dinner for the harvesters in the middle of the day, but a good hot supper at night. Prudent farmers fattened "some old animal" to feed the workers on, and they were given this, or pickled pork, and suet puddings, furnished with plums for the first fortnight but plain after that. According to Ellis they had five meals a day, at least two with meat, and the others chiefly bread and cheese and apple-pie. Tea appears for the first time at their breakfasts, and beer flowed plentifully, "small beer" on the field, and the better quality morning and evening in the house.[2]

In the threshing of the corn curious variations occurred. In Yorkshire a return to (or survival of) the primitive method of treading it out was noted in the case of oats. When mown they were laid out in a circle on the grass, or on a sheet by the stack with the heads inwards, and were trodden by eight or ten men.[3] In Devon and Cornwall, although barley and oats were threshed in the ordinary way with a flail, a different method, probably very ancient, was used for wheat, in order to preserve the straw unbruised for thatching. It was first either lightly threshed with a flail, or else held in bundles and the ears beaten over the edge of a cask by hand. It was then bound in double handfuls just under the ears and hung on a hook, the weeds were raked off the straw with a long toothed comb, and the "reed" laid evenly on a heap. The ears were then threshed again, lightly with a flail or over a cask, and the straw done up into 36-lb. bundles.[4]

But side by side with these primitive methods machines for threshing were beginning to appear, although some mystery is

<hr />

[1] M.G., i. 101.  M.W., i. 14, 166.  M.M., i. 165-8.  M.N., i. 184.

[2] W. Ellis: *Husbandry* (abridged), ii. 35.  Ibid, *Country Housewife's Companion*, 47, 71.  *Tusser Redivivus*, August 9.

[3] M.Y., ii. 20.

[4] M.W., i. 179.

preserved about them. Young mentions one at Belford (Northumberland) worked by a horse, with a boy to drive and a man to feed it, but the details were only obtainable on subscription. Another "thrashing-mill" is vaguely mentioned by Marshall in the West,[1] but the first really successful one was invented in 1786 by the Scotsman Andrew Meikle, and was worked on much the same principle as modern ones. In this the loosened sheaves were passed from a feeding-board between two fluted revolving rollers to a beating cylinder, where the ears were struck by four iron-shod beaters. From this a series of revolving rakes conducted the corn on to a concave sieve beneath, and the straw was afterwards ejected. It was not, however, until the following century that these mechanical devices were widely adopted. The ordinary "tasker" or thresher with a flail wore shoes made out of an old hat so as not to bruise the corn. A machine-fan for winnowing was introduced from Holland about the middle of the century, and improved by Marshall's father. It needed careful regulation, but when properly adjusted saved much time and labour, and seems to have been used a good deal in the North,[2] and was penetrating into other parts, even the far West. In the Midlands and South the sail fan, which seems to have supplanted the wicker fan, was largely used. The former was composed of slats arranged in a cylindrical form, with a piece of sacking attached as a sail to each slat. This needed three men; one to turn the fan, one to heave the corn, and one to fill the riddle. The old method of cleaning the corn by casting was considered by Mortimer superior to any other, and survived in some parts, particularly in Norfolk and the Chilterns. It was dexterously cast by a shovel the whole length of the threshing-floor, against the wind, once or twice, and then sieved on a riddle.[3] In Devon and Cornwall a still more primitive method survived, by which the corn was carried on packhorses to "the summit of some airy swell" and there spread before the wind by women; a "slavish and truly barbarous employment" on a chilly autumn day.[4] At the mill the taking of toll in kind was abolished in 1796, and a money toll made compulsory instead.

Many new sorts of wheat and other grains were used, and the idea of sowing wheat in spring, suggested vainly in the last century, was now to some extent taken up. In Yorkshire and

[1] Y.N., iii. 47.  M.W., ii. 254.
[2] M.Y., i. 26S, 361.
[3] M N., i. 189.  W. Ellis: *Husbandry* (abridged), i. 261.
[4] M.W., i. 182.

Gloucester it did not "gain establishment," but in the Midlands it was very successful. It was sown in April.[1] In some places the ground was levelled with spade or mattock before wheat was sown, and afterwards it was often rolled or sheep-trodden to prevent wireworm. Smut was combated by treating the seed with salt and lime, or arsenic. The practice of eating down too forward wheat with sheep was now condemned.

Common long-eared barley was the most usual variety of that grain, but two important new sorts appear, a six-rowed or winter barley in Yorkshire, which it was admitted was hardly yet understood, and a beardless kind grown in Staffordshire, mentioned by Mortimer as early as 1707.[2] Linnaeus seems to have first hit on the idea of timing sowing by the foliation of the trees instead of by the Calendar, hence the Norfolk rhyme:

> When the oak puts on his gossling grey
> 'Tis time to sow barley night and day.

Oats were the chief crop in moorland districts, such as the wolds of Yorkshire and the higher parts of Devon and Cornwall, but they were not so popular elsewhere.[3] Rye was practically confined to the moorland dales of Yorkshire, and the new white or Dantzic variety had almost entirely replaced the old black. Meslin also was grown in that district, the bread made from it being esteemed there, and it was thought that the rye kept mildew from the wheat. Rye by itself was grown in Kent to feed cart horses, and with tares in West Sussex for sheep food.[4] Dredge seems to have fallen into disuse. Beans were much grown in Kent, being included in the ordinary course of crops, and were drilled in wide rows and horse hoed. Vast quantities were grown round Bedford, but not so many in the other parts of the Midlands. In Gloucester peas and beans were dibbled and hand hoed by women, but in the Sussex Weald and in West Dorset beans were not grown at all, and neither peas nor beans in the hilly districts of Devon and Cornwall. Both were specially well cultivated in Suffolk, separately, and in Yorkshire a mixture of the two, or of beans and fitches (vetches), were often sown together.[5]

The yield of crops seems to have varied so much that it is difficult to arrive at an average, but roughly the usual quantities seem to have been, wheat 3 to 4 quarters, barley 3½ to 4½

[1] M.M., i. 183. M.Y., ii. 44. M.G., i. 113.
[2] M.Y., ii. 15. J. Mortimer: *Whole Art of Husbandry*, 100.
[3] M.Y., ii. 344. M.W., i. 190.
[4] M.Y., ii. 14. M.S., i. 137; ii. 188.
[5] Y.N., i. 41. M.Y., i. 24. M.G., i. 90, 144.

quarters, oats 3 to 5 quarters, rye about 3 quarters, peas 2½ to 3½ quarters, and beans 3 to 4 quarters. On some of the improved farms the yield was much higher.[1]

Hay meadows, in parts which still followed the old order of things, were left open to the cattle until April, or even May in the North, which hardly gave the hay a fair chance; however, the ordinary yield there was 1½ to 2 tons per acre. The after-grass was sometimes left as late as Michaelmas before the cattle were turned on to it. But now, instead of trusting to natural growth for this kind of food, farmers began to sow "permanent leys" with a suitable selection of clover and grass seeds. A common mixture for this purpose was red and white, or Dutch clover; trefoil (hop clover) or occasionally burnet; and rye grass (darnel), rib grass, or dwarf meadow grass.[2] Experiments were made with other sorts of grass, and Coke of Holkham (at the very end of the century) was particularly addicted to cocksfoot grass, the seeds of which he caused to be collected by women and children, paying them 3d. a bushel for the ears. Mr. Blomfield, in the same county, introduced the practice of "inoculating" pasture, by which a tract of land was dotted over in winter with small pieces of turf, carefully rammed down, and a few grass seeds sown in the intervals in spring, the turf thus spreading over the whole.[3]

Clover, most thoroughly cultivated in East Anglia, spread from there across the centre of England, was much cultivated in Kent, and to some extent in the West, but was far from being universally grown.[4]

The other "artificial grasses," sainfoin, lucerne and burnet, were still in the experimental stage. The two former did well in the North and in Kent, but lucerne particularly was little cultivated elsewhere. Burnet met with little success except in Rutland, for although it should have been very useful as an early spring food which would stand frost, opinions about it were contradictory, and as cattle would not eat it too near seeding it was considered best mixed with other foods.[5]

Cabbages as a field crop were favoured a good deal in the North, but not so much elsewhere. It was realized that they kept the cows in good milk through the winter with less expense than on hay, and that they filled a gap in March and April, between the end of the turnips and the beginning of the spring

[1] See YS., 242 on.
[2] M.Y., ii. 88.  M.N., i. 301.  Y.N., ii. 239, 374, 490; i. 31.
[3] *Pamphleteer*, xiii. 458, 467.
[4] Y.S., 281.  M.S., i. 145.
[5] Y.N., iv. 218, 222; i. 79.  M.S., i. 153, 155.  Y.E., iv. 141, 148.

grass, but at that season it was necessary to remove carefully the rotten outside leaves, as these made the butter "absolutely stink." Used for sheep they had not this disadvantage. The large Scotch cabbage answered the best, and after that the red. The Anjou did not do so well, and the Russia and Savoys would not stand the winter.

The cabbage-turnip or kohl-rabi was introduced as a field crop in 1767 by Mr. Reynolds of Addisham, Kent, and was also valuable as a late spring food. The tops were fed off by cows and the roots dug up for sheep or pigs.[1] A few growers of "turnip-cabbages" are mentioned in Yorkshire, but there they were not considered as good as the ordinary Scotch cabbage. Borecole or kale was sometimes grown.[2]

That invaluable root, the common turnip, so inseparably connected with the name of Lord Townshend, was as a matter of fact cradled in East Anglia considerably before his day, although to him belongs the credit of developing its culture on the lines laid down by Tull.[3] Over the rest of England they were somewhat sparsely cultivated, reaching the West about the middle of the century, and Yorkshire and Kent even later. Some counties, notably Dorset, did not favour them at all, and even where they were adopted they were often insufficiently hoed, or not at all, thus doing away with their value as a fallow crop. The seed was sown from June to August, and four sorts were principally used, the common white stock, or Norfolk, the purple, the green, and the pudding stock or "tankard" of the Midlands. In Norfolk the seed for sowing was carefully raised in plots of old transplanted roots. When the seed was ripening the birds were scared off either by boys or by the ingenious expedient of fixing a bell on a post in the centre of the patch and running a cord from it to the kitchen. The bell pull was hung in the most frequented part, so that mistress and maids passing kept it well going, and the birds "having no respite from alarms, forsook their prey."[4]

When the turnip crop was grown it was either eaten in the ground by sheep, which was wasteful as part of the roots were uneaten, or pulled and eaten on the spot by cattle, or carted to the yard or other fields where they were wanted. A machine for slicing turnips was used at Belford (Northumberland) which consisted of a blade fixed in a horizontal frame, pulled back-

[1] Y.E., iii. 60.
[2] Y.N., ii. 373; iv. 201. Y.E., i. 76.
[3] Ernle: *English Farming*, 166. The culture was established there at the end of the seventeenth century.
[4] M.N., i. 257, 283.

wards and forwards on rollers under a hopper, by two men, seated at each end.[1] The young turnip plants suffered a good deal from fly, and a preventive expedient was to burn heaps of weed and straw on the windward side of the field so that the smoke blew across it. Tull advocated drilling the seed at two different depths, and with mixed old and new seeds, so that it would come up at four different times and some of it would be sure to escape the fly.[2]

The Swedish turnip or swede made its appearance in England towards the close of this century, and the mangel-wurzel is said to have been introduced from France in 1786, only the leaves being used at first, and the roots thrown away, until it was noticed that swine relished them. The introduction to Suffolk of these two very valuable additions to cattle food is attributed to Arthur Young,[3] and a little later were carefully cultivated by Coke of Holkham, swedes forming his principal and most valued root crop. To keep them he took them up in November, cut off the tails and stacked them in an orchard or on old turf land, tops upward, and gave them a thin covering of litter in severe weather. They were sliced for feeding sheep.[4]

Carrots were a root crop for which Suffolk had long been famous, but elsewhere they were very little cultivated except experimentally on model farms. They were given to cows and other cattle, but were specially useful for fattening pigs.[5]

Potatoes flourished in Yorkshire, Lancashire and the North, a good many were grown in Essex where they were considered a suitable preparation for any kind of grain, as their shade mellowed the ground. In Norfolk, Kent and the West they were introduced late in the century, and Ellis in 1772 refers to them as a crop little cultivated. Large crops of from 200 to 300 bushels per acre were obtained in Yorkshire and the Midlands, and they were used, boiled, principally to feed pigs, but all sorts of cattle and poultry as well. They were stored in shallow pits lined with straw, ridged up and covered with straw and earth. The disease of curlytops prevailed for a time, but was got rid of by using new seed.[6]

Lentils, formerly a fairly common crop in Yorkshire, had now almost disappeared, being found only in a few places in the North. Buckwheat, too, had fallen into disuse in most parts.

[1] Y.N., iii. 49.
[2] T. Tull, op. cit., 112.   Y.E., ii. 387.
[3] J. Caird: *English Agriculture*, 144.
[4] *Pamphleteer*, xiii. 461-8.
[5] Y.E., ii. 170.
[6] Y.N., iv. 155.   Y.S., i. 263.   M.S., i. 142.

Some was grown in Norfolk and a little in Yorkshire, the common use being to plough it in as manure when in flower, or sometimes to fatten hogs on the grain.[1] The culture of flax and hemp was declining. A quantity of hemp was grown at the beginning of the century in the Ely fens and surrounding districts, and flax still flourished in the Vale of Pickering, Yorks, in 1787, but even there was growing less, and except for a few spots on the south coast was hardly found elsewhere.[2]

Rape was very popular in Yorkshire, especially round Pickering. It was sown in July, one gallon to an acre, transplanted to fill gaps in October, and reaped the following July or August, yielding about four quarters an acre, the seed being sent to the oil mills. A delightful description of the busy scene at a public rape-threshing, to which all the countryside collected, is given by Marshall. Cloths of "hessen" twenty yards square were laid upon a smooth piece of ground, and the rape conveyed from the fields where it had been drying to this spot in "carrying cloths" slung on poles and carried by two men on their shoulders. On the spread "hessen" the threshers, marching in pairs, dealt with the crop, the straw of "almost silver brightness" being thrown aside, and the seed raked up into bags, the whole making a most animated scene: "The two divisions of threshers, moving in close phalanx, with flails nimbly brandishing, sometimes in open view, sometimes partially hid among the piles of straw; the clothmen busy and attentive to their various employments; the team drawing off the loaded seed; the carriers from every hand pressing to the threshing-floor with their seemingly cumbrous loads; and the distant groups of fillers, scattered on every side of the foreground, could not fail of affording matter interesting to the painter."[3]

The great hop-growing districts were Kent, and the neighbourhood of Farnham in Surrey. Then as now an army of hop pickers descended upon the gardens early in September, and were accommodated in any kind of place: "If superior happiness belong to the cottage," observes Marshall, "how supreme must be that of a hopper's hut!" There was a "hop-supper" at the end of the picking, and at Farnham this was extended to revelry, in which decorated women paraded the streets singing and shouting, which was followed by dancing and drinking. Oasthouses with their peculiar cowls were naturally a feature of the countryside, each containing receiving rooms for the green hops.

[1] M.N., i. 254.
[2] M.Y., ii. 64. Defoe's *Tours*, i. 76.
[3] M.Y., ii. 27-42.

several kilns for the drying, storerooms for the hops when dried, a packing place, and a warehouse below for the completed packages. The kiln was usually cube-shaped, the hops being laid on a platform of wooden bars covered with hair cloth, and the heat supplied either by an open stove underneath, or at the side with gratings, or otherwise by an iron cylinder or brick flues passing through the space. Coal was the fuel used in these closed ones, but otherwise charcoal, or coke and charcoal mixed. The dried and cooled hops were poured through a hole in the floor into bags suspended beneath, the unfortunate "bagster" standing in the bag and treading them down with the hops showering on top of him, a "filthy and laborious job." The usual produce was half a ton per acre.[1]

Occasional crops in particular districts were saffron in Essex, woad and weld in Northants, madder in Surrey and Kent, and tobacco in the Vales of Pickering and York. It was not legal to grow more than half a rod of tobacco, and that for purposes of "physic and chirurgy," and it was used for cutaneous disorders of cattle and sheep.[2]

The favourite dressing for all kinds of crops during this century was lime, and almost as popular was the method of paring the surface of the ground, burning the sods and sprinkling the ashes, as described in the last chapter. Suffolk had its special dressing of crag, a soil composed of crumbled shells, and in Norfolk wonders were worked with marl, under the vigorous leadership of Lord Townshend. Folding was neglected in the North and West, but was practised to a certain extent in other parts. Ordinary farmyard manure was comparatively little used except on pasture lands, but some farmers made good compost by laying a thick deposit of litter or earth over their yards at the beginning of winter, when the cattle were kept at home. Some of the strange mixtures of the last century were also used, and occasional experiments tried with rape dust, oil cake, and bone dust. Soot was sometimes used as a top dressing.[3]

Proper drainage of the fields did not receive due attention over the greater part of England, but in Suffolk and Essex hollow drains two feet deep were carefully constructed, the bottoms being filled with small wood and straw, earthed over. In a few parts of the Midlands and North there were drains on the same principle but filled in the bottom with stones, and on one or two model farms actual tunnels were constructed, with brick arches.

[1] M.S., i. 190-284, 407; ii. 53-77.
[2] M.Y., ii. 75.
[3] Y.N., iv. 482. Y.E., ii. 173. M.W., ii. 21, etc.

Up-to-date farmers such as Arbuthnot were careful to have a good hollow drain in each of the furrows of their ridged lands, but the ordinary farmer was content with letting off the surface water. Early in the century Bradley advocated the use of a Persian wheel for raising and carrying away the water from the draining trenches of a "piece of ground annoyed with water," and this system seems to have been in use in Holderness in Young's day. Another idea of Bradley's was to carry water over a hill through an arch of elm pipes, having the discharging end lower than the other, but probably this new principle was little grasped. Joseph Elkington, in the latter half of the century, was the first to practise "sink-hole drainage," by which, after cutting a deep drain, holes were sunk into the porous sub-soil, and the water pent up there released and carried away.[1]

The method of watering meadows underwent a slight change during this century. Water was still directed over them from the nearest stream, with sluice-gates and trenches, but instead of damming up the water at the bottom of the field and letting it lie stagnant on the ground, which was called "floating," the dam was now placed at the top, and the water allowed to run over the meadow when desired and drained off at the bottom, with the idea that running water was less chilly to the ground than stagnant.[2]

In the vegetable garden few new plants were introduced, except apparently salsify and celery.[3] Large market gardens were springing up round London, notably at Sandy and Lewisham, to supply the capital with vegetables. As an addition to the fruit garden the modern large red strawberry was evolved from large varieties introduced from Carolina, and from Chile by way of Holland early in the century,[4] the previous sorts having been derived from the little wild English strawberry transplanted. The Dutch red currant was now esteemed above the English sorts, and many sorts of gooseberries were grown, white, green, yellow, red, black and striped. Raspberries were "layered garden," white, or the common wild sort.[5]

Orchards, according to one writer, were usually little enclosures fenced off near the house or garden regardless of aspect,

[1] R. Bradley: *Complete Body of Husbandry*, 115-17. Y.N., i. 241; ii. 254, 415. Y.S., i. 83.

[2] M.M., ii. 56; i. 223.

[3] Y.N., ii. 414. J. Mortimer, op. cit., 482.

[4] Fair-sized strawberries from Virginia were introduced in the seventeenth century.

[5] J. Mortimer, *Whole Art of Husbandry*, 455-88.

but in Gloucestershire care was taken to face them south-east and shelter them from the north, and all over the South and West of England orchards flourished. The Ribston pippin was introduced by Sir Henry Goodriche of Ribston, who raised the first trees from the pips of a particularly good apple he ate in Rouen in 1709.[1] In Kent apples, cherries, filberts, some pears and a few walnuts and chestnuts were grown, the filberts being trained into the form of a drinking cup with a short pedicle, and the stems of all the trees being whitewashed, to keep off sheep and hares. In Hereford, where liquor was the chief object, the favourite apple was the "stire," and after that the Hagloe crab and Golden pippin, with a few of the old Redstreak. The "squash" pear was most favoured, and also the Oldfield, Barland and Red. The trees there were planted at least 10 to 20 yards apart, and often 22 yards, pear-trees outliving two or three generations. In Devon and Cornwall apples were grown for cider, and pears, cherries and walnuts for the market, the trees there being trained particularly low, only 3 or 4 feet high, and 5½ yards apart. The height increased towards Somerset, and round Taunton they were 5 or 6 feet high. In Devon only horses and calves were allowed to feed in the orchards, but in Gloucester cattle fed between the widely separate, well-guarded trees, and in Hereford and sometimes in Kent crops were grown between them. In Kent only the surplus apple crop was made into cider, but in the West liquor was the primary consideration. A drink of mixed apples and pears. or of wild crabs and sweet pears was made in Hereford for family consumption, as well as proper cider and perry for the market.

The mill-house, or cider-house, which was inevitably attached to a fruit-growing farm, contained an upper chamber where the fruit was matured (and usually left too long), a mill below for breaking up the fruit, and a press. In the West the mill was usually a circular one like a quern, turned by a horse, but in the South it was a wooden roller stuck with hobnails, working against an upright board also nailed. In some parts of Cornwall the old method of pounding the fruit in a tub with club-shaped wooden pestles set with nail-heads, still survived. After milling the "pomage" or "must" was pressed, either with a double-lever and suspended weight, or the more modern screw-press. The juice was run in pipes from the millhouse into casks to ferment in the West of its own accord, but in the South with the help of a little yeast toast. It was observed that in the soft climate of

[1] Rider Haggard, *Rural England*, ii. 290

the West cider seemed to unbrace and relax rather than to give energy and cheerfulness.[1]

Farm servants were a good deal simplified from the old days, consisting commonly of a ploughman, who looked after the horses, unless he had a horsekeeper to help him, and often shod them or even mended their harness at a pinch; a shepherd, a tasker or thresher, and an odd man, whose duties were so extensive that, like the women, his work was never done. He waited on everybody, helped with the cows, calves and hogs, and the dairy, fetched fuel and water and heated the oven, helped at haytime and harvest, and ran errands for master, mistress and maids. He may well have been glad of a little relaxation, but one writer recommends that a grindstone be kept at every farmyard, to keep the servants from always gadding to the smithy, "which with the mill is the seat of news." It was often the custom in summer, both indoors and out, to rise to work with the dawn and have a good interval for rest in the middle of the day, from 11 to 1.30 or 2 o'clock, after which work on the farm was resumed again until 7 or 7.30.[2]

Malt was now made entirely by maltsters and not at home, which would relieve the housewife in one of her departments, but she still usually superintended the brewing. Otherwise she was busy enough with the pigs, poultry and dairy.

Now that beef was killed all the year round, instead of the old-time slaughtering for "hung beef" at Martinmas, the farmer ate much more fresh meat, at any rate where neighbours were numerous and near enough to share a bullock or a sheep: in isolated spots much still had to be salted. In many parts meat was given three times a day, but in more frugal districts, such as Yorkshire, milk foods were still the staple diet with meat once a day, and broth was still a standby in the North. In the Cotswolds bacon and vegetables were still mostly eaten, and in Devon and Cornwall bacon with potatoes, skim-milk cheese and barley bread. Pickled pork was still favoured, and apple-pies and pasties had become a main part of the farm labourer's fare, particularly for taking out to the fields. Wheat bread by the end of the century had become the staple fare in most parts, but other sorts survived in the North and West. In Durham rye or meslin was commonly eaten, in Lancashire oaten bread, in Westmorland rye and barley, in Cumberland, Devon and Cornwall barley

[1] M.G., ii. 217-358. M.W., i. 209-31. M.S., i. 304-17.
[2] D.H.: *Tusser Redivivus*, December 9. W. Ellis: *Husbandry* (abridged), i 336-43.

bread, and in Wales oats and barley.[1]   Oatcake of the kind still associated with Yorkshire was made there, the oatmeal being mixed into a batter and poured out on a hot stone to cook. Occasionally in the North lentils and oatmeal were mixed for bread; or bean and pea flour, which made a "rank hearty bread" which would be little appreciated in these days.   Home-made cheese was of course a great deal eaten, and in Gloucestershire and other cheese-making districts whey butter was largely consumed, being preferred to poor quality milk butter. Cheese, however, was too expensive for the poor labourer.

Beer (rather than the true old ale) seems still to have been a good deal brewed and drunk at home, especially at harvest-time, until the tax on malt made it too expensive, and in the West large quantities of cider were consumed, two gallons a day being allowed on some farms.   But the great innovation of this century was the spread of tea-drinking, not a beneficial change in some ways, for good tea was still extremely expensive, and what the labourer had learned by the last quarter of the century to drink in such quantities though not intoxicating was very inferior and probably unwholesome stuff.   Wines were made from all kinds of fruits by the more skilled, or leisured, housewives, and in Kent a very good black-currant wine called "Gazle" was made.   The old-fashioned mead or metheglin, and birch wine, made in March, are mentioned early in the century, together with a concoction called "Mum," composed of malt, oatmeal, beans, flowers and herbs, finished off with half a score of new-laid eggs, and closed to mature for two years;[2] but probably in the more bustling times of what is known as the "Agricultural Revolution" these more elaborate drinks passed out of fashion.

---

[1] James Caird: *English Agriculture*, 283, 367.   E. Davies: *Case of Labourers in Husbandry*. 133-5.
[2] I.M.: *Whole Art of Husbandry*, 602-8.   M.S., i. 318.

# CHAPTER XII

## THE NINETEENTH CENTURY AND AFTER

THE beginning of the nineteenth century saw the new order of things established in the agricultural world, and the Enclosure Acts of its early years brought about, with a few isolated exceptions, the final extinction of the mediaeval communal village farm. It was, of course, intended in this process that everyone should have his compact little holding instead of a number of scattered strips in the village fields, and that he should be no worse off than before, but in practice it unfortunately worked out very differently. Men who were merely tenants, without right to the soil, were often turned out; the peasant, hampered by the necessity of fencing his new holding and harassed by changed conditions and rising expenses, often sold his acres to the landlord. But the great stumbling-block was the loss of the commons, those lands so invaluable to the poor, of which the simple uses are so well set forth in the poem quoted by Lord Ernle:

> *Thomas.* Why, 'tis a handy thing
> To have a bit of common, I do know
> To put a little cow upon in spring,
> The while woone's bit ov orchard grass do grow.
>
> *John.* Aye, that's the thing, you zee. Now I do mow
> My bit o' grass, and mëake a little rick;
> An' in the zummer, while do grow,
> My cow do run in common vor to pick
> A blëade or two o' grass, if she can vind 'em.
> Vor tother cattle don't leave much behind 'em.
> An' then, bezides the cow, why we do let
> Our geese run out among the emmet hills;
> An' then, when we do pluck 'em, we do get
> Vor zëale some veathers and some quills;
> An' in the winter we do fat 'em well,
> An' car 'em to the market vor to zell
> To gentle-volks. . . .
> An' then, when I ha' nothin else to do
> Why, I can tëake my hook an' gloves an' goo
> To cut a lot o' vuzz and briars
> Vor hetèn ovens or vor lightèn viers;
> An' when the children be too young to earn
> A penny, they can g'out in zunny weather,
> An' run about, an' get together,
> A bag o' cow-dung vor to burn.[1]

---

[1] Ernle: *English Farming*, 306, quoting Barnes, the Dorsetshire poet.

Under the old conditions a co-operative concern, run with the prestige and experience of centuries, although antiquated in method and unsuited to the developments of higher farming did, from the point of view of giving support to the mass of the villagers, succeed better than the efforts of detached units, and consequently when the ruthless march of progress overthrew the old system, disaster ensued to many of the poorer inhabitants. Farming under the new system could produce meat of good quality, plump poultry, varied and well-tended crops, in quantities to satisfy an abundant market, and against this it was hopeless for the miscellaneous and scraggy beasts of the village commons, and the limited and hazardous crops of the open fields to compete.   The process of enclosure, already far advanced, inevitably pushed on to completion.  But the bewildered peasant, accustomed from time immemorial to mutual working arrangements with his neighbours, and the communal rights of pasture, woodland, and fuel, can hardly have known what to do when suddenly thrown on his own resources and obliged to act on his own initiative.  His plot of exchanged and consolidated land was often insufficient to maintain his cattle and poultry, and gave him little or no fuel or wood for repairs; he had no common on which to run his geese and pigs, the new factories in the towns deprived him of his little domestic industries, and unless he had unusual resources of mind and pocket he often came to grief.  Many, as we have already seen, became that unhappy thing the landless agricultural labourer.

On the other hand, the well-to-do peasants and the small farmers who had already established a measure of prosperity had an opportunity of rising.  Many of these sold their land and sank their capital in large farms and did very well, if they could afford to do things on a large enough scale.  General conditions were not in favour of small farms, yet many of them sprang up round large towns, which created a special market for their produce.  Good enclosed land could produce far better stock than in the old days, and within limits could do well in small lines such as calf-rearing, chicken fattening, or vegetable-growing, but with the increase of population and labour-supply which took place in the eighteenth century, and with improving agricultural conditions, the scale was turning in favour of the large farmer, who not only could produce large quantities of corn and rear good stock on a generous scale, but with his big turnover could afford to bide his time for selling, whereas the small man must perforce sell continually to pay his debts, and not when prices were the best.

Yet in this country of odd survivals one or two villages worked on the ancient open-field system still remain, notably Laxton in Nottinghamshire, and possibly some others.[1] But by the nineteenth century England had long ceased to be a land of peasant farmers, where every country-dweller had his own few acres to till; now the classes concerned with agriculture had come to be, most usually, large landlords farming but little if any of their own land, many substantial tenant farmers, and a few independent ones, and below them the cottage labourers with no stake in the land. The practice of having the farm hands living in the farm-house began to fall out of use early in the nineteenth century, which would increase the cottage population.

The fortunes of these different sections of society, now unhappily become separate in their interests, may be followed in any Economic History, but we must return to tracing the progress made during this period in the farm-yard.

The nineteenth century, as we all know, was a time of marvellous scientific discovery and invention in every direction, and to the assistance of the capitalist farmer all these things were turned, until the earliest and humblest of occupations came to be organized on a scientific basis and assisted by every kind of mechanical power.

In the opinion of some these changes were not the product of the farmer's brain working out improvements by personal experience, as in the case of Jethro Tull and others in the eighteenth century, but were rather the result of scientific experiment and invention in other spheres, now applied to the purposes of agriculture.

The farmer's house continued comfortable enough, but the habitations of his labourers for a long time went from bad to worse. In the West conditions were particularly bad, the ordinary dwelling there, until well on in the second half of the century, being a wattle and mud building of two or three rooms with a thatched roof. In some parts dwellings were better than this, and notably on such model estates as that of the Duke of Bedford, where ample accommodation was provided, well built of brick, with copper, kitchen range, bedroom fireplace, and a common oven for each block. But it was not until the 'seventies that a universal conviction of the unseemliness of the mediaeval style of accommodation, with a large family of several generations crowded into a tiny dwelling, without any attempt at sanitation, with no water supply except a probably polluted

[1] *The Observer*, 17.9.33.

stream or well, and the pigsty outside adding to the general savouriness of the surroundings, really seemed to dawn on the community.

The prosperous farmer landlords were obliged to take some steps for improvement, as they already had in some parts, and from that time the homes of the labourers steadily improved. The general run of farm buildings also showed but slow improvement. In 1850 the majority were still built of rough wood and thatch, very combustible material needing constant repair, although the thatch had the advantage of being warm in winter and cool in summer. In the West the walls were often of wattled furze or mud ("cob"), but rarely of stone or other good material except where enlightened landlords had erected more substantial buildings. In Northumberland at that date primitive dwellings could still be found where the cow and pig lived under the same roof as the family, merely separated from the living and sleeping-room by a partition, and entering by the same door. One improvement, however, which was met with sometimes even in the wood and thatch cattle sheds was a slatted wood flooring raised above the watertight floor underneath, so that the liquid manure drained off and was not lost.[1] The discovery of Liebig that cattle when kept warm needed less food led a little later in the century to the erection of more solid and weather-proof houses by intelligent farmers.

Nowadays the main principle for the housing of cattle is that buildings should be light, airy and warm, sunny if possible but free from draughts, and easily kept clean. All animals should have warm dry beds, therefore houses and yards must be well drained with good litter or sometimes duck-boards over a sloping concrete floor. Half covered yards are much used.

With the coming of threshing and winnowing machines, which dealt with the whole crop rapidly out in the open, the necessity no longer existed for the great barn, with its long threshing-floor, high wide doors to receive the loaded wagons, opening towards the prevailing wind for the convenience of winnowing; so that this picturesque feature of the old farm-yard, with its fine timbering and great sweep of roof, tended to become a mere storage place instead of a scene of activity. With this shrinkage of office much less space was required, and a modern writer on farming recommends as the sole accommodation for a small farm one barn 48 feet by 60 feet, with a partition across the middle, one half being used as a stackyard and the other for the housing of all kinds of beasts, with a granary and mixing

[1] J. Caird: *English Agriculture*, 69, 141, 389, 490.

floor for foods over it, and tanks to receive liquid manure and rainwater underneath.   Thus has the mediaeval establishment with its courtyards and buildings for every kind of stock and crop, and its array of bakehouse, brewhouse, salting-house, and other domestic offices, tended to shrink back to almost primitive proportions, though for very different reasons.

In material for construction the modern farmer has derived much benefit from concrete for well-drained floors, wire for fencing, and the useful but deplorably unsightly corrugated iron for all kinds of purposes, notably roofing.   Thus the row of neatly thatched ricks and haystacks is often replaced by a large Dutch barn with curved iron roof.   Thatching however has not died out, and between the two World Wars was fostered by Local Education Committees.

As the nineteenth century advanced the farmer's *stock*[1] became of increasing importance to him owing to the declining profits of corn-growing.   Once more, as in patriarchal times, cattle, especially dairy cattle, bid fair to constitute the country-man's chief wealth.   But even in this direction the developments of the new age soon brought a serious handicap. The invention of steamships bringing about rapid communication with foreign lands, cheap freightage and still more the process of "chilling' perishable foods, so that meat, butter and cheese could be poured in from more favoured lands where these things could be more cheaply produced than in England, was a serious blow to the home farmer, the effects of which were felt increasingly until the situation created by two World Wars made the home-production of food a vital necessity.

The science of stock-breeding was, as we have seen, well estab-lished during the eighteenth century, though far from being generally disseminated, but during the nineteenth century much was done towards further improvement and still more towards the universal prevalence of a good type of all sorts of beasts, and "herd books," or records of every variety of pedigree animal were established.

Throughout the century many cottagers, although deprived of the means of maintaining a cow, managed to keep a *pig* or two, and the Board of Trade Reports at the beginning of the twentieth century show that it was still a pretty general practice in the Midlands and the West, in Northumberland, Lancashire, Lincoln and the East Riding of Yorkshire, in Essex and to some

[1] The main authorities for this section are: R. Wallace and J. A. Scott Watson, *Farm Livestock of Great Britain*, 5th edit.; F. H. Garner, *Cattle of Britain;* J. F. H. Thomas, *Sheep.*

extent in Suffolk; but not in the Home Counties or South, except to a small extent in Kent.[1] Cobbett, in riding through Sussex and Hants between 1820 and 1830, noted large numbers of cottagers' pigs, which he considered an "infallible mark of a happy people," but that good fortune must have departed from them later. About 1880 a village in Wiltshire had a thriving Pig Insurance Scheme, to which each pig-keeper paid 6d. entrance fee and a weekly premium of ½d. benefit being received on the loss of a pig after three months insurance.[2]

Breeds of pigs underwent a good deal of change. The Berkshire, so popular at the end of the eighteenth century, has continued up to the present day under that name, but has changed in colour from the original sandy with black spots. By 1845, owing to much crossing with Chinese breeds, many of them were nearly black, and some black and white; and now the correct Berkshire pig is black with white face, feet and tail tip, the original hue only occurring in an occasional cross.

FIG. 72.—BERKSHIRE SOW (1847)

Sandy or red pigs, however, have continued in the Tamworth or Staffordshire, a breed not noticed under that name until late in the nineteenth century (being registered in 1885), but obviously of ancient origin, its ancestors probably being the "all

[1] Quoted by F. G. Heath: *British Rural Life and Labour*, 64-73.
[2] *British Rural Life and Labour*, 288-9.

sanded" pigs of the seventeenth century.[1] It seems likely that the Berkshire were really an offshoot of this breed and that both were originally classed under the latter name, for Parkinson in 1810 stated that the true breed of Berkshire pigs were rough-haired and feather-eared which, he considered, "looked rather unseemly,"[2] but which are both still characteristics of the Tamworth, as are also, occasionally, black spots. Low in 1845 gives the Berkshire as the prevailing breed of Staffordshire, as if they had not then been differentiated. It is suggested, therefore, that those which were crossed with the Chinese and thus acquired the characteristic very retroussé countenance and subsequent change of colour, continued under the name of Berkshire, and that those retaining the red-brown hairy coat and original long straight nose became later known as the Staffordshire or Tamworth. Both have erect ears.

The white pigs of Yorkshire and the East Coast were divided at first into large and small whites, the latter being the result of a cross with the Chinese and thus gaining the characteristic face, which when large and small were crossed was retained as a feature of the Middle White, a breed registered in 1885 and popular at the present day. Large Whites are now the most popular breed for bacon, and are sometimes crossed with Large Black Essex or Wessex Saddleback sows, producing what are called Blue Pigs. Middle Whites and Berkshires are favoured for the London Pork trade. Both Middle and Large Whites have erect ears. Other varieties, white but lop-eared, now bred are the Lincolnshire Curly-coated or Baston, the Cumberland, and the Devonshire or Long Lop-eared White. This last variety has displaced, or perhaps improved out of recognition, the old, gaunt, white pigs of the West, which at the beginning of the nineteenth century in Cornwall were a "wolf-shaped" breed with sharp head and ears.[3] The white breeds of Gloucester, Shropshire, and Northants seem to have died out before the middle of the century, but the existing Gloucester Old Spot is a survival of the spotted pigs of that district, and is hairy and lop-eared. In 1845 Hampshire hogs were noted for their size and suitability for bacon, and the Rudgwick, from the borders of Surrey and Sussex, was the largest of all;[4] but these breeds are not now kept. The old black and white pigs of Essex, which possibly got their

[1] See above, p. 252.
[2] R. Parkinson: *Treatise on Livestock*, 228.
[3] J. Lawrence: *General Treatise on Cattle*, 523.
[4] Low: *Domesticated Animals*, 431. In 1805 the Hampshire were dark-spotted or black.

mixture of colour from a Chinese cross, survive, much improved, having white shoulders, legs, nose and tail-tip. A breed exactly similar except for the white nose and tail-tip is the Wessex Saddleback. Neither of these breeds, although ancient, was registered until 1919.

Black pigs received little attention until the middle of last century, since when they have been carefully bred in Devon and Cornwall, and in Suffolk and Essex, but the "Large Black Pig Society" was not instituted until 1899, after which they became for a while very popular all over the South, but have since declined. They were of impressive size with large lop ears.

The farmer's pigs are now kept in clean warm dry houses, or sometimes—particularly breeding sows—on pasture, or in portable huts on grass land. The cottager's pig of course is not so fortunate but often lives in a shed akin to the old odorous pigsty.

*Cattle* are now either "dual-purpose," i.e., good both for the dairy and as beef-producers, or "special purpose", i.e., one of those two, the third purpose of farm draught being no longer considered. Up to 1914 a few ox-teams lingered in the Cotswolds and on the Sussex Downs, hilly districts where their powers of steady endurance were specially useful. The last Sussex team, at Birling Manor, Eastdean, was given up in 1929, and the sole survivor now seems to be Earl Bathurst's team of six Herefords near Cirencester. These wear collar and harness and not the ancient yoke. The Sussex farmers for some reason latterly rejected their own excellent breed in favour of long-horned black Welsh runts. In other parts draught cattle had long disappeared before the demand for fresh young beef, which made the sale of their more elderly carcases unprofitable.

The Longhorn breed of cattle, the original inhabitants of the North-west and a good part of the Midlands as far south as the Severn, in spite of their improvement by Bakewell of Dishley shrank in popularity before the shorthorns, chiefly owing to their slowness in maturing, and in the middle of the nineteenth century were almost in danger of extinction. There was a certain inconvenience about beasts with horns so long that they could not get through the cow-house door without turning their heads first to one side and then to the other, besides other dangers, but their hardiness of constitution was unrivalled, they were easy to feed and gave rich milk, and consequently demand for them revived in the North, mainly as a useful cross for other breeds. They are usually dark red, often brindled with black spots, and should have a white line along the back and other white patches. This colouring was already established, with some variations, in

Fig 73.—The Last Sussex Ox Team

*Owners.*—Major and Mrs. Harding, Birling Manor, East Dean, near Eastbourne
*Trainer and Ploughman.*—William Henry Burton
*Ox-boy.*—William Stephen Burton (son)

1800, and the breed was then considered an excellent one for producing butter.[1]

The shorthorns have undergone various developments. Well advertised by Collings' new method of sending the famous "Durham Ox" round the country in a cart, animals of this sort became very popular, being adaptable to climate and of all-round use. The most famous successors of the Collings brothers were the Booth family of Killerby and Warlaby, and Bates of Kirklevington (Yorks), while the Scottish herd of Cruikshank introduced a good cross with these.

The Lincolnshire Red Shorthorns, although ancient inhabitants, were not recorded in a separate herd book until 1895. This variety is all red, but other shorthorns can be red, white, or any mixture of the two, including roan. Special Dairy Shorthorns (of all these hues) were differentiated in 1905 and are a very popular dual-purpose breed. White Shorthorn bulls are sometimes crossed with black Angus or Galloway cows to produce a blue-grey shade. There is also a beef-shorthorn.

The ancient red cattle of the South (formerly known as Middlehorns), the breeds of Hereford and Gloucester, Devon and Sussex, are still much valued. The Herefords, considered by Cobbett the finest and most beautiful of cattle, still retain their white faces, spines and under parts, probably the result of a mixture with the similarly coloured cattle of Glamorgan, now extinct, or the old white Welsh cattle. They are a hardy race, not given to tuberculosis, but have only in recent years become good milkers. The old Gloucester breed, of which a few remain, have a white streak along the rump and underneath the body, with a white tail, but are otherwise red. They are good dairy cows.

The Devon cattle are now divided into the original or North Devons, and South Devons (or South Hams). The latter are larger and lighter in colour, possibly as a result of crossing with Channel Island and Normandy cattle. They received a herd book in 1891, and are good dual-purpose beasts but give specially rich milk. The North Devons are a beef breed.

The dark red Sussex cattle are large and solid creatures, excellent beef-producers, the bullocks being obviously fitted for their previous use as draught animals.

Polled (hornless) red cattle were known in various parts at the beginning of the nineteenth century; a Northern or Yorkshire polled variety is mentioned by Lawrence in 1805, which

[1] J. Lawrence: *New Farmer's Calendar* (5th edit.), 506. *General Treatise on Cattle*, 34.

would probably be an offshoot of the shorthorns, and they seem to have existed on the borders of Suffolk much earlier.[1] The modern Red Poll, however, is the result of a cross begun early in the century between the old Norfolk red horned breed and the Suffolk Dun, both of which are now extinct, the result being an unsurpassed type of dual-purpose cattle combining the good qualities of both. The Red Poll Society was founded in 1888.

A breed from Derbyshire, of unknown origin, which has lately come to the fore as a dual-purpose animal, is the Blue Albion. It is much of the shorthorn type but blue-roan and white in colour. The herd book was published in 1920.

Channel Islands cattle, for a long time generically referred to as "Alderney," were increasingly imported during the nineteenth century, and both Jerseys and Guernseys are now bred in England, the Jersey Society being formed in 1878. These beautiful animals are of great merit as particularly rich milkers but of no use in any other way.

Various Scotch breeds have become popular in England. The rough-haired polled black Galloways, as we have seen, were brought by drovers to East Anglia in the eighteenth century or before, and are now valued as "prime Scots beef," and there is also a sheeted or belted variety, so called from the white band round the centre of the body, which came chiefly from Tyneside, and is now represented by a few herds in England.

The Ayrshire, the famous Scots dairy cow red and white in colour with lyre-shaped horns, has become the most popular of dairy breeds in England. The Aberdeen Angus, a black-polled breed, is pre-eminent as a beef-producer owing to its unsurpassed fineness of bone, an English Society being formed in 1900. The Welsh black cattle are also a good hardy dual-purpose breed, but are often crossed with Herefords.

Of Irish cattle the little black or red Dexter for dual purpose, and the black Kerry for the dairy have recently become popular.

Lastly the Friesian, a handsome black and white shorthorn of great dairy qualities, was imported in considerable numbers during the latter part of the nineeenth century and the British Friesian Society was formed in 1909. This breed has a wonderful milking capacity, some cows having been known to give as much as 3,000 gallons a year.[2] Their milk, however, is not particularly rich in fat, but is being improved.

Since the early days of Smithfield Market (first mentioned in 1732) "Drover's cattle" were driven all the way from Scotland to

---

[1] Wallace and Watson: *Farm Livestock of Great Britain* (5th edit.), 91.
[2] The annual yield of the average cow is 595 gallons (1950).

East Anglia to be fattened, along a settled route where fields
were regularly used by them for rest and grazing, and when
fattened they proceeded by shorter stages to London.  But as
transport developed this practice naturally diminished in favour
first of rail and then of lorry.  "Drover's cattle" came to be
looked on with some doubt, as it was impossible to discover at
first sight what injuries they might have received in transit, or
what unsuitable food they might have indulged in.

*Calves* on the farm are either brought up on the pail or
suckled by a foster mother, one cow being kept for this purpose
to feed a number of calves.

Dairy operations have received much attention from
mechanical inventors, and what is still more important the
necessity of extreme cleanliness throughout has been realized
and is fairly well enforced. All cow sheds must now have concrete floors.  On dairy-farms there is usually a milking-shed used
for that purpose only, and thus easily kept clean.  Cows, hands
and clothes of cowmen, buildings, buckets, and churns are kept
scrupulously clean.  Hand-milking is still widespread especially
on small farms with not more than ten cows, but milking-
machines are used on very many.  These either deliver the milk
into a bucket beside the cow, or the milk is carried by an over-
head pipeline direct to the cooler, and in some cases passes from
that into bottles.  Milk is tested for cleanliness, and the cows of
accredited herds are tuberculin tested.   In small dairies the milk
may be merely strained and run over a cooler, but in large
dairies it is often pasteurised to destroy germs. It is now sold in
sealed bottles or cartons except in some country districts.

Milk-recording was started officially in England in 1913, and
in 1933 the Milk Marketing Board was instituted to control the
milk industry.

For butter-making cream-separators on the centrifugal
principle worked by hand, horse or steam power, began to come
into use about 1879 and did their work much more quickly
and thoroughly than hand skimming, though with rather a
detrimental effect on the pigs' diet.  Churns of course can be
operated by machinery,[1] and also butter-workers, but little butter
is now made on the farm, as the Milk Marketing Board give a
higher price for liquid milk; consequently butter is more and
more made in factories.

As early as 1850 a small manufactory of concentrated milk
was established in Staffordshire.  The milk was steamed in a

[1] Horse-power churns were already known in Bucks in 1850.

large copper pan, and patiently stirred for four hours by persons slowly walking round it, after which it was soldered up in tins and lowered into boiling water for a while.[1]

Cheese-making was able to make a great advance in general standard of excellence, chiefly owing to the services of chemical and bacteriological research. The nature of the all-important ferments used was discovered, and tests invented by which the amount of acidity present in the milk and whey could be determined, and the right moment for proceeding with the delicate process of cheese-making fixed. Thus first-class cheese (and butter) can now be produced with some certainty. The making of whey butter, although still customary in cheese-making districts in the middle of the nineteenth century, is very rare nowadays, as the improved methods of separating the curd leave too little cream in the whey to make it worth while.

Cheddar cheese, which made its name during the nineteenth century, was no new thing, being known and esteemed in Stewart times, but it seems to have been generally classed as a Gloucester cheese, that name extending over Somerset and the neighbouring counties, and at the end of the eighteenth century it was sold in London as "double Gloucester," owing to the enormous size of the cheeses. It was not until 1856 that a definite formula for its manufacture was issued by Joseph Harding, as the result of a deputation from Scotland.[2] Since then this variety of cheese seems to have become the most popular of all.

Stilton cheese, formerly made with cream but now as a rule merely of full milk, has the peculiarity of not being put in a press. Wensleydale, another local cheese that has gained repute, is of similar nature but is pressed like other cheeses.

Cheeses were not always made solely in the locality of their place name, thus both Caerphilly and Cheddar cheeses were made and sold in Sussex, and doubtless in other counties equally remote from those of their origin. But cheese-making also has been largely transferred to factories: only in Somerset the Cheddar cheese industry is still maintained on the farms, being encouraged by the M.M.B., but in other districts it has been almost discontinued. Cheshire, Cheddar, Lancashire, Wensleydale, Caerphilly and Stilton are still manufactured.

*Sheep* declined in numbers during both World Wars, though shewing some recovery up to 1939. Their numbers have remained steady in highland districts where arable and dairy farming do not thrive, but by the end of the last war their total numbers

[1] J. Caird, *English Agriculture*, 234.
[2] *Victoria Counties' History*, Somerset, ii. 538-9.

had sunk to eighteen million. They are now kept mainly to produce mutton, but with the rising price of wool it may become worth while to keep them for their fleeces also. Unfortunately few young men seem likely to be attracted by the life of a shepherd.

Sheep are of three sorts: mountain sheep (whose coarse wool is used for carpet-making), green land sheep, such as the Romney Marsh, and arable land sheep kept for the value of their manure. They are indeed invaluable for the manuring and consolidating of light land, besides needing less water and shelter than cattle.

We now have in this country between 30 and 40 recognized breeds. Of the old long-wooled breeds remain the Cotswold, the sheep which anciently gave its name to the hills—the wolds of the sheepcotes; the Romney Marsh, described by Cobbett as "white as a piece of writing paper"; and also the Leicester, so invaluable for improving other strains. The Border Leicester, evolved during the closing years of the eighteenth century by the Collings brothers, originally by a cross with the Cheviot, is very popular in the North, and the ram is an impressive animal as he appears at North Country Shows, his broad, stout carcase enveloped in a fleece of long, beautifully dressed curls, from which emerge four thin white legs, rather conspicuous white ears, and a long pale aquiline countenance, with a pair of very wide-awake golden eyes; there is nothing of the stolid stupidity of the southern sheep in his expression.

FIG. 74.—GLOUCESTER SHEEP

The old Lincoln is now an improved breed by crossing with the Leicester, and is known as the Lincoln Longwool; it is the largest sheep in England.

The Wensleydale breed of blue-faced sheep are the successors of the old Yorkshire "mugs" or Teeswater, crossed with the Leicester. In the West the Devon Longwool is the outcome of a cross between the Bampton Nott and the Southern Nott,[1] with an infusion of the useful Leicester blood. The South Devon, a still heavier sheep, is the commonest breed in Cornwall and Mid-Devon, and is sometimes crossed with Oxford Downs. There is also the Improved Dartmoor.

The Southdown has been as useful to short-wooled sheep as the Leicester to longwools, and has enjoyed unimpaired popularity. This sheep's stout, compact little carcase with its short close wool, short legs and discreet grey face, has a neat and comfortable appearance comparable to a well-upholstered seat. Early in the nineteenth century a blending of the old Wiltshire and Berkshire breeds with the Southdown produced the Hampshire Down, which still remains, and by a later cross with the Southdown has produced the Dorset Down, registered in 1904. The old Norfolk horned breed by crossing with Southdown rams produced early in the century the hornless Suffolk, and the original stock of Shropshire was much improved by the same admixture. The Oxford Down originated about 1830 by the crossing of Cotswold rams with Hampshire ewes, or sometimes with Southdowns. The Ryeland, the old short-wooled breed of Herefordshire, remains, also the horned but short-wooled breeds of Dorset and Wiltshire, the last now called the Western. This last breed grows very little wool and it peels off naturally during the summer. The Devon Closewool, developed from the Devon Longwool and the Exmoor Horn, was registered in 1923. Of these sheep all have black or dark-brown faces except the Ryeland, Dorset Horn, Devon Closewool and Western, which have white, and the Southdown grey faces.

Of mountain breeds, the Cheviot has maintained unbroken popularity in the North, hornless except for an occasional ram, and the Herdwick is still the sheep of the Lake District and the surrounding country. The black-faced (or mottled) mountain sheep, of which the Highland sheep are the Scotch variety, in England became divided into six branches; the Rough-Fell of Westmorland; the Lonk of North Lancashire, West Yorkshire and part of Derby; the Dale sheep of Yorkshire of which the chief is the Swaledale; and the Derbyshire Gritstone without horns. The Penistone and the Limestone, having white faces and legs, are now practically extinct.

Of the ancient Welsh sheep two breeds are now prominent,

[1] Nott or Natt = hornless

the Welsh mountain, and the Kerry Hill; both are short-wooled, and the latter are hornless, as are most usually the mountain ewes.  The Radnor and the Clun Forest are still kept in their localities.  In the West the Exmoor Horn remains, a sheep with fairly long wool.

The average weight of fleeces varies from 2 lb. (the Welsh Mountain) to 13 lb., or as much as 15 lb., in the long-wooled breeds.  The Downs breeds average 5½ lb., the Blackface 4½ lb., Cheviots 4¼ lb., and Swaledale 3¼ lb.[1]

At the beginning of the nineteenth century ewes were milked in Wales for three months in the year, giving about a quart a day,[2] but the practice now seems to have entirely died out.

Sheep-shearing, which takes place between May and August, is no longer an occasion of festal celebration except perhaps in the Lake District.  Shearing machines, driven either by an engine or by electricity, are now in general use, but it remains a skilled art whether done by hand or machine. Washing, where it is still done, takes place about ten days beforehand, and the washer can now be provided with a water-tight box to stand in to keep him comparatively dry.  But as a rule nowadays the manufacturers prefer to wash the wool themselves.  The preparation of sheep-dips for scab and other parasites is a benefit to the farmer provided by modern chemistry and enforced by law, in some districts once and in others twice; taking place usually a few days after shearing.

Coloured sheep-dips are now discouraged as injuring the wool for light-coloured dyes.  A sheep-marking fluid is now sold which does not harm the wool and does away with the dangers of boiling pitch to the marker.  Foot rot, that ancient scourge of sheep, is to a large extent prevented by careful paring of the hoofs to prevent rough in-turning edges.  The shepherd's smock and his chiming sheep-bells have died out, even from the Sussex Downs, where they lingered until early in the present century.

*Goats*, after suffering a period of almost total eclipse, began to return to favour during the last quarter of the nineteenth century, the establishment of the British Goat Society in 1879 doing much to revive interest in them.  During the war their milking properties were so valuable that a good many were kept, and at the present day there are goat societies in many counties. To the common English goat, however, has been added some very popular foreign breeds.  Crossed with the Indian goat, or later with the imported Nubian, an animal of a similar type,

[1] Ministry of Agriculture and Fisheries, 1940-49.
[2] J. Lawrence: *General Treatise on Cattle*, 420.

it has produced a variety called the Anglo-Nubian, a strange-looking beast with long, drooping ears, a Roman nose, and an undershot jaw. The other popular type, of more ordinary appearance, is the Swiss, represented by the drab or chocolate Toggenburg, the British Alpine, and the white or cream Saanen. British Alpine are the best milkers, giving as much as 2½ gallons a day.

*Horses* have undergone a sad eclipse on the farm before the advance of machinery, but have not been entirely superseded, and are not likely to disappear, especially on small farms. Tractors relieve them of the heaviest work, but horses are economical when carting for short distances and are better for working small or irregular-shaped fields and for hoeing. They can also work on much softer land than can a tractor and can be kept on farm produce, whereas tractors are a considerable expense to run and keep in order. Some large farms buy horses for the autumn work of carting root crops and sell them again later. But the fine teams of wagon-horses which were once such a feature at hay time and harvest have disappeared, and their bright brass ornaments, once handed down in carters' families have become "antiques." The great Shire horses, the heaviest breed in existence, yield their largest specimens for occasional dray-work in the towns, but some of these are first hardened on the land. In the middle of the nineteenth century they were still as a rule black with white points,[1] but nowadays that is an exceptional colour. A mare of this species should be over sixteen hands, and a stallion over seventeen hands. The defect of the breed is still its slowness of action, but for massive strength it is unrivalled, and the great proud stallions with their wide, arched necks, gleaming hides polished to the last hair, and silky fringes round their ponderous feet, are a wonderful sight, though rarely seen now.

The Clydesdales, the famous Scottish breed of cart horses, began to spread into the North of England in some numbers early in the nineteenth century. Originating in Lanarkshire, where a good breed of draught horses had long existed, they were improved by English horses brought back to that district by drovers who had gone south with the herds of cattle. Upon this foundation John Paterson of Lochlyon built up the true Clydesdale breed by the importation of a black Flemish stallion from England in 1715 or 1720. Impetus to their breeding was given by the encouragement of the Highland Society from 1827 on, but in them the black colour disappeared even sooner

[1] Low: *Domesticated Animals*, 611.

than in the Shire, brown or black bays being considered the correct colour at that date.  Grey or chestnut are now permitted, but are not so popular as the others.  In frame they are rather less massive than the Shires, have a better action, and their working life is longer, while the Shires come more rapidly to maturity.  They became the prevailing type of working-horse in the four northern counties of England, and are largely bred in Cumberland.

The Suffolk Punches of the present day are all descended from a horse foaled in 1768, the property of Crisp of Ufford. They retain their short legs with clean feet, round body, longevity and vitality, and in colour are one of seven different shades of chestnut, their average height being 16¼ hands.

After 1918 another breed of cart horse which came into favour is the Percheron, taking its name from the district of La Perche in Normandy.  This is a rather smaller horse than the three preceding breeds, the mares averaging 15 to 16 hands, but of a powerful build, good workers, strong, active and enduring.  Their colour is generally grey, sometimes black, and occasionally bay or brown.

The Cleveland Bay, described by Rider Haggard as "a long horse standing over a great deal of ground," was considered very useful for lighter work on the farm, and was used mainly in Yorkshire.  Its origin is thought to have been the crossing of Yorkshire cart mares of the "Great horse" type with an imported Barb, and the colour is light or dark bay with clean black legs.

The importance of good breeding, or at least of descent from a good working family, was recognized with regard to the shepherd's *dog*, as well as other animals on the farm.  Two main breeds were early utilised, and are still; the collie and the "old English" sheep-dog.  The latter belongs to the south and is shaggy all over, bob-tailed, and generally some shade of grey or blue-grey mixed with white; powerful and pleasant dogs, but apt to be noisy if not well trained in that respect, as was noted by Low in 1845.  The collies of the North, named, according to the same writer, from the Celtic "Coillean," a little dog, are of various sorts, both rough and smooth, and are sometimes naturally bob-tailed.  The ordinary working collie is very often black and white, light and swift, and of extraordinary intelligence and patience in rounding up numbers of scattered sheep on a rough hillside, directed only by an occasional whistle or cry from the shepherd.  Sheep-dog trials have become a feature of many county shows.  Between north and south a variety of sheep-dogs is found, but these are the two orthodox breeds.  A cross between

v

a shepherd's dog and a terrier used to be known as a cur, and that between the former and a greyhound a lurcher, and these were often used as watchdogs on the farm, on account of their alertness.[1]   A "true vermin-bred cur" was early recommended for a weekly rat hunt on the farm, with the help of ferrets, as already in 1800 the grey rats were a scourge, and they and the mice were calculated to do 50s. worth of damage a week.   John Lawrence, the writer of that date, urged that rat-traps should be of the cage shape, so as not to endanger the *cats*, "a most useful species of domestics fully entitled to our care and kindness," and he had the perspicacity to see that a good cat should hunt from love of sport and not from hunger, as the eating of rats is not good for them.[2]   Even modern writers on farming consider that no farm is complete without a few cats.

The keeping of the barn-door *fowl*, like other small fry, languished during the time of the farmer's general prosperity and was revived later.   At the beginning of this century there was a great increase in the numbers kept, and the handful of indiscriminate poultry picking round the farmyard was in most cases replaced by a carefully kept flock of well-bred birds.   They are now recognized as an important part of the farmyard economy and are no longer the exclusive perquisite of the housewife.   Many poultry farms have sprung up, although hampered latterly by the rationing of chicken foods.

The chicken-cramming industry of Surrey and Sussex became almost extinct again during the two wars owing to the high cost of the foodstuffs.   The same fate overtook the fattening of Aylesbury ducks, but this has revived a little.   Many fresh breeds of fowls have been introduced.   The chief sorts at the beginning of the nineteenth century were the game fowl (the nearest to the original Bankiva Jungle fowl), the Dorking, and the Poland, a name erroneously given to hens of the crested type.   The Malay or Chittagong is also mentioned, but the eggs, though large, were declared "only fit for soup."[3]   To these, as the century went on, were added fowls of the Cochin breed, including Brahmas and Langshans, from which have sprung the Plymouth Rocks, the Wyandottes and the Orpingtons; the Buff Orpington being a cross between the Cochin and the Dorking.   Another important new species was the Spanish fowl, said to have been introduced after the Peninsular War, to which are closely related the Andalusians, Minorcas, Leghorns, and

[1] Low: *Domesticated Animals*, 711.

[2] J. Lawrence: *New Farmer's Calendar*, 196-8.

[3] J. Lawrence: *New Farmer's Calendar* (5th edit.), 552.

the Anconas.   The pert, trim little Bantams have also gained a footing in spite of the smallness of their eggs.   Since the beginning of the twentieth century there has been a rapid evolution of utility breeds of hens, and "laying trials" to test their capacity for producing eggs have become the custom.   At the beginning of the century one hundred and twenty-seven eggs per hen was considered a good laying average: now a well-bred hen should lay two hundred.   The most popular breeds at present are Light Sussex, Rhode Island Red (introduced comparatively recently from America) and White Wyandottes, of the heavy breeds, and the Leghorns (white, black and brown) light birds for laying.   Crosses of all these are favoured, and various methods of keeping them are followed.   They may have a fixed house with two alternative runs, or a movable house with free range in the field selected.   Others are folded like sheep in small movable runs to manure the ground.   It is now realized that they improve pasture in this way, and on stubble break up and spread the straw ready for ploughing in, at the same time picking up a lot of food.   They are also kept intensively in indoor runs well littered and even in separate cages so that their individual egg-production can be checked.   Incubating has been developed, and improved "foster-mothers," gently heated, provided for the chicks on leaving the incubator.

*Geese* have not undergone a very great change, the three sorts kept being the Roman white, the large grey or Toulouse, and the white or Embden, or a useful cross between the last two.

During the nineteenth century the most common breeds of *ducks* were the Rouen (doubtless the same as the "Normandy" of the eighteenth century) the Aylesbury, or Old English White, the Peking, and some black breeds. During the last fifteen years of the century the Indian Runner duck became popular, owing to its small need of water and its good laying properties, and these are still much kept, another breed, popular since 1921, being the Khaki Campbell and more recently the Muscovy.

*Turkeys* may be of the bronze or Cambridge breed, the black or Norfolk, fawn, and white, the American Mammoth bronze being the most popular.   All of these colours except the fawn are mentioned at the beginning of the nineteenth century, but the white was considered the most delicate.   Cobbett, about 1826, noticed a farm near Burghclere (Hants) where the fowls, ducks and turkeys were all white.[1]

Guinea-fowls are kept on some farms, in spite of their

[1] W. Cobbett: *Rural Rides*, 468 (1853 edit.).

raucous voices, but peacocks have become an occasional ornament of stately gardens.

Thousands of *pigeon* houses still existed in 1813, but with
more efficient ideas of agriculture they died out.   The birds
were favoured by Cobbett not so much for profit as because
they were a good training for children in the care of tame things,
and would sit on a shelf in the cowshed and give little trouble;[1]
but they are hardly ever kept now.

*Rabbits,* like other home-produced food, came into favour
during wartime and were bred in immense numbers for the
table.   The Belgian Hare and Flemish Giant were found large
and useful animals for this purpose, and also a good cross for
the ordinary English.   A profitable branch was developed in
the keeping of these animals for their fur, and foreign breeds
were imported for this, the most popular sorts at present being
Beverens, Havana, Angora, and the Rex-coated varieties. The
Chinchilla is the most useful as a dual-purpose animal, and the
pelts of all these command good prices and make beautiful
furs.   The Angora has a long woolly coat which is shorn every
three months and spun like any other wool.

Public attention was recalled to *Bee*-keeping by the formation
in 1874 of the British Bee-keepers' Association, the first President
being the distinguished antiquary, Lord Avebury.    British
honey was recognized as second to none in quality and the
industry once more began to flourish, branch societies, not for
professional bee-keepers, but for those who kept a few hives in
addition to other activities, being formed in many counties.
Progress in this art was made possible by the invention of new
types of hives, which, unlike the old straw skep, could be opened
when required and the work of the inhabitants supervised.
Much more knowledge of the habits of bees was obtained in
this way, movable frames could be inserted for the making of
honeycombs, much larger returns of honey could be obtained,
and, still more important, the ancient practice of "burning up"
the bees to clear out the hive, was done away with.   The pretty
old-fashioned straw skep is now going out of use, except for
such purposes as hiving and carrying swarms, and a neat but
unpicturesque wooden structure, suggestive of a pile of boxes
with a slightly gabled roof, has taken its place.   Their value
as fertilizers to keepers of orchards and market gardens is now
well recognized, and many are kept in Kent and the Midlands
for this reason.

In 1906 the plague known as "Isle of Wight disease"

[1] W. Cobbett: *Cottage Economy,* 181.

appeared and caused much destruction, and much trouble is being taken at the present day to find strains of bees which are immune from it.

Much damage was caused among the stock during the nineteenth century by various diseases; with cattle foot-and-mouth disease was a great trouble from 1839 on, pleuro-pneumonia appeared in the following year and reached its worst in 1872; the rinderpest or cattle plague descended upon the farm in 1865 and caused enormous loss, but was stamped out by 1877. The drastic measures taken to combat this plague, making the slaughter of diseased animals compulsory, happily resulted in almost extinguishing the two former as well, but in the black year of 1879 millions of sheep, as well as other cattle and even hares, rabbits and deer, died of the rot. Yet in spite of all these misfortunes cattle-breeding flourished and great improvement took place, aided by all the resources of modern veterinary science. The Veterinary College was founded in 1791, but as usual it was a long time before its enlightened practices were taken advantage of on the ordinary farm. Horses seem to have benefited first, but the ancient practice of bleeding them twice a year continued in some parts until the middle of the century, and for the other animals the local "leech" or cow doctor long continued his ministrations, often worse than useless.[1]

The improvements of modern farming, and particularly the lavish provision of winter foods, made a great difference to the comfort of the animals, for instead of having to fend for themselves through the winter or exist half-starved on what fodder remained for them, they now lived in good condition all the year round; in fact the position was so much reversed from the days of our ancestors that it became customary to fatten them during the winter for sale in the spring, a very different practice from feeding up the thin winter beasts on "a handful of grass" in the summer and selling them off before herbage got short. Modern cattle foods are many and various, backed by careful scientific investigation into the value of the foodstuffs and the needs of the animal. Linseed cake, cotton cake and other similar foods were evolved, and after the first World War a balanced concentrated cake was provided. This was later called "National Cattle-food" and was issued mainly in nuts. Skilled attention was turned to the production of home-grown foods, and a careful succession of crops was planned, to leave as little need as possible for the purchase of expensive foods. Thus the

[1] Sir Walter Gilbey: *Farm Stock One Hundred Years Ago*, 2-4.

animals can, for instance, start on maize in August, continuing until the first frosts, carrying on with marrow-stem kale until the January frosts, with swedes throughout the winter; thousand-headed kale will then take them through to March, with mangolds from February to April, and "hungry-gap" to supplement. These, with rye, will keep them until the mixed grasses are ready in May, and they go through a succession of these, with the aftermath of the mowing, until autumn comes again. These fresh crops, combined with hay, silage, and perhaps dried grass constitute a modern economical round of crops. The Tithe Commutation Act of 1836, by which the farmer gave a money payment, varying with the price of corn, instead of having a tenth of the produce of his cattle and crops actually removed by the tithe-owner, did away with a cumbrous system which had often hampered him in the gathering in of his harvest. Aided by all the new resources of science and their general dissemination the farmers flourished on the growing of wheat and other grain crops during the first part of Victoria's reign, until the great change of general conditions in the second half of the century. With the expansion of the Middle West in America and Canada, the development of Australian wheat-growing, and the commercial exploitation of South Russia, corn poured into the country from lands favoured with a more reliable climate, and the English farmer was unable to stand against such competition. A series of bad harvests in the 'seventies, culminating in the disastrous season of 1879, sounded the knell of England as a corn-growing country. Before the first War only 42 per cent. of all our food was home-produced. But during the second World War especially the acreage of cornland enormously increased again owing to the pressing needs of the time and has not much decreased since.

In 1850 the eastern half of the country, from north to south as far as Middlesex, and below that the southern counties as far west as Berkshire, South Wilts and Dorset, were still corn-growing, the rest being chiefly given over to grazing and dairy-farming. The principle of the four-course rotation, that of alternate corn and cattle or "fallow" crops, was generally adhered to, with variations to suit the situation, the point aimed at being to keep the land dry, clean and unimpoverished. Caird gives the average yield of wheat per acre at that time as $26\frac{2}{3}$ bushels, a crop we have improved since.[1] The average yield of crops for the years 1940-49 was: Wheat 19.1 cwt. (about 34 bushels) per acre, barley 17.8 cwt. ($35\frac{1}{2}$ bushels), oats 16.8 cwt.

[1] J. Caird: *English Agriculture*, Map, 501, 521.

(43 bushels), beans 14.3 cwt. (22½ bushels), and peas 13.3 cwt. (28 bushels).

Root crops were universally established before the middle of the nineteenth century, and mangolds, at first rather neglected, in the end became even more popular than swedes owing to their drought-resisting and good keeping qualities. With the increased need for sugar the cultivation of sugar-beet has come much to the fore. Not only is the root valuable for that purpose, but the green tops, and the pulp returned by the factories, form a very useful cattle food. Among green crops lucerne and sainfoin became increasingly popular; Italian rye-grass was introduced, in addition to the sorts already known, and in 1850 produced great crops in Yorkshire and Middlesex, mowings being taken from it six to nine times during the year. In addition to the ordinary red and white clovers the handsome crimson Italian variety was introduced, with its long-shaped brilliant-coloured head, commonly called "trifolium," although this name properly belongs to all clovers. Another useful new variety was alsike, a hybrid clover with pinkish-white flowers. Lupins are occasionally grown for forage; and Indian corn, a crop which was cultivated in Essex in 1880,[1] and still seems to flourish there. Cobbett in 1821 recommended the ears for pigs and poultry, but nowadays it is more valued as green food. Much attention is now given to the selection of grass-seed for leys, the most usual mixture to be sown being ryegrass of various sorts, red and white clover, timothy, and sometimes cocksfoot. The cultivation of flax and hemp died out when large factories were concentrated in towns, though flax was grown near Bridport and prepared in the ancient way up to 1914. Woad and madder have long disappeared, but the last English woad farm, at Boston, Lincs, carried on until 1930.[2]

Hops still flourish in parts of Hereford and Worcester, and the hop-growing districts of Kent, Surrey and West Sussex, the maximum acreage being reached in 1878.

The necessary processes of husbandry were made much quicker and easier during the nineteenth century by the progress of mechanical invention, and also by the discovery of new means of power, notably steam, by which to propel the machinery. Crops could thus be saved in much better condition, and the field-work of women was rendered unnecessary. Motor traction and electric power have been the contribution of recent years.

[1] J. Lawrence: *New Farmer's Calendar* (5th edit.), 395.
[2] T. Hennell: *Change in the Farm*, 191, 194.

The plough, being a simple form of implement, did not undergo much alteration.  The heavy old wooden plough with iron parts lingered on stiff soils, and in Sussex in 1850 even the mould-board was still a flat wooden structure, although arranged as a turnwrest; indeed solid wooden plough-frames can still be seen in use in that county.  In Buckinghamshire the farmers used a wooden plough in winter, and took to a more modern iron one in spring "with the cuckoo."  By the middle of the century the "Bedford" light iron plough with two small wheels was already in use in the Midlands,[1] and this material, presently improved by steel cutting parts, became increasingly popular. The one great improvement in method was in the turning of the furrow.  Up to the latter part of the nineteenth century mould-boards had a long gradual curve which turned the soil over gently, giving a long, unbroken furrow slice, which looked neat, and gave a certain amount of natural drainage in the hollow; this is still used, but it is now considered better in many cases to have a short, steeply curved mould-board which turns the slice sharply over so as to break it and pulverize the soil, thus giving a better tillage.  This form is called a "digging plough," because it gives much the same effect as a spade.  Now, however, all these horse-drawn implements have been largely superseded by the many-bladed tractor-plough.  Two- or three-furrow ploughs are the most popular, but those with as many as eight blades are occasionally seen.

Cultivators were known at the very beginning of the nineteenth century, and although in those days probably not more than a strong horserake, were used for cleaning fallows at less expense than ploughing.[2]  Nowadays they are used either to break up new land or to stir it after ploughing, to work in dressings and prepare it for sowing.  They are fitted with arrow-shaped blades like a horse-hoe, or sharp tines like a rake.

The drilling machines, for sowing all kinds of seed, which had begun to appear at the end of the eighteenth century were developed and perfected in many ways, and various elaborate types are now used.  Seed is now bought from the seedsman ready-dressed with the appropriate chemicals to prevent disease. The ancient method of broadcasting seed is still occasionally seen in the North and West, but sowing now is almost always done by machinery; large areas being sometimes sown with a combination of plough, roller, seed-box and harrows in one

[1] J. Caird: *English Agriculture*, 10, 102, 127, 159, 239.
[2] J. Lawrence: *New Farmer's Calendar* (5th edit.), 345.

operation.[1]   Drills of all sorts are now used for a great variety of crops, and also for drilling and spreading artificial manures. The good old practice of dibbling beans by hand, five to a hole,

> One for the pigeon, one for the crow,
> One to die, and two to grow.

went out as child labour was prohibited and adult labour grew scarce or expensive, and now they are often drilled or ploughed in.   Mangels and potatoes can also be set by a drill, and a special machine is made for unearthing the latter when ready to gather up.   The advance of science applied the theory of heredity even to plants, and pedigree strains of wheat and other grains can be obtained, with improved characteristics. The study of entomology too has resulted in much fresh knowledge of insect pests and means of preventing their ravages.

Harrows have not undergone any radical change except in material of construction.   Iron rollers early replaced the primitive slightly shaped tree-trunk, a cast-iron one being noted in Staffordshire as early as 1800.[2]   They were made at first with smooth cylinders, and later another variety was added, with discs, producing a ribbed surface when rolling.   The heaviest sort, for breaking up the soil, are called clod crushers. Fallowing, "the miserable substitute of former times for manure and the hoe culture," and which occupied so much of the time and energy of the mediæval peasant, is now, with the assistance of root crops carefully cleaned by the horse-hoe, a brief business of weeding and preparing the ground, instead of a matter of repeated ploughings spread over a whole year.   Some soils, however, still need an occasional bare fallow.   Weeds among the crops are now being dealt with by spraying with chemicals, but as this is still in the experimental stage it is not without danger both to the sprayer and the consumer.

The process of reaping the crops was the last to be successfully treated by mechanical invention, and owing to the specially elaborate machinery required was only slowly adopted, indeed in 1895 the old-fashioned saw-edged shearing hook was still in use in the wolds of Yorkshire.   The smooth-bladed "fagging or swapping hook," with which the corn was slashed close to the roots, had already a good deal superseded it, and the scythe is still sometimes used for beaten-down crops, and to cut a strip round the field before the machine cutting begins.   Various

---

[1] The quantities of seed sown remain much the same—2 to 3 bushels of wheat are drilled to the acre, and barley the same; oats 3 to 4 bushels per acre.

[2] J. Lawrence: *New Farmer's Calendar* (5th edit.), 227.

attempts at reaping machines were made early in the century. James Smith of Deanston, whose name is usually associated with drainage, put one forward in 1811, pushed from behind by one horse, the cutting being done by a horizontally revolving cylinder; but it was not a success.   The first notable machine was one on the principle of a series of scissors set on a board, invented by the Rev. Patrick Bell of Carmylie, Fife, in 1826. The beginning of the present form came from America, evolved by C. H. McCormick in 1831, having a straight shearing action provided by a series of triangular knives on a bar, travelling backwards and forwards in slots.   Even this simple apparatus was still unused in England in the middle of the century, though it began to gain ground soon after.

Three classes of reaping machines were successively evolved from this.   The first, called the manual delivery, between 1860 and 1870 had a board added behind the cutting apparatus, which held enough corn for a sheaf, and this was cast off at intervals by the driver, but fell in the track of the machine, so that it had to be gathered up and bound before the next round.   The second, the self-delivery reaper, was fitted with windmill-like arms which swept up the corn behind the cutter onto a platform, and when sufficient cast it off to one side. The third and most elaborate machine does all this, and also by the addition of a binder gathers up the corn, ties the sheaves, and neatly cuts off the string afterwards.   This astonishing degree of efficiency was only brought about after many trials, at first with wire, which was found unsatisfactory, and the ingenious form of knot later used, with cord, was hit upon by J. E. Appleby, an English engineer, many of the other ideas coming from America.   These machines cut about an acre an hour, whereas a band of five mediæval reapers with the sickle only accomplished two acres in a day.

An early objection to mechanical reapers was that they bruised the straw and thus impaired its lasting qualities for thatching.

The latest invention, within the last few years, is the Combine Harvester, which reaps (even beaten-down corn), threshes, winnows and delivers the straw in one place and the dressed corn in another, a truly marvellous condensation of labour.   It does not, however, deal satisfactorily with unripe or damp corn, and if fed too rapidly with straw is not thorough in threshing. The working of many agricultural machines is more or less obvious to the naked eye, but the processes of combines and threshers are mysteriously conducted within the confines of a closed case and seem little short of magical.    Could such a demonstration have been given in the Middle Ages the inventor would have had a short life.

The principle of the modern threshing-machine had already been invented in Scotland during the last century, as we have seen, and all that was needed was to improve and develop it. The revolving drum with projecting beaters which rubbed the corn against the concave frame surrounding it remained throughout.    At one time pegs were inserted into the concave frame to help the process of beating, and this was called a "peg-mill," but the rubbing principle was soon reverted to in England, the "concave" being made adjustable to the size of the seed being threshed.    In 1800 threshing machines although common in Scotland were only used here and there in England, and were far from general for some time afterwards.    Like most labour-saving mechanism they met with great opposition from the farm workers, and these machines more than others because they did away with the steady work through the winter of threshing out the corn in the barns.    In the disturbances of 1830 many threshing machines were smashed by gangs of discontented labourers.    By 1850 they were largely used for wheat, but barley was more often threshed by the flail owing to the idea that the machines injured it for malting, and in some counties the flail was still used for all corn.[1]    The machines were worked at first by water- or horse-power, but by the middle of the century steam engines were used on large and up-to-date farms; Caird describes a farm in Dorset where already in 1850 the steam engine was adapted to thresh and winnow the corn, cut the chaff, grind the cattle-cakes and work the bone-crusher for manure, while the damp corn and beans were dried on a loft over the engine as in a kiln.[2]

A winnowing-machine was first added to the thresher in 1800, and a machine was noted in Essex in that year, worked by one bullock, which threshed and dressed the corn and cut the chaff.[3]    This, however, was unusual, and hand winnowing survived for a long time, the sail-fan still being found in Wilts and Surrey in 1850.    Separate winnowing and dressing machines, and rotary screens or sieves for separating the grain are now provided, for the final processes supplementary to the thresh-

[1] J. Lawrence: *New Farmer's Calendar*, 209.    J. Caird: *English Agriculture*, 21, 67, 159, 468, etc.

[2] J. Caird, op. cit., 67.

[3] J. Lawrence: *New Farmer's Calendar*, xxxi.

ing, but in elaborate threshing-machines all these processes are included.[1]

To complete the process at the rick-yard end the emerging straw is run on to an elevator, a revolving belt or endless chain, set with tines, which hoists it to the stack.

Other modern machines dealing with the harvested crops are the clover-huller, which threshes and dresses small seeds, and the maize-sheller for Indian corn, besides all sorts of chaff-cutters and root-slicers.

These and many other farm appliances can be run by electric current, or in default of that, by small motor or oil engines.

Gleaning has died out, but at Farnham in Essex the gleaning bell was rung up to 1931.[2]

Haymaking, curiously enough, was very late in receiving help from the mechanical inventor, and it was not until the last quarter of the nineteenth century that the scythe, rake and pitchfork were superseded to any large extent. The mowing-machine which then came into use had a row of little blades set on a bar running along the ground in a similar way to the reaper, but without any of the other intricate machinery. A mower cutting a wide swathe of 5 or 6 feet cut about 10 acres a day; one man could cut with a scythe 1½ acres of level meadow. When the cutting was completed the grass could be turned over by a machine with circulating tines called a "tedder," or merely tossed in the air by a set of double forks moving backwards and forwards, very aptly called a "hay-kicker." Clover was dealt with very similarly by a "swathe-turner." When dry it was gathered up by the horse-rake, which was of course a much earlier invention, but was later improved by levers to raise the rake and free the hay at regular intervals, and was increased in width to 16 or 18 feet. Another invention, introduced from America in 1894, was the huge sweep- or buck-rake for gathering up the hay and carrying it to the stack, long iron-tipped teeth going along the ground scooping up the hay and carrying the whole load to the elevator which, as with threshed straw, hoisted it to the rick. In the North, where the drying process is slower, the hay was made into cocks and little tipping platforms were often used, on to

---

[1] According to Markham (1620) threshers with the flail completed in a *day* 4 bushels of wheat or rye, 6 bushels of oats or barley, and 5 of peas or beans. These quantities may have improved later, but were still less in mediaeval times.

[2] T. Hennell: *Change in the Farm*, 134.

which the cock was bodily removed and drawn by a horse to the rick.

The latest device is a pick-up baler, which gathers up hay or straw and compresses it into neat bales ready to be removed, after seasoning awhile, by tractor and trailer and piled in a Dutch barn.  Care however is necessary to see that the material is thoroughly dry or the bales are apt to be found mouldy in the middle.

With the adoption of all these mechanical devices the need of extra hands for haymaking and harvest is obviated.  No longer, except on very small farms, does every available member of the household turn out merrily to toss the hay, with any other help that can be got, and the cheerful bustle of harvest, with its feast at the end when the last sheaf is safely gathered in, is unfortunately a thing of the past.  With the increased speed of the operations the farmer is at any rate not so dependent on the holding up of the weather, but a pleasant feature of country life has disappeared.

A valuable expedient for the storing of winter forage, which although hinted at as early as the seventeenth century, was not really practised to any wide extent until the last quarter of the nineteenth century, is *silage*.  This is the process of storing green fodder in its fresh state instead of drying it for hay, and it is effected by the exclusion of air.  Silos may be cylindrical buildings of wood, concrete or asbestos, or even of wire and paper, or more often a pit in the ground.  The green stuff is tightly packed and pressed down by the tractor and its load, each layer being sprayed with treacle diluted with an equal quantity of water.  Young grass, oats, peas, beans and vetches are treated in this way, also (without treacle) kale and sugar-beet tops.

The most recent innovation is dried grass, by which grass is cut young and dried by heat in some kind of kiln.  This grass gives a highly concentrated food to be mixed with other cattle feed, but the plant is unfortunately very costly.

*Carts* from the application of new mechanical knowledge received an improvement in the wheel and its adjustment.  The wheel was often made "dished," or receded in the middle like a plate, and the axle-end to which the hub of the wheel was attached was sloped slightly downwards, so that the spoke coming between it and the ground attained a vertical position and was thus able to take the weight of the load.  The upper part of the wheel being then sloped outwards the dirt fell off it away from the cart.

The Norfolk variety of cart with an additional pair of wheels attachable in front, noticed by Young in 1771, was still in use there at the beginning of the twentieth century, and had come to be called a "maffie," as an abbreviation of "hermaphrodite," to denote a vehicle neither cart nor wagon.[1]

But farm carts are rare sights now, and the great gaily-painted hay-wains and harvest wagons, the pattern varying in each county, are things of the past. The chime of horse-bells which in old days gave warning of their approach in narrow lanes, is now replaced by the hideous clatter of the tractor.

Among the *farm servants* cattlemen naturally tended to increase as grazing supplanted corn-growing, and as cattle unfortunately need feeding and milking on Sundays as well as other days, this has brought into prominence the natural objection to Sunday work which has added to the increasing difficulty of getting farm labour during the last few decades. The shepherd is still employed on some farms, but nowadays, instead of being given a lamb or a bell-wether's fleece, he often receives a small money allowance on each lamb, this being increased if the number of lambs born exceeds the number of ewes kept. Lambs' tails at tailing time, are often considered the shepherd's perquisite, and are esteemed a great delicacy. A few shepherds in Northumberland were still paid entirely in kind at the beginning of this century, being supplied with a house and garden and an allowance of grain or meal, potatoes, straw, fuel, with cartage, and the keeping of sheep, sometimes of a cow also.[2]

Payments in kind to any servants have become uncommon, and early in the nineteenth century the practice of lodging them in the farm-houses began to go out. The houses were built with less accommodation than in the old spacious days, moreover the farmers' wives began to fancy themselves and to grow house-proud and did not care to have the farm hands living there, or to do so much cooking and providing for them. Incidentally, with this control removed the youth of the village became less well-behaved, and the licence of their manners caused much searching of heart to the authorities. As usual this innovation began in the South; in the North, where hiring fairs also survive, many farm hands still live in, and also to some extent in the West.

The "general utility" or odd-jobs man is still in request on small farms, but his accomplishments are far less varied and skilled than of old. The "tasker" or thresher has disappeared

---

[1] H. Rider Haggard: *A Farmer's Year*, 312.
[2] F. G. Heath: *British Rural Life and Labour*, 13, 14, 28, 37.

but most farms still need a cowman or stockman.    A new factotum is the tractor-driver or engineman, who needs a very different form of knowledge from the old farm hand.

Working hours for all farm labourers, including land girls, are now 47 hours a week, beginning usually at 6.30 in summer and 7 in winter.    But many jobs are done on a piece-work basis. Carters, where they still exist, work with their horses until 2.30, and after that are busy about the stable or yard.    Work on Sundays and Bank Holidays is treated as overtime and paid accordingly, and all have holidays with pay.

Cobbett in 1821 regarded with contempt a household servant who could not bake or brew but only "dawdle about with a bucket and a broom," and considered her merely a consumer of food, but these and other useful domestic arts went out with the nineteenth century, and the farm-house maid's life although perhaps less hard must also be less interesting.

In no department of farm work did scientific discovery do more useful work than in the *manuring* and dressing of the soil.    As far back as the seventeenth century men such as Evelyn had groped after the idea of the different nature of soils and their complements by different kinds of manure, but now by the aid of chemistry the actual constituents of the earth could be analysed, those that were necessary to the growth of plant life discovered, and those elements exhausted in the soil by each kind of crop observed.    Hence when the nature of manures was also analysed the ground could be supplied with whatever constituents it lacked, either in some natural form, such as farm-yard manure, or in the concentrated and easily portable form of prepared chemical compounds.    In England Sir Humphry Davy's eight lectures on agricultural chemistry laid the foundations of this science in 1812; the work of the great German chemist, Justus Liebig, tracing the connection between the nutrition of plants and the compostion of the soil, was published in English in 1840; and his pupil Sir John Lawes, with Dr. Gilbert, founded in 1843 the agricultural experiment station at Rothamsted, which has done so much wonderful work.    The elements most often lacking in the soil, so the scientists inform us, are nitrogen, phosphorus, potash and some-times lime.    The first of these, nitrogen, was well supplied by the expedients of our forefathers, farmyard manure, seaweed, wool clippings, parings of horns and hoofs, dried blood, and suchlike, and a little later by soot, refuse cakes and meals, and sewage sludge.    In addition to these nitrate of soda was obtained from Chile and Peru after about 1830, later sulphate of ammonia was extracted from the products of gasworks and is still much used for grassland.   As a modern expedient calcium cyanomide and calcium nitrate are obtained from electrical sources, all of which give the soil its needed nitrogen.

The production of phosphorus needed more enterprise. Bones, which are rich in this material, had been utilized in the eighteenth century, but crushed by a hammer or a horse-mill were laborious to produce in any quantity, and it was not until about 1840, by which time they could be dealt with by iron rollers worked by steam, that their use became at all universal.   About 1840 Liebig (whose name is more familiar to us in connection with extract of beef) suggested the treatment of bones with oil of vitriol (sulphuric acid), and Lawes took this up and applied it to the dissolution of coprolite and other mineral phosphates in 1843.   Superphosphates in a convenient form were thus obtained.   Recently a useful form of this kind of fertilizer has been discovered in basic slag, the product of steel manufacture by the Thomas-Gilchrist or "basic" process.

Bones contain both nitrogen and phosphates, as do guanos. Guano proper consists of the droppings of sea-birds, obtained from islands off the coast of Peru, and was first brought from there by von Humbolt in 1804, but the name is improperly applied to preparations of fish and meat refuse.   By 1850 the use of bones and guano had become widespread and superphosphates were also used to a certain extent.[1]

Potash was supplied in the past chiefly by wood ashes, but after 1861 the products of the potash mines of Strassfurt in Prussia were mainly used.   The best known of these are kainite, sulphate of potash and muriate of potash.

Lime is not properly a manure but improves a heavy soil by making it less sticky, prevents acidity, and releases food properties in the soil which would otherwise lie dormant. Nitrate of lime and chalk are used.   Salt has the useful property of liberating the potash in the soil.

Marling, even in Norfolk, has been discontinued as too costly.

Besides all these applications to the ground it is still found beneficial, especially on light soils, to grow a green crop, which collects nitrogen by its leaves, and communicates it to the soil.   Red and crimson clover, trefoil, tares and lupins are found best for this, and mustard, buckwheat and rape for ploughing in.

Farmers in the vicinity of large towns, especially those in

[1] J. Caird: *English Agriculture*, 35, 81, 92, 188, 337, etc.

Hertfordshire, within reach of London, were in the habit of getting large quantities of manure from the stables there, but since horses have been so much superseded by motors the supply has practically ceased. Even with all the fertilizing resources just described the ancient practice of folding sheep on the land is still practised with benefit, and as one modern writer observes sheep are the best manure cart for the farm.

The value of liquid manure is now realized (as on the Continent) and is used profitably on arable land. A manure-loader for farmyard manure has been invented which saves much heavy and disagreeable work.

Along with this great advance in the science of manuring it was realized that all dressings would fail of their proper effect without sufficient *drainage* of the land, and attention was necessarily turned in this direction also. A well-drained field can be worked on many more days in the year than a wet one, it can be sown earlier, and the crops on it are sooner ready for harvesting and of a better quality. In porous land with a wet sub-soil the rule was laid down that the "water table" or level of saturation must not rise too near the roots of the plants, and the moisture must be drawn off by deep drainage. On clay soils on the contrary, where the water is held on the surface, the old practice of "ridging up" was not a bad one, and would have been effectual had good surface drains been provided in the furrows. It often had the good effect, too, of mixing the subsoil with the clay and making it more porous, which can now be done by other applications, such as burnt clay or lime.

The great pioneer of deep and thorough drainage was James Smith of Deanston, who began to experiment in 1823, making a return to the seventeenth-century suggestions of Blith and Worlidge, which although not entirely forgotten had only been practised by the more enlightened farmers. This method was to drain a field by means of a complete system of parallel underground channels, about 30 inches deep and from 16 to 20 feet apart, running in the line of the greatest slope, and with a main receiving drain at the lowest part. Broken stones were laid in the bottom to the depth of a foot and the trenches were then filled in. Smith's successor, Josiah Parkes, advocated deeper and wider trenches, sometimes as deep as 4 feet. In some parts stubble, fern, or brushwood, was substituted for the stones, and bush draining is still used on heavy lands where a great expense is not worth while, but early in the century tile-draining was invented as a substitute. At first horseshoe-shaped tiles standing on flat ones were used, but about 1843 small cylindrical pipes

w

began to be made for this purpose, at first only 1½ to 2 inches in diameter, but increasing in size later with experience. By 1850 tile-draining was pretty general, and pipes were spreading, a very common practice being for the landlord to supply the tiles and the tenant the labour of putting them in. The bore-hole system of Elkington also continued in use.[1]

Trenching and laying of pipe-drains is still much resorted to, and a machine can be had for carrying out both these processes.

A useful expedient for stiff soils which will retain a hollow for some time without falling in is mole-draining. For this process a steam-engine or tractor is used which draws through the ground a pointed steel "mole," shaped like an explosive shell, attached to the base of a steel plate. Mole-ploughs are mentioned in 1800 and 1850,[2] and since they can hardly have been an apparatus of this sort were probably a kind of deep trenching-plough, possibly with a boring share as in modern ones.

The practice of *irrigation* of meadow-land has not been found generally necessary in a climate so liberally endowed with rainfall, but is occasionally resorted to for the stimulation of pasture, during the winter months. Caird in 1850 noticed its success in Devon, in which hilly country it was easy to direct water from any little stream over a warm slope, and the system was copied by Mr. Pusey in Berkshire, and with more elaboration at Clipstone Park in Nottinghamshire, where the value of the stream was much augmented by the sewage from the town of Mansfield.[3]

Various systems are now used. In the first, known as "bedwork" irrigation, a system of trenches is made connected with the water source, regularly distributed over the field on slight ridges, with a carefully corresponding drainage system in parallel hollows to carry the water off again and prevent it from standing stagnant and damaging the grass roots. By the "catchwork" system a meadow on a lower level is irrigated by the water from the drainage system of the upper one. Another and quite modern method is the "subterraneous" or upward, by which drain-pipes are laid underground with vertical shafts to the surface at intervals. Water is supplied from trenches at the top of the field, and by stopping the drainage outlet at the bottom by strong sluices the water is forced up the vertical

---

[1] J. Caird: *English Agriculture*, 17, 187, 230, 354, etc.

[2] J. Caird: *English Agriculture*, 135. L. Lawrence, op. cit., 178.

[3] J. Caird, op. cit., 55, 110, 206.

shafts to flood the meadow, and can be drawn off again when
desired by opening the sluices.

Warping, or the utilization of tidal waters to deposit a coat
of mud on arable land, is practised only in Lincolnshire and
Yorkshire, round the Humber estuary and the rivers of Trent,
Ouse and Don flowing into it.  It is worked by a system of
canals and sluices, with beneficial results, and is said to have
been the custom there since about 1760.[1]

"Inoculating"[2] of pasture was observed in Herefordshire in
1850 but seems to have died out since then.

The growth of fruit and vegetables for the market received
great stimulus from the increase of population in the towns and
the lessening prosperity of ordinary farming, and market-
gardening particularly is now a profitable industry.  All sorts
of vegetables have been immensely improved by scientific culture
and experiment but few new plants have been introduced, the
most important newcomer being the tomato.  Watercress was
promoted from its wild state to cultivated beds at the beginning
of the nineteenth century.[3]

The farm-house garden profited like the fields from new
knowledge of manures, and in these days a few pence expended
on packets of seeds will add to the garden any new vegetable or
flower that may be desired.  The old neighbourly habit of inter-
change of seed was pleasant but necessarily restricted in its
range.

As regards food, the farmer continued to fare well, but the
farm hand with very varying comfort according to his wages.
In 1850 porridge was still common for breakfast as far south
as Derby, and a thin kind of broth was customary in the western
counties.  In both parts cheese was sometimes added.  Food
was best where the hands were boarded in the farm-house, and
there meat or bacon, with broth or pudding, were given for
dinner.  In Cardigan in the early part of the century pudding
was so much appreciated that it is said to have become customary
to begin with it, since the day when one of the farm hands had
had the misfortune to expire before reaching that course in the
usual order.  In the West, notably Wiltshire, which long continued the poorest part, the men often had to be content with
bread and cheese, or in Devon with a barley dumpling containing a small piece of bacon.  For supper milk-porridge, sometimes with cheese, was given in the northern farm houses, in

[1] J. Lawrence: *New Farmer's Calendar*, 171.
[2] See above, p. 313.
[3] G. Henslow: *Uses of British Plants*, 16.

Wilts there was bread and water, or potatoes with occasional bacon, and in Devon the same (with barley bread) varied by salt fish. The Midland counties fared much better than the West.[1]

By 1880 the food of Devon and Wilts had slightly improved, the Cornishmen did pretty well on pasties, and in Dorset and Somerset there was meat or bacon for dinner every day, and often for breakfast and supper as well.[2]

The Board of Trade reports at the beginning of the present century presumably give a good idea of the farm worker's fare all over the kingdom at that time, but it should be regarded as a general impression rather than a series of fixed menus. Breakfast commonly consisted of bread and butter and tea, with or without milk, sometimes replaced by coffee in the Midlands. Bacon, cold or hot, was often added, in Devon with fried potatoes. The bloater appears as an occasional variant in Cambridge, Hants and Norfolk, and porridge in a few districts, but not, as one would expect, in the North. For dinner cold meat left from Sunday, and for the rest of the week pork or bacon was usually eaten, often accompanied by potatoes and other vegetables, and fairly often by pudding. Soup seems peculiar to Norfolk, and only in Essex was a meatless dinner of bread and cheese, potatoes, and tea usual. This was, however, to some extent made up for by eating cold meat for supper. Hot, fresh meat on Sundays was a universal custom, often accompanied by fruit pie. In Norfolk, East Suffolk, West Cambridge, Warwick, Yorkshire, Hereford, Worcester and Devon it was not customary to have tea (as a meal) as well as supper, and the evening meal in those counties consisted of bread and butter and tea supplemented by jam or syrup, cold meat or bacon, cheese, or fruit tart in Yorkshire and pasties in Devon. In other counties, where bread and butter and tea were taken in the afternoon, sometimes with cheese, jam or treacle, supper consisted of bread and cheese and tea, or in Wiltshire an unappetizing dish called "kettle-broth," consisting of boiling water poured on bread and butter or dripping, with sometimes a scrap of bacon. This was also eaten in Wales, under the name of "browis," and in that country porridge and broth were more common than in England.[3] "Flummery," a preparation of

[1] J. Caird: *English Agriculture*, 84, 358, 395. F. G. Heath: *Rural Life and Labour*, 287.

[2] F. G. Heath, op. cit., 286-7.

[3] F. G. Heath: *British Rural Life and Labour*, 59, 64-74, 88-90. The Reports quoted, however, do not include all counties; Bedford, Berks, Bucks, Herts, Middlesex, Surrey, Sussex, Hants, Leicester, Rutland, Shropshire, Cumberland, Westmorland and Cornwall being omitted.

soaked and strained oatmeal, boiled to a blancmange-like consistency, was still eaten in Wales in 1896.   More meat was eaten in the north of England and in the Midlands than in the South, as wages were higher and pigs more commonly kept.   Present-day rationing has imposed considerable restrictions on diet, the limitation of tea and cheese being perhaps most felt, although agricultural labourers get a little extra of the latter.

Drink has also undergone a good deal of change.   Cobbett complained between 1820 and 1830 that labourers no longer did their own brewing unless actually given the malt, this being apparently due to the heavy tax on malt and hops, though he says there was a little malt-making in "unkent places" which escaped the eye of the authorities.   With home-brewing falling into disuse many farm workers resorted to the inferior beer of the public-house, a place which now, instead of the mill or smithy, became "the seat of news."   Deprived of the milk of the cow they were no longer able to keep, as a home drink, in the later eighteenth century they took to poor quality tea.   Cobbett becomes wonderfully eloquent on the disadvantages of this beverage and the terrible amount of time wasted in making and drinking it,[1] but in spite of all his efforts to stimulate the making of wholesome beer at home the practice steadily declined and is now extinct, although an instance could be quoted of a labourer's wife who 25 years ago boiled up the family supply of beer twice a week in a fish kettle.

In 1850 a daily allowance of 3 pints to 2 quarters of cider was usually made on Devon farms, and doubtless on others in the West, and at haytime and harvest there was a liberal supply of beer in most parts.   It is not to be wondered at that water has never been a popular drink, since up to the last quarter of the nineteenth century that liquid was always apt to have a richer taste and odour than was at all wholesome.

The farmer's wife was relieved from her old-time occupation of spinning and weaving of linen and wool, for as early as 1821 Cobbett remarks on the rare sight of a woman bleaching home-spun and home-woven linen,[2] and as the century went on more and more of these ancient employments dropped from her, the activities of the still-room with its home-made remedies, the making of homely fruit wines and other drinks, and malting and brewing, all fell into disuse.   Home-baking fortunately remained the custom longer, but even that is rarely done nowadays, though not extinct in the North and East.   In cheese-

[1] W. Cobbett: *Cottage Economy*, 20.
[2] W. Cobbett: *Rural Rides*, 172.

making districts, however, she remained the most important person in the household as the superintendent of that intricate process. At the beginning of the nineteenth century candles were taxed and might not be made at home, but rush lights could be made by peeling rushes, all but one little strip to hold the pith together, soaking them in a long vessel of melted fat, and laying them to dry on a hanging shelf of bark. Cobbett did his best to introduce straw-plaiting for hats, and this continued as a home employment for women and young girls in Hertfordshire and Bedfordshire to the end of the century. The same writer recommended the housewife to have nothing in her house of deal or crockery but everything of good wood or pewter,[1] but few were probably able or willing to take his advice.

In the house the farmer's wife still has to bestir herself, even though her duties are now mainly restricted to housekeeping, except on small family farms, where she may still be called upon to undertake many jobs. But there is little dairy work now, and on the larger farms there is often a land-girl to look after the poultry and calves. In the house she has less help than of old, for active young damsels are hard to come by, and with more complicated possessions and a higher standard of living there is more to do. However, her cooking facilities were improved about the middle of the nineteenth century by the introduction of the closed kitchen range, and now in many places she is able to cook by gas or electricity, which current will also run other labour-saving devices. It is unusual now for her to work on the land except in market-gardening districts.

Here we come to the end of our survey of the development of the farm-yard, a long, slow process from the rude flint knife to the modern combine, from the first meagre little domesticated cattle to the great well-fed beasts of to-day. Many-sided and rapid has been the progress during the last one hundred and fifty years, a progress first in the hands of individual farming pioneers, and more recently taken over by the State. The establishment of the Royal Agricultural Society and its kindred associations in each county, aided by printed journals to disseminate its discoveries and ideas, has made knowledge of better farming widespread, with a consequent rise in the general level of attainment.

The first farmer's thought was to feed his family, next he took his part in the self-supporting village, then he learnt to extend his efforts to supplying the needs of his country, but now he has to take his place in the world's economy. What this

[1] W. Cobbett: *Cottage Economy*, 193-200.

place is to be, and what changes and readjustments it may bring about in English farming the future must decide.

In this restless age, the lads who should be growing up into skilled farmers and farm workers drift away to the alluring towns, where there are higher wages and more obvious pleasures. Rarely does the villager own land of his own to give his sons the true incentive to follow in his footsteps. Little enough, in spite of the outcry over shortage of labour, is done to keep his interest rooted in the land. The system of education he receives is not often adapted to the needs of the country; hardly any effort is made to unfold to him the fascination of the many-sided store of knowledge about all the elements of farming and its plant and animal life; he does not leave school at haytime and harvest for practical experience as continental children do. Too late, when he is just at the age when the bustling life of the town is most attractive, is the possibility of farm work put before him, and naturally he rejects it as dull.

History cannot go back on her steps; we cannot return, even if we would, to the simple days when man worked, not for commercial gain, but to supply directly the needs of his own family; but still the fields endure, the most ancient occupation of mankind continues, and the natural wholesome life of the farm is there for any who will pursue it.

# BIBLIOGRAPHY

## GENERAL

ADDY, S. O.   Evolution of the English House.
AMHERST, A.   History of Gardening in England.
ASHLEY, Prof.   Place of Rye in the History of English Food.   (*Econ. Journal*, vol. xxxi, 1921.)
BENNETT, R., and ELTON, J.   History of Corn Milling.
CANDOLLE, A.   Origin of Cultivated Plants.
CUNNINGHAM, W.   Growth of English Industry and Commerce.
CURTLER, W. H. R.   Short History of English Agriculture.
DARWIN, C. R.   Variation of Animals and Plants under Domestication.
ERNLE, Lord.   English Farming, Past and Present.
GARNIER, R. M.   Annals of the British Peasantry.
GILBEY, Sir W.   The Great Horse.
GOMME, G. L.   The Village Community.
GOTCH, J. A.   Growth of the English House.
GRAS, N. S. B.   Evolution of the English Corn Market.
HASBACH, W.   History of the English Agricultural Labourer.
HUGHES, T. McKENNY.   Breeds of Cattle in the British Isles.
(*Archaeologia*, vol. lv.)
JOHNSON, W.   Byways in British Archaeology.
PEAKE, H.   The English Village.
POCOCK, R.   Horses.
RIDGEWAY, W.   Origin and Influence of the Thoroughbred Horse.
ROGERS, J. E. THOROLD.
Six Centuries of Work and Wages.
History of Agriculture and Prices in England.
SEEBOHM, F.
The English Village Community.
Customary Acres.
SLATER, G.   English Peasantry and the Enclosure of Common Fields.
TRAILL, H. D.   Social England.
VINOGRADOFF, Sir P.   Growth of the Manor.
YOUATT, W.
The Dog.
The Horse.
The Sheep.
The Pig.

## CHAPTERS I AND II.   STONE AND BRONZE AGES

AVEBURY, LORD.   Prehistoric Times (7th edition).
BOYD DAWKINS, W.
Early Man in Britain.
Cave Hunting.
British Pleistocene Mammalia. (Palaeontographical Soc.)
ITISH MUSEUM GUIDES.
Stone Age.
Bronze Age.
DECHELETTE, J.   Manuel d'Archéologie Préhistorique.
CHILDE, V. GORDON.   Prehistoric Communities of the British Isles.

CLARKE, G.   Prehistoric England.
CURWEN, E. C.   Plough and Pasture.
ELTON, C. I.   Origins of English History.
EVANS, Sir. J.
    Ancient Stone Implements of Great Britain.
    Ancient Bronze Implements of Great Britain.
GREENWELL and ROLLESTON.   British Barrows.
HAWKES, J.   Early Britain.
KEITH, A.   Bronze Age Invaders of Britain.
KENDRICK and J. HAWKES.   Archaeology in England and Wales.
JERROLD, D.   Introduction to the History of England.
MONTELIUS, O.   Chronology of the British Bronze Age.   (*Archaeological*
    vol. lxi, Pt. I.)
PAGE, J. W.   From Hunter to Husbandman.
PEAKE, H.   Bronze Age and the Celtic World.
REID, C.   Submerged Forests.
RICE HOLMES, T.   Ancient Britain and the Invasions of Julius Caesar.
TYLER, J. M. New Stone Age in Northern Europe.
TYLOR, E. B.   On the Origin of the Plough and Wheel Carriage.
    (*Journal of the Anthropological Institute,* vol. x.)
WINDLE, B. C. A.
    Life in Early Britain.
    Remains of the Prehistoric Age in Britain.


                CHAPTER III.   CELTIC AND EARLY IRON AGE
*Contemporary.*
    ADAMNAN.   Life of St. Columba.   Ed. Reeves.
    ANCIENT LAWS AND INSTITUTIONS OF WALES.   (Rec. Com.)
    ANCIENT LAWS OF WALES.   H. Lewis.
    CAESAR.   De Bello Gallico.   Bk. V.
    DIODORUS SICULUS.   Mon. Brit.   Excerpta ii.
    GIRALDUS CAMBRENSIS.
        Itinerary through Wales.
        Description of Wales.
    MABINOGION.   Trans. Lady Charlotte Guest.
    PLINY.   Natural History, xi, xvii, xviii.

*Modern.*
    BRITISH MUSEUM GUIDES.   Early Iron Age.
    BULLEID, A., and GRAY, H. ST. GEORGE.   The Glastonbury Lake
        Village.
    GUEST, E.   Origines Celticae.
    LLOYD, J. E.   Early Welsh Agriculture.
    RHYS, J.   Celtic Britain.
    RHYS, J., and BRYNMOR JONES, D.   The Welsh People.
    SEEBOHM, F.   The Tribal System in Wales.
    SQUIRE, C.   Mythology of Ancient Britain and Ireland.


                        CHAPTER IV.   ROMAN
COLLINGWOOD, R. G. and MYRES, J. N. L.   Roman Britain and the
        English Settlements.   Second Edition.
CONYBEARE, E.   Roman Britain.
COOTE, H. C.   Romans of Britain.
FOX-PITT-RIVERS, A. H. L.   Excavations at Cranbourne Chase.

HAVERFIELD, F.   Romanization of Roman Britain.
HAVERFIELD and MACDONALD.   Roman Occupation of Britain.
PALLADIUS.   On Husbondrie.   E.E.T.S.
SCARTH, H. M.   Roman Britain.
SILCHESTER EXCAVATIONS.   *Archaeologia*, vols. liii, lx, etc
SUMNER, H.   Ancient Earthworks of Cranbourne Chase.
WARD, J.   Roman Era in Britain.
WRIGHT, T.   Celt, Roman, and Briton.

### CHAPTER V.   SAXON

*Contemporary*
ATTENBOROUGH, F. L.   Laws of the Earliest English Kings.
LIEBERMANN, FELIX.   Die Gesetze der Angelsachsen, vol. i.
    (Gerefa.)
THORPE, BENJ.
    Ancient Laws and Institutions of England.
    Analecta Anglo-Saxonica.

*Modern.*
FILES, G. T.   The Anglo-Saxon House.
KEMBLE, J.   The Saxons in England.
SEEBOHM, F.   Tribal Custom in Anglo-Saxon Law.
THRUPP, J.   Domestication of Animals in England.   (*Transactions
    of Ethnological Society*, N.S. vol. iv.)
VINOGRADOFF, Sir P.   English Society in the Eleventh Century

*Danish.*
DASENT, G. W.   The Story of Burnt Njal.
KEYSER, P.   Private Life of the Old Northmen.
STENTON, F. M.   Types of Manorial Structure in the Northern
    Danelaw.   (Oxford Studies, vol. ii.)
VIGFUSSON and POWELL.   Origines Icelandicae.
WRIGHT, T.   Homes of Other Days.

### CHAPTERS VI and VII.   1066 TO 1348.

*Contemporary.*
ANGLICUS, BARTHOLOMEW.   Mediaeval Lore.   Ed. R. Steele.
BATTLE ABBEY, CUSTUMALS OF.   (Camden Soc.)
DOMESDAY BOOK.   (Rec. Com.)
DOMESDAY OF ST. PAUL'S.   (Camden Soc.)
FLETA, vol. ii.
GLASTONIAE, RENTALIA ET CUSTUMALIA ABBATIAE.   (Somerset
    Record Soc.)
HENLEY, WALTER OF.   Husbandry.   Ed. E. Lamond.
NECKHAM, ALEXANDER.
    Treatise De Utensilibus.   Ed. J. Meyer.
    De Naturis Rerum.
S. PETRI DE BURGO, LIBER NIGER MONASTERII.   (Chronicon
    Petroburgense, Camden Soc.)
WORCESTER PRIORY, REGISTER OF.   (Camden Soc.)

*Modern.*
BALLARD, A. Domesday Inquest.
BATESON, M.   Mediaeval England.
DAVENPORT, F. C.   Economic Development of a Norfolk Manor.

*Modern*—continued.

HONE, N.   The Manor and Manorial Records.
INMAN, A. H.   Domesday and Feudal Statistics.
MAITLAND, F. W.
    Domesday Book and Beyond.
    Select Pleas from Manorial Courts.
SMYTH, J.   Lives of the Berkeleys.
TURNER, T. HUDSON.   State of Horticulture in Early Times
    (*Archaeological Journal*, vol. v.)
VINOGRADOFF, Sir P.   Villeinage in England

### CHAPTER VIII.   1348 TO 1500

*Contemporary.*

BABEE'S BOOK, etc.   Ed. Furnivall.   E.E.T.S.
CHAUCER.   Canterbury Tales.
DURHAM HALMOTE ROLLS.   (Surtees Soc.)
FORTESCUE, Sir JOHN.   Works.
FIFTY EARLIEST ENGLISH WILLS.   E.E.T.S.
GARDENER, JOHN.   The Feate of Gardening.   (*Archaeologia*, vol. liv.)
HEXHAM, PRIORY OF.   (Surtees Soc.)
ITALIAN RELATION OF ENGLAND.   (Camden Soc.)
PIERS PLOWMAN.   Ed. Skeat.
POLITICAL, RELIGIOUS AND LOVE POEMS.   Ed. Furnivall.   E.E.T.S.
SCOTTER, LINCS, COURT ROLLS OF.   (*Archaeologia*, vol. xlvi.)

*Modern.*

ABRAM, A.
    English Life and Manners in the later Middle Ages.
    Social England in the Fifteenth Century.
BRADLEY, HARRIETT.   The Enclosures in England.   (*Studies in
    History, Economics and Public Law*, vol. lxxx. No. 2, Columbia
    University.)
CAPES, W. W.   Rural Life in Hants.
DENTON, W.   England in the Fifteenth Century.
DUNKIN, J.   History of Bicester.
GASQUET, F. A.   The Great Pestilence.

### CHAPTER IX.   SIXTEENTH CENTURY

*Contemporary.*

BOORDE, ANDREW.
    Introduction of Knowledge.   Ed. Furnivall.   E.E.T.S.
    Dyetary.   Ed. Furnivall.   E.E.T.S.
BARCLAY, ALEX.   Eclogues.
FITZHERBERT.
    Boke of Husbondrye.   1523
    Surveyinge.   1523.
GERARD, JOHN.   The Herball, 1597.
HERESBACH, C. (Barnabe Googe).   Foure Bookes of Husbandry.
    1577.
HARRISON, W.   Description of England in Shakespeare's Youth.
    Ed. Furnivall.   1577.
HALL, Bishop.   Satires.

*Contemporary*—continued.

> MASCALL, LEONARD.
>> Countreyman's Jewel or the Government of Cattell.   1591.
>> Husbandeye Ordring of Poultrie.   1581.
>
> MORE, SIR THOMAS.   Utopia.   1516.
> SPENSER, EDMUND.   Shepheardes Calendar.   1579.
> STARKEY, T.   Life and Letters.   (England in the Reign of Henry VIII.)
> TUSSER, THOS.   Five Hundred Points of Good Husbandry.   1573.
> WHITFORD, RICHARD.   A Werke for Householders.   1533.

*Modern.*

> HALL, H.   Society in the Elizabethan Age.
> PROTHERO, R. E. (Lord Ernle).   Shakespeare's England, section xii.   Agriculture and Gardening.
> TAWNEY, R. H.   Agrarian Problem in the Sixteenth Century.

## CHAPTER X.   SEVENTEENTH CENTURY

ATKINSON, J. A. C.   Quarter Sessions Records of Yorkshire.
BLITH, WALTER.   The English Improver Improved.   1652.
BEST, HENRY.   Rural Economy in Yorkshire in 1641.   (Surtees Soc.)
CLARK, ALICE.   Working Life of Women in the Seventeenth Century.
COX, O. C.   Quarter Sessions Records of Derby.
EVELYN, JOHN.   Terra.   1675.
GENT, A. S.   The Husbandman, Farmer and Grasier's Compleat Instructor.   1679.
HAMILTON, A. H. A.   Quarter Sessions Records of Devon.
HARTLIB, SAMUEL.
> Compleat Husbandman.   1659.
> His Legacie.   1651.

LAWSON, WILLIAM.
> A New Orchard and Garden.   1618.
> The Country Housewife's Garden.
> The Husbandman's Fruitful Orchard.

LEVETT, JOHN.   The Ordering of Bees.   1634.
MARKHAM, GERVASE.
> Cheape and Good Husbandry.   1614.
> Country Contentments, and the English Huswife.   1615.
> Farewell to Husbandry.   1620.
> Inrichment of the Weald of Kent.   1625.
> A Way to get Wealth.   1628.

MAXEY, EDWARD.   A New Instruction of Plowing and Setting of Corn.   1601.
MEAGER, LEONARD.   Mystery of Husbandry.   1697.
NORDEN, JOHN.   Surveyor's Dialogue.   1607.
PARKYNS, SIR THOS.   Servants in Husbandry.
PLATT, SIR HUGH.
> Jewell House of Art and Nature.   1594.
> New and Admirable Art of Setting Corne.   1600.

PLATTES, GABRIEL.   Discovery of Infinite Treasure.   1639.
SPEED, ADOLPHUS.   Adam out of Eden.   1659.
STEVENSON, MATTHEW.   The Twelve Moneths.   1661.
TOPSELL, EDWARD.   Historie of Foure-Footed Beastes.   1607

WORLIDGE, J. (I.W.), Systema Agriculturae. 1668.
YARRANTON, A. Great Improvement of Lands by Clover. 1663.


CHAPTER XI. EIGHTEENTH CENTURY.

BRADLEY, RICHARD.
  Complete Body of Husbandry. 1727.
  Country Housewife and Lady's Director. 1727.
  Country Housewife and Lady's Director. Part II. 1732.
CULLEY, GEORGE. Observations on Livestock. 1786.
DAVIES, E. Case of Labourers in Husbandry. 1795.
DEFOE, D. Tours.
ELLIS, WILLIAM.
  Complete System of Experienced Improvements on Sheep. 1749.
  Country Housewife's Companion. 1750.
  Husbandry (abridged). 1772.
JACOB, GILES. Country Gentleman's Vade Mecum. 1717.
MARSHALL, WILLIAM.
  Rural Economy of Yorkshire. 1787.
  Rural Economy of Norfolk. 1787.
  Rural Economy of Gloucester. 1788-9.
  Rural Economy of the Midlands. 1789.
  Rural Economy of the West of England. 1791-6.
  Rural Economy of the Southern Counties. 1797.
MORTIMER, J. Whole Art of Husbandry. 1707.
NOURSE, TIM. Campania Foelix. 1700.
RIGBY, Dr. Agriculture of Holkham. (*Pamphleteer*, vol. xiii.)
TULL, JETHRO. Horse-hoeing Husbandry. 1731.
TUSSER, REDIVIVUS. (D. Hillman.) 1710.
YOUNG, ARTHUR.
  Essay on the Management of Hogs. 1765.
  Northern Tour. 1770.
  Eastern Tour. 1771.
  Six Weeks Tour of the Southern Counties. 1772.
  Farmer's Calendar. 1804.
  On the Husbandry of Three Celebrated Farmers. 1811.


CHAPTER XII. NINETEENTH CENTURY. CONCLUSION.

CAIRD, JAMES. English Agriculture in 1850.
COBBETT, W.
  Cottage Economy. 1821.
  Rural Rides. 1821-32.
GARNER, F. H.
  Cattle of Britain.
  The Farmer's Animals.
GUNSTON, J. Farming Today and Tomorrow.
GILBEY, Sir W. Farmstock a Hundred Years Ago. 1910.
HEATH, F. G. British Rural Life and Labour. 1911.
HENNELL, T. Change in the Farm.
LAWRENCE, JOHN.
  New Farmer's Calendar. 1800.
  General Treatise on Cattle. 1805.

Low, David. On the Domesticated Animals of the British Islands. 1845.

Parkinson, Richard.    Treatise on Breeding and Management of Livestock. 1810.

Rider Haggard, H.
    A Farmer's Year.
    Rural England.

Thomas, J. F. H.   Sheep.

Tod, W. M.   Farming.

Wallace, Robert, and Watson, J. A. S.   Farm Livestock of Great Britain (5th edition)

# INDEX

Abbreviations showing Period of reference: Ne. = Neolithic Age; Br. = Bronze Age; El. = Early Iron Age; C. = Celtic; R. = Roman; S. = Saxon; N. = Norman; Ch. = 1348-1500; T. = 16th Century; St. = 17th Century; xviii. = 18th Century; M. = 19th Century and after.